KNITTING

FUNDAMENTALS, MACHINES, STRUCTURES AND DEVELOPMENTS

KNITTING

FUNDAMENTALS, MACHINES, STRUCTURES AND DEVELOPMENTS

N Anbumani
Assistant Professor
Department of Textile Technology
PSG College of Technology
Coimbatore, Tamil Nadu
India

Committed to Educate the Nation

Publishing Globally

IN INDIA

NEW AGE INTERNATIONAL (P) LIMITED, PUBLISHERS

LONDON • NEW DELHI • NAIROBI

Bangalore • Chennai • Cochin • Guwahati • Hyderabad • Kolkata • Lucknow • Mumbai

Visit us at **www.newagepublishers.com**

Published by New Age International (P) Ltd., Publishers
First Edition: 2007
Reprint: 2021

GLOBAL OFFICES

- **New Delhi** **NEW AGE INTERNATIONAL (P) LIMITED, PUBLISHERS**
  7/30 A, Daryaganj, New Delhi-110002, (INDIA)
  Tel.: (011) 23253771, 23253472, **Telefax:** 23267437, 43551305
  E-mail: contactus@newagepublishers.com • Visit us at www.newagepublishers.com
- **London** **NEW AGE INTERNATIONAL (UK) LTD.**
  27 Old Gloucester Street, London, WC1N 3AX, UK
  E-mail: info@newacademicscience.co.uk • Visit us at www.newacademicscience.co.uk
- **Nairobi** **NEW AGE GOLDEN (EAST AFRICA) LTD.**
  Ground Floor, Westlands Arcade, Chiromo Road (Next to Naivas Supermarket), Westlands, Nairobi, KENYA
  Tel.: 00-254-713848772, 00-254-725700286
  E-mail: kenya@newagepublishers.com

BRANCHES

- **Bangalore** 37/10, 8th Cross (Near Hanuman Temple), Azad Nagar, Chamarajpet, Bangalore- 560 018
  Tel.: (080) 26756823, **Telefax:** 26756820, **E-mail: bangalore@newagepublishers.com**
- **Chennai** 26, Damodaran Street, T. Nagar, Chennai-600 017, **Tel.:** (044) 24353401, **Telefax:** 24351463
  E-mail: chennai@newagepublishers.com
- **Cochin** CC-39/1016, Carrier Station Road, Ernakulam South, Cochin-682 016, **Tel.:** (0484) 2377303, **Telefax:** 4051304
  E-mail: cochin@newagepublishers.com
- **Guwahati** Hemsen Complex, Mohd. Shah Road, Paltan Bazar, Near Starline Hotel, Guwahati-781 008,
  Tel.: (0361) 2513881, **Telefax:** 2543669, **E-mail: guwahati@newagepublishers.com**
- **Hyderabad** 105, 1st Floor, Madhiray Kaveri Tower, 3-2-19, Azam Jahi Road, Near Kumar Theater, Nimboliadda Kachiguda, Hyderabad-500 027, **Tel.:** (040) 24652456, **Telefax:** 24652457
  E-mail: hyderabad@newagepublishers.com
- **Kolkata** RDB Chambers (Formerly Lotus Cinema) 106A, 1st Floor, S N Banerjee Road, Kolkata-700 014
  Tel.: (033) 22273773, **Telefax:** 22275247, **E-mail: kolkata@newagepublishers.com**
- **Lucknow** 16-A, Jopling Road, Lucknow-226 001, **Tel.:** (0522) 2209578, 4045297, **Telefax:** 2204098
  E-mail: lucknow@newagepublishers.com
- **Mumbai** 142C, Victor House, Ground Floor, N.M. Joshi Marg, Lower Parel, Mumbai-400 013
  Tel.: (022) 24927869, **Telefax:** 24915415, **E-mail: mumbai@newagepublishers.com**
- **New Delhi** 22, Golden House, Daryaganj, New Delhi-110 002, **Tel.:** (011) 23262368, 23262370, **Telefax:** 43551305
  E-mail: sales@newagepublishers.com

ISBN: 978-81-224-1954-2
C-20-11-12563

Printed in India at Glorious Printers, Delhi.
Typeset at Shubham Composers, New Delhi.

NEW AGE INTERNATIONAL (P) LIMITED, PUBLISHERS
7/30 A, Daryaganj, New Delhi-110002
Visit us at **www.newagepublishers.com**
(CIN: U74899DL1966PTC004618)

Preface

The knitting industry is gaining tremendous importance in Indian Textile Industry by its remarkable achievements in earning foreign exchange through substantial export of knitted goods. It is felt for long time to have a comprehensive text book on the subject of Knitting Technology, which will cater both the academic and trade needs. A thorough understanding of the Knitting machines is essential for the proficiency with knitted fabrics and vice versa. As this book is written based on the fifteen years of teaching the subject to the undergraduates and postgraduates, I am sure that the book will exactly meet the academic requirements of the students. And due emphasis is given to make this book as a trouble shooting ready reckoner for the practicing textile technologists, professionals and non-professionals alike involved in this field.

The book starts with knitting elements, stitch formation and knitted structures for both weft and warp knitting. The aspects covered in this book are weft knitting, warp knitting, flat bed knitting and hosiery socks knitting. Needle techniques, maintenance of knitting machines, associated fabric problems and remedies, knitting geometry and dimensional characteristics, knitting advancements, finishing of knitted goods, knitting quality and calculations, flatbed knitting and hosiery socks knitting are some of the highlights of this book. Simple illustrative diagrams and point paper designs of the knitted structures give a clear understanding of the knitting process.

I hope the book will serve for the complete requirements of all the students of diploma, undergraduation and postgraduation courses as well as textile technologists, practicing trainees, engineers and personnel in knitting industry and trade.

N. Anbumani

Acknowledgements

I wish to express my sense of gratitude to all the people who have contributed towards the preparation of this book.

My sincere gratitude is due to my good friend Prof. R.P. Sundaram who has encouraged to take up this venture and guided in the completion of this book with a deep sense of self contentment.

I am extremely thankful to the management of PSG institution for providing all the amenitities towards the preparation of this book.

I whole heartedly acknowledge with a deep sense of gratitude for the multi faceted constant support given to me by all my colleagues for the successful completion of this book.

My special thanks are due to Mr.A.P. Arun, PG Scholar of 2006 Batch, for his relentless involvement in neatly editing and organizing the chapters.

I thankfully place on record the untiring support rendered to me by my lab assistant Mr. P. Mahendran.

And I should mention that it is my family's deep sense of love and understanding on me which has made to complete this work successfully at ease.

Contents

Circular Weft Knitting

1

Weft Knitting Machines and Industry

1.1 INTRODUCTION

The art of knitting has been rapidly progressing in the developed countries of the world. Unlike woven fabrics, knitted fabrics are popular for their shape fitting properties, softer handle, bulkier nature and high extension at low tension. The hand knitting was a much older technology. The application of knitting was popular by the term hose to produce a complete covering of the legs and since then the word hosiery is conveniently used for a series of articles on stockings, socks and knitted leg coverings, and even to the general circular knitted goods. Though the invention of the first knitting machine is attributed to the Reverend William Lee of England in the year 1589, the growth of machine knitting in Britain was not spectacular. The Lee hand stocking machine established the principle of bearded needle with sinkers, knockover bits etc. In nineteenth century, many inventors struggled to convert the hand operated into steam power by which the knitting speed attained 100 percent greater. The invention of latch needle and its introduction to circular knitting (first machine 1808) simplified the knitting action and enabled circular knitting machines, to be built for plain, rib and purl fabrics.

1.2 STRUCTURE OF WEFT KNITTING INDUSTRY

In knitting industry, the end product is generally a clothing article, so that the specification of the finished article depends on its comfort in wear. The Indian knitting industry by its raw material and end products falls into two groups viz., cotton knitting and woolen hosiery industries, for which the southern and northern centres are specialized respectively. Bombay and Delhi are functioning as nylon socks and synthetic knitwear producers.

Though varieties of knitted products emerge out of the multidimensional growth of the knitting industry, the following Table 1.1 serves to understand the overall structure of the weft knitting industry both in apparel and non-apparel sectors.

Table: 1.1 Structure of Weft Knitting Industry

Fabric producers circular/flat machines	*Fabric Producers/ circular machines*	*Hosiery knitting circular machines*	*Fully fashioned flat machines*
Yarn goods in cloth beams 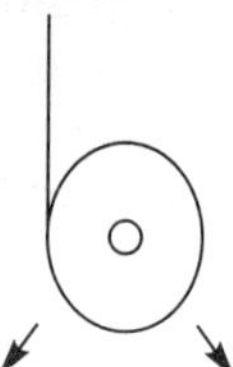 Special purpose applications; Cutting and Sewing → Non-Apparels: Upholstry, Curtains, Technical textiles, Floor coverings, Sports, Packing, Bandages; Garments	Garment length machines for the shape of garments  Cutting and sewing ↓ Garments of vests, Sweaters etc	Panty hose Scarves Caps Hats Socks and stocking in both single and double knit machines  Seamless type	Shaped garments 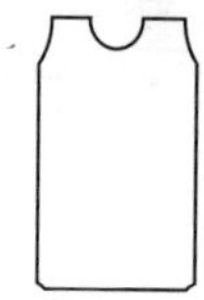 Sewing ↓ Garments ↓ Seemed stockings ↓ Sweaters

1.3 COMPARISON OF WOVEN AND KNITTED FABRICS

Knitting began with wool knitting and expanded fast to cotton and lately into the synthetics including all types of blends. Knitting is comparatively faster and more economical process to convert yarn into fabric, even straight into apparels, socks etc. Piece knitting is possible whereas piece weaving is not yet known. Besides being economical, knitted fabric being stretchable offers more comfort and better fitting in the most type of apparels. Designing possibilities are also larger though at times difficult.The mass production of knits have overriding advantages. The appeal of knitted fabrics lies in part in the very nature of the knitted as compared to the woven structure–its superior draping powers, its ready packability and its ease and excellent comfort.

Besides better appeal, popularity of knits in principles stems from the nature of knitting technology. Basically cloth can be turned out more economically on a knitting unit than on a loom. The knitting machine also has some distinct advantages over the loom in terms of patterning potential, the speed and ease of patterning. Modern living style is becoming more and more informal, go as you like and easy going. Man's life in these days is full of tension and informality helps him relax and ease tension to a great extent. It is for these reasons that knitted casual wear, sportswear, T. Shirts, Ladies tops, skirts, overalls, jackets, suits and what not are catching up fast and replacing woven wear years back.

The following gives the general comparison of woven and knitted fabrics:

- The woven fabrics are produced by the interlacement of warp and weft yarns and thus require at least two yarn systems for their production. In knitted fabric production one yarn is sufficient; however a group of yarns are required for warp knitting.
- Elasticity as well as stretchability of woven fabrics are much less and unless special elastic yarns are used, there is almost no elasticity. Knitted fabrics show high amount of stretch

and elasticity due to their loop structure. Thus they have the ability to wrap the body and conform to the body movement.

- The lesser inherent tensions and good dimensional stability of woven fabrics results in minimum shrinkage and loss of size. The formation of loops by the given yarn causes inner tension which produce higher shrinkage values in knitted fabrics.
- The woven fabrics are more durable than knits as their durability depends on yarn strength, yarn twist, yarn structure and fabric weave. The durability of knitted fabrics is based on yarn strength, loop size, stitch density and the kind of knit. Due to the elasticity character of the loop, tensile and tearing strength tests are not performed for knitted fabrics. Their bursting, wear and other resistance performances are good.
- Moisture absorption of knitted fabrics is more than that of woven because of their comparatively loose and voluminous in construction.
- Tight construction and intersecting of yarns at right angles provide good stability to woven fabrics and hence there are no much problems of slackening and loosening of the fabrics. Because of the loose structure by the loops and also because of their inability to return back to their original form after stretching in knitted fabrics, there is the problem of slackening and loosening after wearing a long time.
- Though the air permeability, heat conductivity and insulation properties are depended on fibre structure, fabric thickness, fabric construction and finishing processes, the open and voluminous structure of knitted fabrics makes them advantageous with respect to the above characteristics.
- As the woven fabrics are more inclined to crease, their ironing and iron retention properties are better than knitted fabrics. Knitted fabrics are more resistant to crease, and their garments can be worn without ironing. Their ironing and ironing retention properties are also poor.

1.4 CLASSIFICATION OF WEFT KNITTING MACHINES

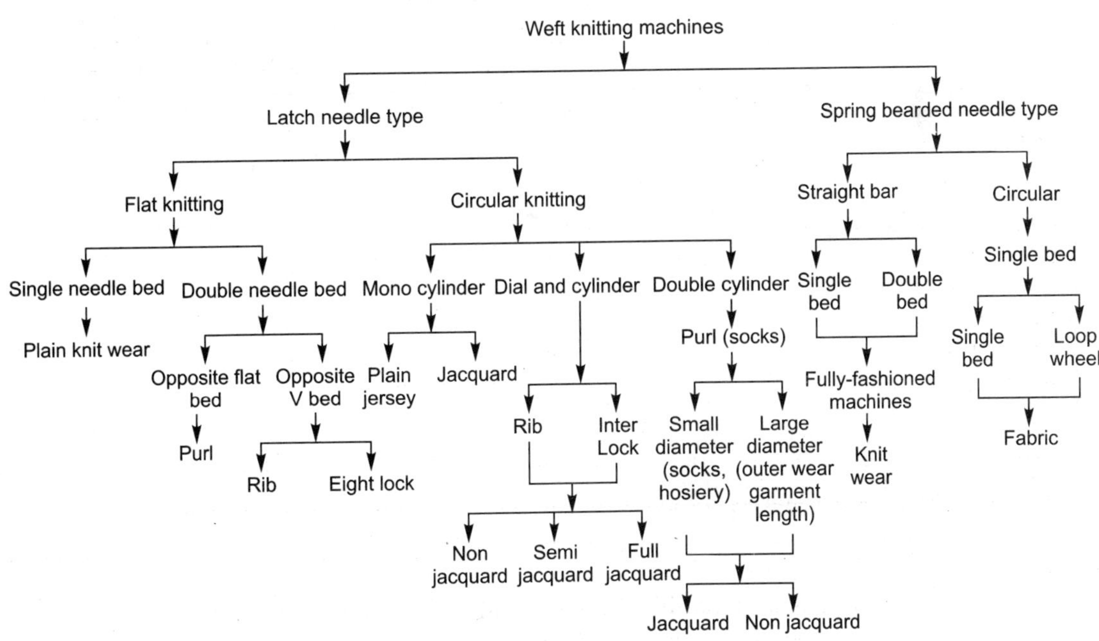

1.5 YARN QUALITY REQUIREMENTS

The yarn characteristics have a major influence on the performance of knitting as well as on the appearance of finished fabric. Improvements in the performance of knitting industry demand improvements in the knitting machines as well as optimisation of yarn properties. Various hosiery yarns are manufactured using the range of fibres available to suit different end uses. In India, majority of the knitted products are made from cotton yarn.

The following are the important yarn quality requirements (given in the order of importance) for efficient running on knitting:

- High work of rupture value
- Substantial elongation at break
- Low flexural, torsional and initial stretch resistance
- Good resiliency
- High ratio of primary to secondary creep
- Reasonable tenacity and loop strength
- Resistance to fatigue due to cyclic stress and strain
- Regularity of physical characteristics such as twist, evenness, moisture content
- Smooth and low frictional yarn properties.

Conditioning at yarn prior to knitting at a standard atmosphere of 65 ± 2% RH at 27°C temperature for 24 hours to 48 hours will give good results in knitting. Moisture content of cotton and cotton blended yarns not only affects knitting performance but also helps in maintaining consistency of fabric weight. Hence a moisture meter should be used so as to quickly establish yarn/fabric moisture. The following Table 1.2 gives a general idea on incoming spun yarn norms and allowances for certain widely used count ranges.

Table 1.2

Parameters	***Values***				
Yarn Count (Ne)	30s & 32	34 & 36	39 & 40	40s combed	Allowances guide line values
Count CV %	2.4	2.2	2.6	2.5 –3	<2
RKM (CN/tex)	15.3	15.0	15.0	16.0	> 10 CN/tex
CV % RKM	5.6	6.0	6.2	5.4	<5.5
Breaking Elongation %	5.4	5.2	5.2	7.0	> 5
U %	11.2	12.0	12.4	10.5	< 10
TPI	18.7	20.4	21.9	20.5	± 2
TPI CV%	2.5	2.8	3.0	2.7	< 4
Hairiness Index	6.4	6.2	7.1	6.5	< 7
Yarn Appearance rating	B	C+	B	B	± ½ grade
Shade Matching	Grey	Grey	Grey	Grey	< 2

To prevent snarling, steam setting is essential for polyester blends. The recommended steam setting conditions are:

Vacuum gauge – 740 mm Hg
Temperature – 95 – 110°C
Time Duration – 15 to 30 min

In high speed knitting, the knitting performance with respect to yarn hairiness is to be considered on the following accounts:

- Excessive yarn hairiness causes excessive yarn friction between yarn and metal, which hinders proper loop formation and particularly with cotton yarn, generates lots of fly.
- It spoils the fabric appearance and causes excess pilling.
- It also gives improper dyeing.

2

Knitting Terms and Functional Elements

2.1 INTRODUCTION

Weaving is a process of fabric forming by the interlacement of warp and weft yarns. Knitting is a process of fabric forming by the intermeshing of loops of yarns. When one loop is drawn through another, loopstitch is formed. Stitches may be formed in horizontal or vertical direction.

The two main forms of knitting [Fig. 2.1] are:

1. Weft knitting, and
2. Warp knitting

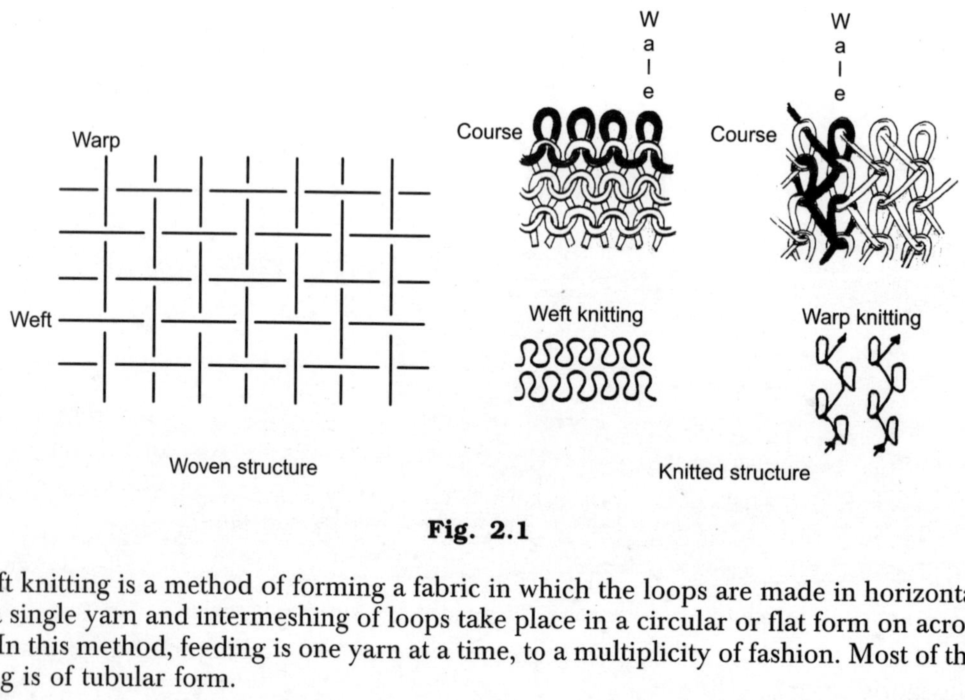

Fig. 2.1

Weft knitting is a method of forming a fabric in which the loops are made in horizontal way from a single yarn and intermeshing of loops take place in a circular or flat form on acrosswise basis. In this method, feeding is one yarn at a time, to a multiplicity of fashion. Most of the weft knitting is of tubular form.

Warp knitting is a method of fabric forming in which the loops are made in a vertical way along the length of the fabric from each warp yarns and intermeshing of loops takes place in a flat form of lengthwise basis. Here, numerous ends of yarns are being fed simultaneously to individual needles placed in a lateral fashion. Most of the knitted structures are flat or open width form.

2.2 TERMS OF KNITTING (Fig. 2.2)

Open Loop: The open loop is one in which the loop forming yarns does not cross at the bottom of the loop.

Closed Loop: In Closed loop, the legs of the loop cross so that the loop closing takes place.

Face Loop: During loop formation, when the new loop emerges through the old loop from back to the face (or front) side, it is called as face loop or weft knit loop.

Back Loop: If the new loop passes from the face side to the back side of old loop, it is called as back loop or weft purl loop.

Technical Face: The side of the knitted fabric that consists all of face or knit loops, is called technical face of the fabric.

Technical Back: The side of the knitted fabric having full of back or purl loops is called needle loop. Normally the reverse side of the technical face is the technical back.

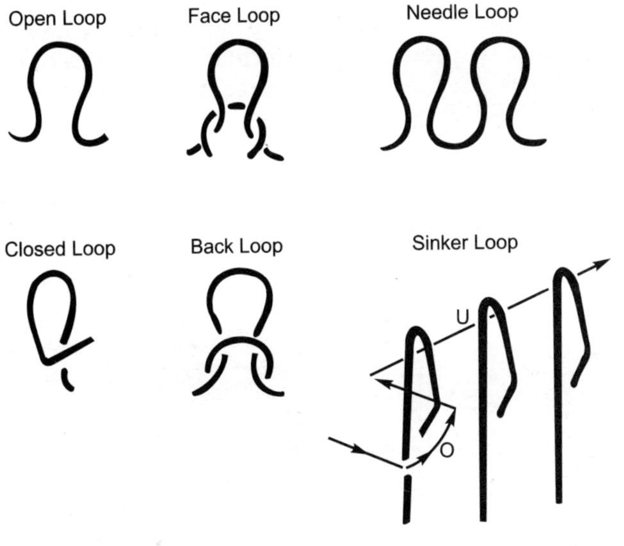

Fig. 2.2

Needle Loop: The upper part of the loop produced by the needle drawing the yarn is called as needle loop.

Sinker Loop: The lower part of the knitted loop is technically referred as sinker loop. It is the connection of two legs belonging to the neighbouring stitches lying laterally.

Course: The series of loops those are connected horizontally, continuously are called as courses.

Wales: The series of loops that intermeshes vertically are known as wales.

Stitch density: It is the number of stitches per unit area of a knitted fabric. It determines the area of the fabric.

Over lap: This term refers mainly to warp knitting. Lateral movements of the guide bars on the front side (or hook side) of the needle is called over lap. This movement is normally restricted to one needle space.

Under lap: This term also refers to warp knitting. Lateral movements of the guidebars at the backside of the needles are called under lap. This movement related to the needle space is variable according to the construction of the fabric.

Single Jersey: The weft knitted fabrics produced with one set of needles (both in tubular or flat forms) are called as single jersey or plain knitted fabrics.

Double Jersey: Weft knitted fabrics (tubular/flat) produced with two sets of needles mounted in two needle beds are called as double jersey or double knit fabrics.

2.3 MACHINE DESCRIPTION

2.3.1 Frame

The circular knitting machine consists three major sections viz., yarn supply, knitting elements and fabric take-down. The Fig. 2.3 shows the machine frame, indicating its various parts. The knitting elements such as needles, sinkers, cylinder, cams and feeders are supported at the centre called as knitting zone. Yarn packages are mounted at the overhead creels and yarns are fed to knitting zone through yarn guides, stop motions and feeders. The knitted fabric goes down inside the cylinder towards the centre of the machine, drawn into the take down device and finally collected on a roll winding mechanism. A fabric spreader gradually converts the tubular fabric into a double layer folded fabric by preventing the formation of pleats or creases. At the knitting zone, single knit plain machines are fitted with a cylinder and sinker ring, whereas the double knit machines have cylinder and dial.

1 – Legs
2 – Cylinder
3 – Dial
4 – Needle
5 – Cam Parts
6 – Feeder guide
7 – Cam
8 – Supply Package
9 – Creel
10 – Top Stop motion
11 – Anti Snarl Device
12 – Tensioner
13 – Positive feeder
14 – Knitted fabric
15 – Fabric spreader
16 – Fabric withdrawal roller
17 – Fabric winding roller

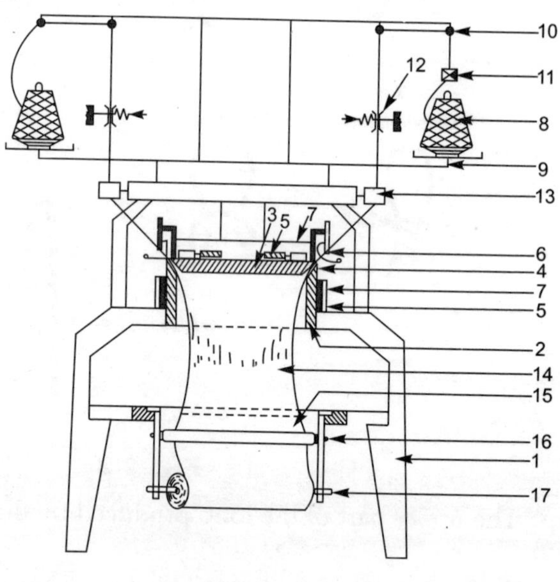

Fig. 2.3

2.3.2 Drive

The drive to the knitting machine is simple and direct. The motor imparts rotary motion to the rotating needle beds, such as cylinder and dial and also to the take-down and cloth winding mechanisms.

2.4 NEEDLES (Figs. 2.4 & 2.5)

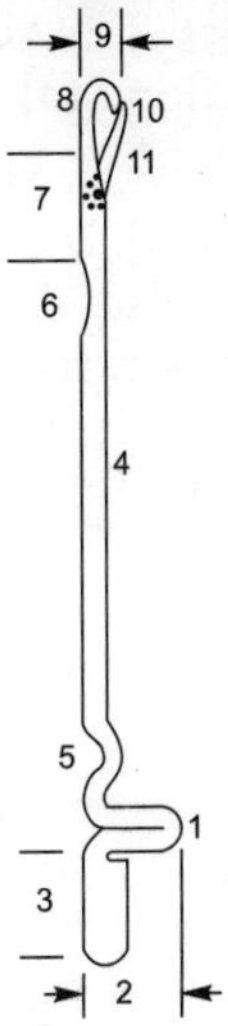

1. Butt
2. Butt height
3. Back shank
4. Stem
5. Crimp
6. Groove
7. Cheek
8. Hook
9. Hook width
10. Latch
11. Rivet

Fig. 2.4

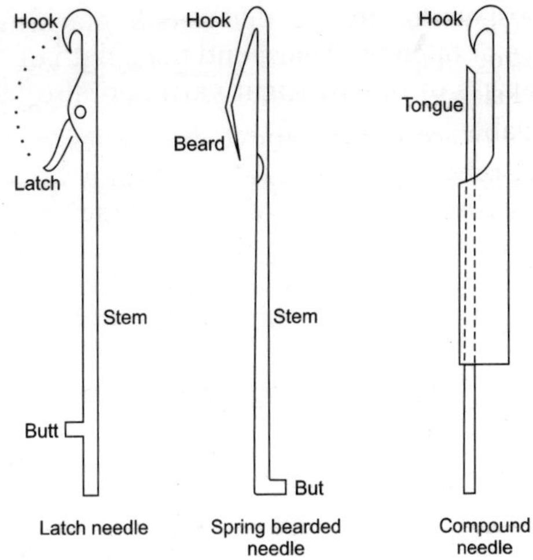

Fig. 2.5

The needles are the most important stitch forming elements. They are displaced vertically up and down and are mounted into the tricks or cuts of the knitting cylinder. There are three types of needles namely:

1. Latch needle
2. Spring bearded needle, and
3. Compound needle.

2.4.1 Latch Needle

It consists of a hook portion at the top and a latch riveted at certain distance from the needle head. Latch needles are given sliding movements in individual grooves called tricks of the cylinder. The latch swing freely and the stem is a straight portion with a protruding butt. Through the butt the reciprocating movement to the needles is given. The latch needles are self acting, and requires only previous loop on the stem and do not require any outside agency to close the hook. The swinging latch has a cup at its end and is riveted to the stem. Therefore, the total thickness of the needle cannot be reduced and hence these needles are coarser in dimensions than the bearded needles.

2.4.2 Spring Bearded Needle

This needle consists of a top hook curved downwards with a finished tip and the downward continuation is called beard. A small eye or groove is cut in the stem, to receive the point of the needle beard when it is closed. It does not require latch, it is finer in cross section and more number of needles per unit space can be and hence finer machine gauge can be achieved (gauge is the number of needles per inch). It requires additional element to close the needle which is done by a presser. The bearded needle produces superior knitted stitches than the latch needle.

2.4.3 Compound Needle

The speed of the bearded needle machine is restricted by two factors. The presser required to close or open the beard and the length of the beard. The swinging action of the latch may cause damages to fine filament yarn and also latch needles are much expensive to manufacture.

To overcome the above difficulties, compound needles are developed. It consists of a hollow steel tube of fine gauge in which a hook-closing element which is also a steel tube of finer gauge, is inserted. To the tip of the upper element, a needle hook is arranged. It claims the advantages of both bearded and latch types. There is no yarn strain and the movement required to form a loop is half to that of other needles. Both pieces have straight movement instead of swinging. Hence the high speed and high productivity.

2.4.4 Loop Forming Sequence of Latch Needle [Fig. 2.6]

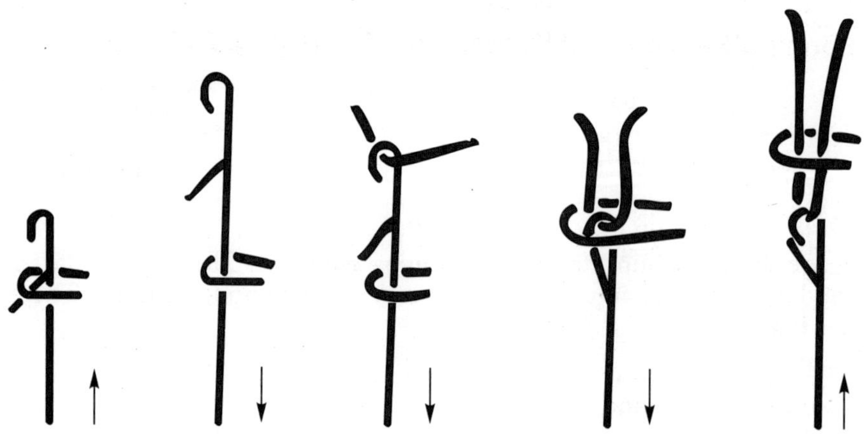

Fig. 2.6

In the running position, the held old loop rests on the top of the open latch. Clearing occurs as the held loop slips of the latch and on to the stem as the needle moves upwards. A downward movement enables the needle hook to engage a new piece of yarn. This is known as feeding. As the needles continue downwards the latch is forced to close under the influence of the held loop. Knock over occurs as the help loop disengages from the needle. Following knock over, the loop pulling occurs and a new knit loop is formed. The needle must return now to the running position to complete the cycle. This is essential, otherwise, cleaning, knock over and loop pulling would not take place. Control of the held loop is usually achieved by the use of sinkers or by the application of the fabric tension.

2.4.5 Loop Forming Sequence of Spring Bearded Needle [Fig. 2.7]

The held loop is initially just below the tip of the beard in the running position. The upward movement by the needle repositions the held loop further down the stem, at which time a new yarn can be fed to the needle i.e., feeding occurs. The needle moves down until the newly fed yarn enters the hook. An auxiliary element known as presser closes off the hook to unable the held loop to leave the needle by deflecting the beard tip into a recess cut in the needle stem. The movement when the held loop is securely located to the outside of the beard is known as the landing position. As the needle continues downwards, knock over and loop pulling take place after which the needle returns to the running position to complete the cycle.

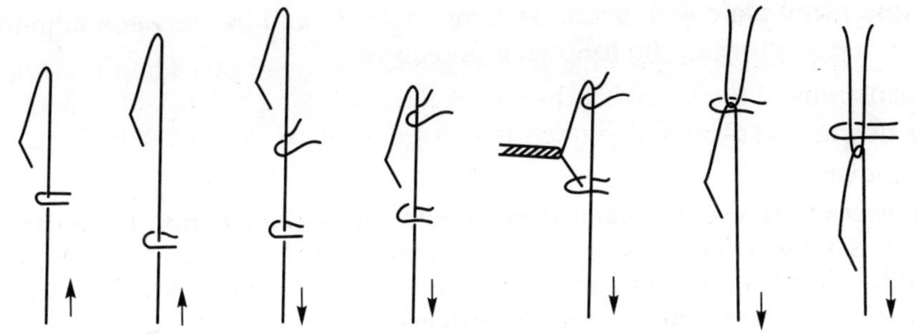

Fig. 2.7

2.4.6 Loop Forming Sequence of Compound Needle [Fig. 2.8]

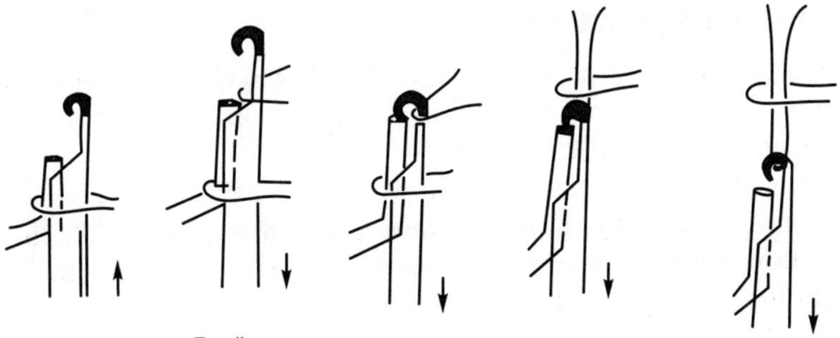

Fig. 2.8

The held loop is shown on the needle stem in the running position. The hook and tongue elements move upwards to that a new yarn may be presented to the hook and hence feeding occurs. Both elements descend, although at different velocities, which causes the tongue to close-off the hook. Thus the heldloop is free to leave the needle, knock over pulling occurs. The needle now returns to the running to complete the cycle. As the needle returns to its starting position, the hook and tongue elements once again move at different velocities resulting in the opening of the hook.

2.5 SINKERS [Fig. 2.9]

1 – Butt
2 – Butt breadth
3 – Height of shank
4 – Buldge
5 – Neb
6 – Length of Neb
7 – Throat Angle
8 – Sinker Platform Height
9 – Breadth of Lower Shank
10 – Clearance
11 – Throat

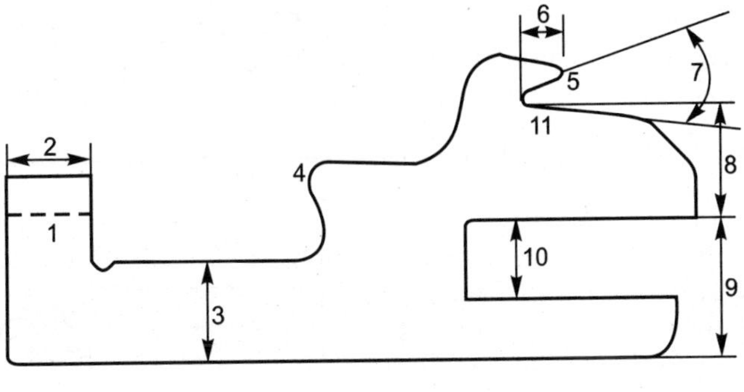

Fig. 2.9

Sinker is a thin metal plate with action at right angles to and fro between adjoining needles. It may perform one or more of the following functions:

1. Loop formation
2. Holding down, and
3. Knocking over.

The main object of a sinker is to assist the needles in loop formation by sinking or knitting the newly laid yarns into a loop. As its forward edges of catch [c] advantages between the two adjoining needles. This is only for bearded needle where as in case of latch needles and warp knitting loop formation is not the function of sinkers.

It also holds down the loops at a lower level of the needle stems and prevents the old loops from being lifted as the needles raise to clear them for their hooks.

The third function is a knocking over at which the needle passes through the old loop by drawing a new loop.

2.5.1 Sinkers Operation [Fig. 2.10]

1. The held loop is positioned in the throat of the sinker when the sinker moves forward and the needle moves upward for clearing. The held loop is held by the throat and hence its movement along the needle is restricted.
2. The sinker remains at its forward position when the needle attains its clearing position.
3. The sinker retracts when the needle comes down after feeding. At this stage, due to sinkers retraction, fabric or held loop is eased out. Also the sinker belly supported the fabric or held loop and hence its movements along the needle is prevented.
4. Sinker remains in backward position and the needle descends to its lowest position drawing the new loop through the old one.
5. Before the needle ascends, the sinker moves forward to push the knitted fabric a little and to hold the old loop away from the head of the needle and to be in a position to control the fabric.

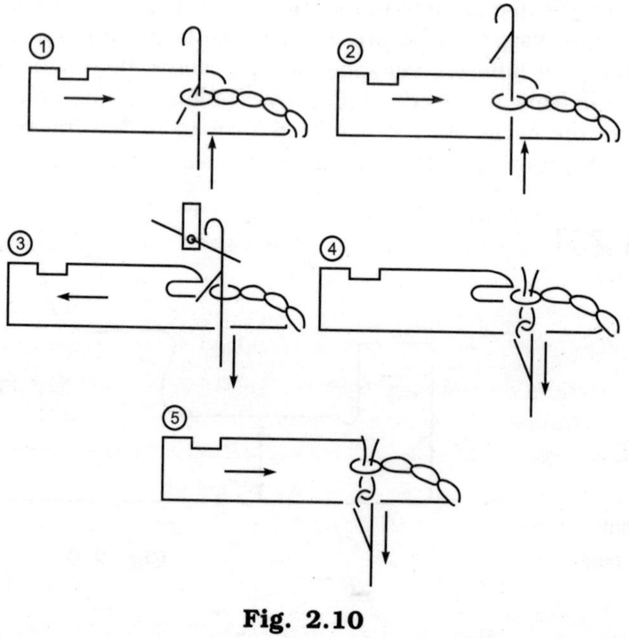

Fig. 2.10

2.6 CYLINDER (Fig. 2.11)

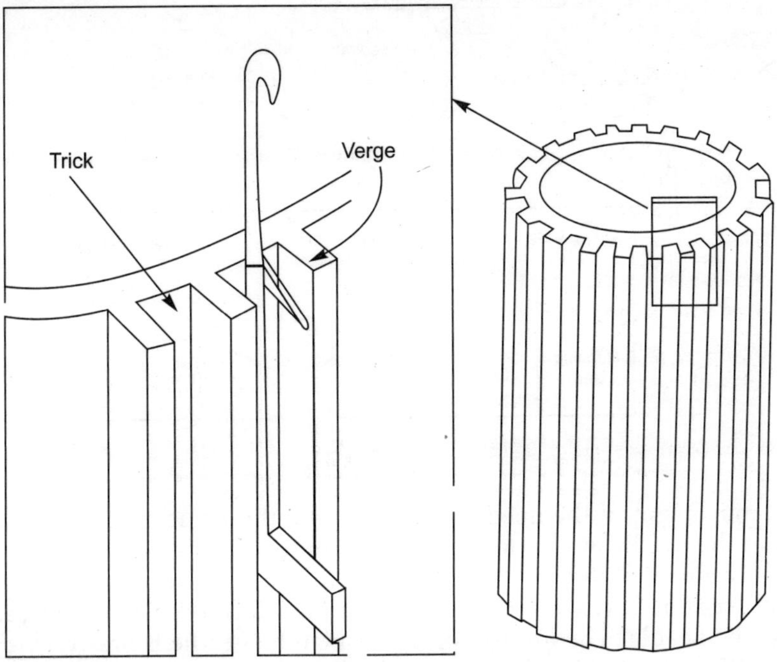

Fig. 2.11

The cylinder is a steel circular bed having grooves/tricks/cuts on its outer periphery into which the needles are mounted. With reference to the tricks, the needles move vertically up and down by their butt being in contact with the cam track. The number of tricks per inch i.e., number of needles per inch decides the gauge of the machine. Machines are built as low as 4 NPI to as high as 32 NPI. Based on the machine gauge, the fineness of the yarn to be knitted can be varied. The diameter of the cylinder also varied based on the type and width of the fabric, and a maximum of 75 cm diameter machines are available.

2.6.1 Dial

Dial is the upper steel needle bed used in double knit machines. Into the grooves of the dial, the needles are mounted horizontally and are allowed to move radially in and out by their dial cams. The number of grooves per unit space conforms with the cylinder gauge in most of the cases.

2.7 CAMS

The knitting cams are hardened steels and they are the assembly of different cam plates so that a track for butt can be arranged. Each needle movement is obtained by means of cams acting on the needle butts. The Fig. (2.12) shows the simplest cam design.

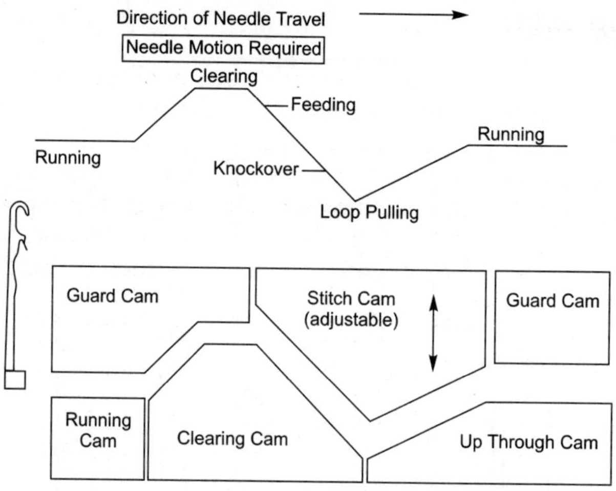

Fig. 2.12

The upward movement of the needle is obtained by the rising cams or clearing cams. The rising cam places the needle at a certain level as it approaches the yarn area. Cams controlling the downward movement of the needles are called stitch cams. The stitch cam draws the needle down below the knitting level, thereby drawing a loop formed by the fed yarn through the loop already on the needle. The lowest point to which the needle is drawn by the stitch cam is called the "cast-off" position. They are screwed to the cylindrical cam ring and are adjustable in vertical direction. If the stitch cam is raised, then shorter loop is drawn below the sinker level and a tighter fabric will result. With lowering of stitch cam, a reverse result is obtained.

Guard cams keep the needle butts in their race-way.

Running cams or [up throw cams] the needle butts at a low level until they meet the next rising cam.

Sometimes, swing cams (Fig. 2.13) or auxillary cams are placed between the rising cams and the stitch cams to change the path of the needle butts to form a raceway and the needle butts travel in this restricted path according to different stitch requirements.

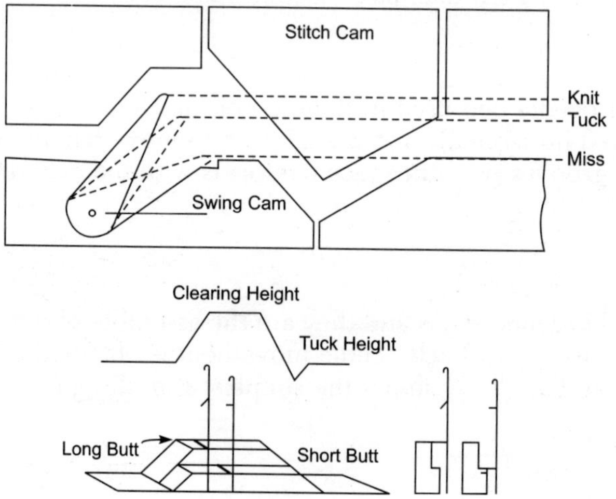

Fig. 2.13

2.8 CREELS (Fig. 2.14)

Creels are the knitting elements which carry the yarn packages to feed the knitting zone. The number of feeders in the knitting machine is equal to the number of cones in the creels. Reserve cones also can be placed in each position. In accordance with the number of feeders, two cone supports per feed are equidistantly fixed on a ring over the circumference of the machine is mounted in the lower part. On machines having yarn meters or yarn storage devices, an additional ring is mounted in the lower part. Fan blade with motors, oiling system for needles and sinkers, knot catchers and stop motion are also mounted on creel section.

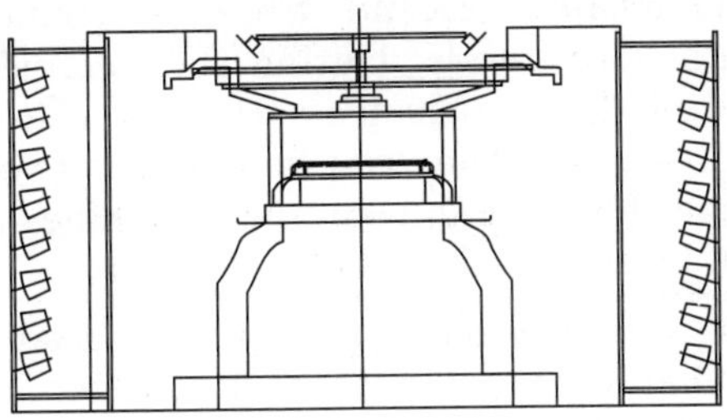

Fig. 2.14

Apart from the over head ring creels, depending on the number of feeders and space availability, creels are arranged on two or three sides of the machine. These construction keep off fibre fly from the machine and are more suitable for high speed circular knitting machines. Differences in yarn tension due to variation in yarn length from the cone to the feeder can be eliminated with these side creels compared to overhead creels. Yarn guiding via yarn guide tubes normally reduce the yarn tension than the open guiding system. Closed yarn guiding also reduces fibre fly and yarn entanglements. Side creels also make easier threading in operation.

2.9 FEEDERS/STRIPERS

Feeders are the yarn guides (Fig. 2.15) placed close to the needles to the full circumference of the knitting zone. The feeders feed the yarn into the needle hooks and control the needle latches in their open position while the needle attain their clearing position. They consist of a yarn guiding hole and a bevel edge to guard the latches of the approaching needles. They are slightly curved to the corresponding curvature of the needle bed. Feeders may have two holes also for the purpose of plating.

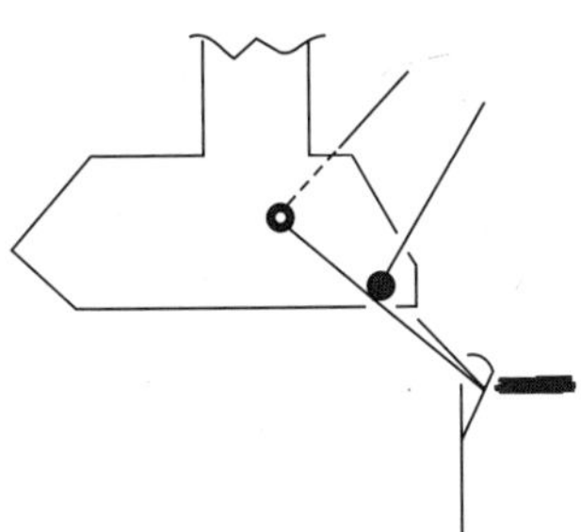

Fig. 2.15

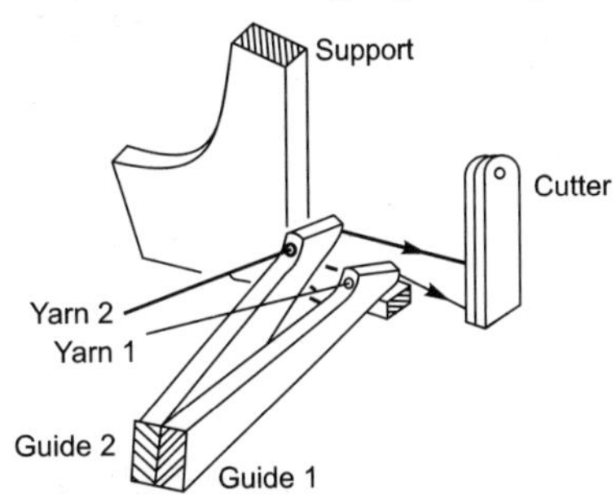

Fig. 2.16

The feeder brackets can be adjusted to set their distance from the needle and to ensure yarn feed into needle hooks.

Stripers are the feeders designed to deliver two or more yarns individually to the same feed. They can be considered as moving guide replacing the holes of fixed guides. In a two colour striper (Fig. 2.16), two different coloured yarns are supplied by two stripper fingers and their engagement is controlled by an endless control chain which governs the guide change at the appropriate feeds. At each revolution, a counter may select the movement of all the striper chains. The stripers are used on both single and double bed machines.

2.10 FABRIC SPREADER

In circular knitting, the tubular fabric descends from the needles is converged into flattened double layer and rolled into yard goods. During such conversion, irregular tensions are developed throughout the fabric due to distance variations which leads to unwanted wrinkles, crease marks and stitch deformations. They can be only removed with great difficulty during fabric finishing. To overcome this problem knitting machines are equipped with spreaders which apply a more uniform tension to the fabric.

A simple fabric spreader (Fig. 2.17) is a solid wooden plate, placed above the takedown rollers and held in position by the fabric tube.Another type of fabric spreader (Fig. 2.18) consists of two curved metal bars with adjustment for their length. The device is placed secured inside the fabric roll.

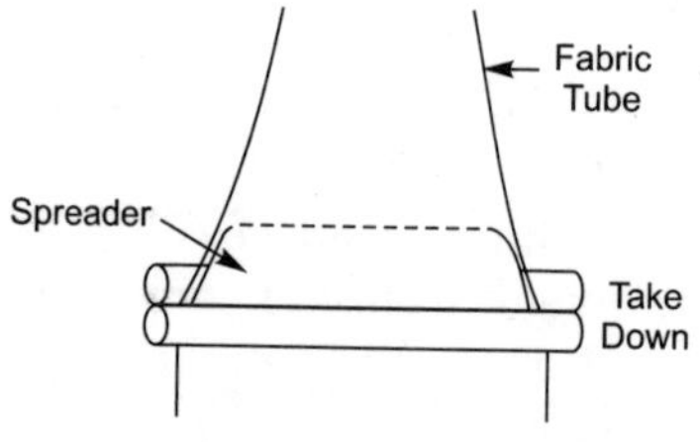

Fig. 2.17

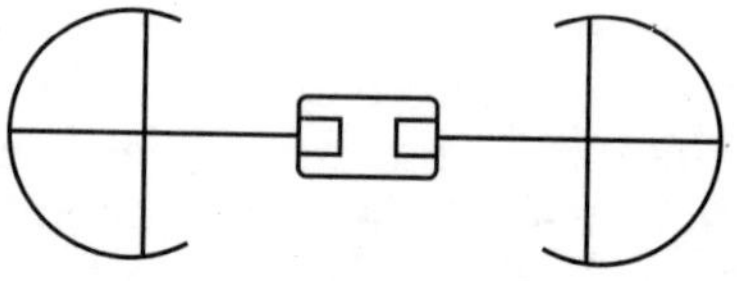

Fig. 2.18

It is important that the spreaders should not bend the courses and increase the takedown tension. Heavy tension on the fabric affects the life span of the knitting elements such as needles, cams and needle bed.

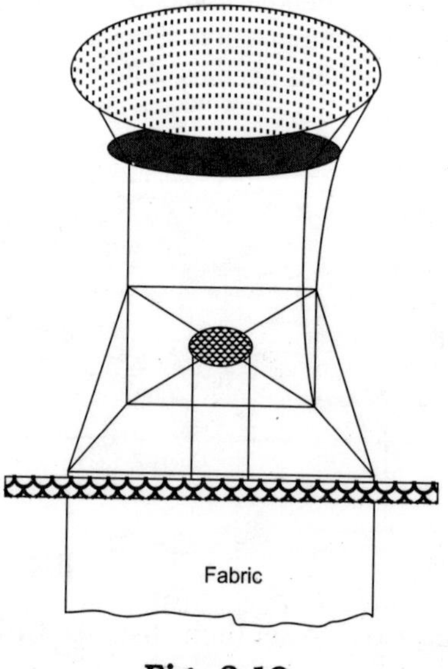

Fig. 2.19

Cadratex (Fig. 2.19) an advanced spreader consists of an internal and external guide clamp. Because of the rectangular shape of this spreader, the circumference of the fabric tube, obtained directly after knitting, is not altered over the entire distance upto the take down rollers. This results in the required evenness of the take down tension of the knitting machines and leads to a uniform stitch shape across the entire fabric width.

2.11 TAKE DOWN AND WINDING MECHANISM

This mechanism withdraws the fabric from the needles and winds them into the cloth roll. The takedown mechanism consists of grooved rollers through which the fabric is nipped and pulled downwards. The roll winding mechanism facilitate the convenient storage of fabric in which the flat, double ply fabric is wound on a revolving square bar. The take-down and winding devices are either driven mechanically by the main drive of the motor or through an additional direct current motor.

2.11.1 Mechanical Drive (Fig. 2.20)

At the take down, the fabric is nipped between two rollers or with some devices having three rollers. The pressure between the rollers is readily adjustable. The lower part of the machine frame is fitted with a circular rail or ring with the same central axis as the cylinder, but positioned inclined to the horizontal. There are two rollers or followers placed diagonally opposite and ride over the rail as the cylinder rotates. The running of the follower over the rail gives oscillating motion to the lever to which they are fixed. The levers inturn drives a toothed ratchet wheel by means of a pawl. This ratchet wheel is fixed to the grooved takedown roller and thus drives the same. A fixed pawl prevents the ratchet wheel from the reversing while the lever oscillates. The extent of the rail inclination governs the withdrawal rate of fabric through the take down rollers. Exactly similar mechanism is extended to the cloth roll through connecting levers from the takedown oscillating lever. Similar kind of ratchet wheel with driving pawl and fixed pawl rotate the cloth roll and wind the fabric.

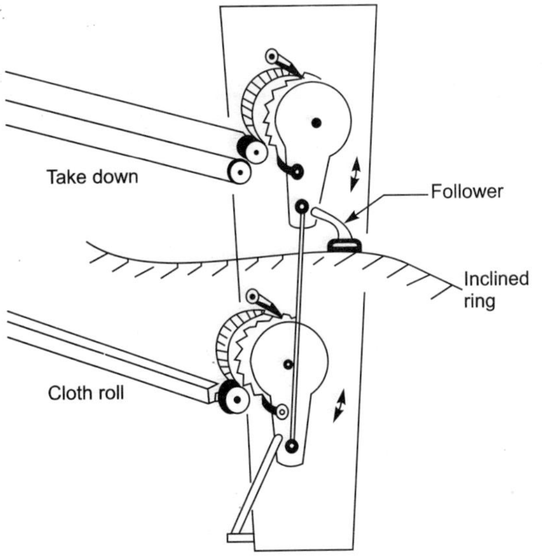

Fig. 2.20

2.11.2 Electrical Drive (Fig. 2.21)

A separate motor drives the take down and winding devices. The fabric takes down and fabric wind on form a compact unit and are as such attached to the rotating base plate in the lower machine frame. The unit consists of two lateral supports, between which the three take down rollers, the winding roll and the fabric winding spindle are provided.

The fabric is led through the take-down rollers which are driven by a gearing motor over a chain gear. The take-down tension can be infinitely and sensitively set on the control panel with a potentiometer without disturbing the machine running. The fabric encircles the take down rollers and is then led down to fabric winder and wound into fabric roll.

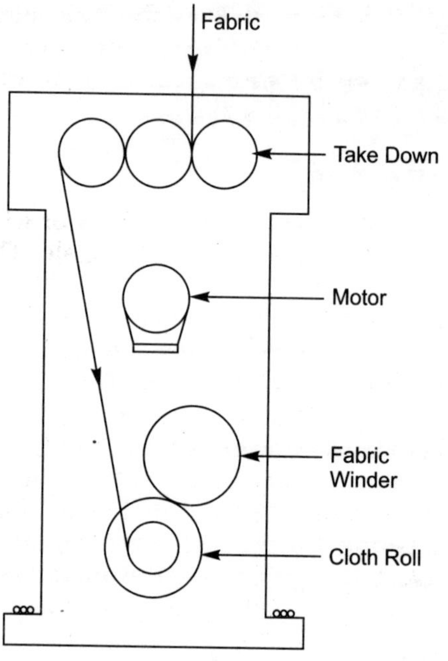

Fig. 2.21

3

Selection Criteria in Weft Knitting

The growth of the knitting industry in the recent years is tremendous in respect of both the product range and the knitting machine to manufacture the same. The growing trend towards modern machines enables wide range of product designs, use of finer yarns, as well as the use of fibres from both natural and synthetic origins.

3.1 SELECTION OF KNITTING MACHINES

The knitting machines selection is purely based on fabric type and style that should satisfy the demand of the market.

Machine productivity is the important factor to be considered in selecting a knitting machine which is decided by the circumferential speed, number of feeders of the machine, diameter of the cylinder and machine efficiency. Though the machine speed can be increased, certain factors such as cam design, cam angle, needles and sinkers fineness, their lubrication, yarn path simplicity are to be considered. Speeds have increased now to a maximum of 45-50 rpm on single knit machine of 30 inch diameter. Higher speeds require a premium quality on yarn, yarn package, the heating of the cylinder and the ability to start and stop motion. Computer designed cam tracks with an optimum angle of 58 to 59^0 enhance the machine speed significantly. The feeder density is restricted to construction versatility, yarn type and fabric spirality. High feed densities greatly reduce the stitch length range, as the steep cam angles put tremendous stress on the yarn and knitting elements. Normally a feed density of 3.2 feeders per diameter inch is found to be optimum.

The flexibility of the knitting machine to incorporate the patterning devices for the ever changing demands of fabric varieties is the other important factor to be considered in machine selection. The machines should offer a wider range of patterning possibilities along with rapid resetting facility. With the advent of design software and electronic controller for needle selection, the patterning flexibility is enhanced. Changeover time requirement for the machine from one pattern to another, adaptability for a wide range of yarn counts and yarn furnishing installations, minimum down time and minimum fault rate per fabric piece are the other factors governing the machine flexibility.

Fabric quality is the next important factor in the selection of knitting machines. Doubtlessly the production of direct quality fabric is the prime objective of every knitter and the accessories that control the yarn feed, lubrication, fabric take down and stop motions play vital role in this respect. Positive yarn feeders are indispensable today for stitch control according to the stitch varieties there by maintain uniformity in stitch length, fabric quality and minimum skewness of the fabric. Effective control over lubrication routine and any rise in temperature is to be taken care by the lubricating system. Positive driven fabric take down mechanism are preferable for minimum skew and less fabric tension according to the machine speed. Package enclosed side creels, effective stop motions, fabric inspection device are also to be considered while selecting a knitting machine to maintain the fabric quality.

3.2 SELECTION OF FIBRES

Wool and cotton fibres were the conventional fibres used traditionally. The increasing popularity of knitwear and its demand towards appearance, characteristics, dimensional stability and fashion requirements leads to the use of synthetic fibres and their blends with natural fibres. In certain sector of the knitting industry, the bulked yarns are also popular. In general, the selection of fibres and their respective yarns depends on the fabric end use, climatic conditions, market demands and economic viability.

Long staple cottons of S4, MCU5, Suvin, and Varalakshi are more suitable for the manufacture of nep free and soft twisted good hosiery yarns. They have good moisture absorption characteristics, better wearing comfort and durability. The products of cotton knits are such as vests, briefs, T- Shirts, underpants, singlets and shirts.

The best wool Merino from Australia, cross breed wools from Argentina, New Zealand, United States and United Kingdom and Cashmere of India are noted for their softness, fineness, strength and elasticity. Good quality virgin wools are used for knitting pullovers, cardigans, jumpers, socks, gloves, caps and mufflers.

The normal viscose fibres and high density and high strength polynosic fibers are used for the production of knitted articles such as petticoats, T-Shirts, sports shirts and vests, either independently or blended with cottons and synthetics.

Heat setting to shape and size is the most important property of nylon fibres in addition to its light weight, high strength, excellent drape, good abrasion, quick drying properties. The nylon fibers are extensively used in knitting industry. They find their suitability in producing socks, blouses, slacks, trousers, baby suits, nightwear, swim wear, sports wear and gloves.

The polyester fibre is known for its high strength, durability, resistance towards chemicals, resilience and drapability. Its excellent wrinkle resistance, crease resistance and shape retention properties leads to its progressive utilization in the production of knitted goods. The textured polyester knitted goods are also emerging out in knitting industry. Double knits, Cardigans, pullovers, socks, shirts and jumpers are some of the knitted products using polyester fibres.

High bulk, lightness, warmth, low stretch, low pilling and softness are the attractive characteristics of acrylic fibres for their use in the production of knitted fabrics. Their light and warm ch acters made thus as replacement for wool knits, thus they are mainly used in sweater manufacture. Also the other products from acrylics are cardigans, pullovers, socks, shirts and jumpers.

Spandex is a synthetic elastomeric fibre having extreme elongation properties. By their rubber like character these fibres when they are mixed with other fibres, possesses sufficient elasticity and holding power for knitted fabrics.

3.3 SELECTION OF YARNS

For the efficient production of knitted fabrics, proper selection of knitting yarn is very essential. Characteristics of raw materials, count of yarn, twist imparted over the yarn, its evenness, knittability, winding behaviour and humidity are the factors to be considered while selecting a knitting yarn.

Raw material characteristics such as fibre staple length, shining, lustre and colour are to be given due weightage for uniformity for cotton yarns. Normally two or three varieties of cotton are blended to get the desired yarn results.

Finer the count of yarn more would be the yield in meter per kilogram. The choice of count depends on the knittability of yarn on the given type of machines. The count choice is also based on the price range within which the fabric is to be produced.

Twist is the most important factor for knitting yarn, as higher the number of twist, suitability of yarn for knitting purposes reduced and vice versa. In order to attain a given breaking strength for a given count of yarn, better cotton mix would be required if the TPI is reduced.

The approximate TPI factors for different yarns are as given below:

Hosiery yarn – $2.5\sqrt{Ne}$

Warp yarn – $3\sqrt{Ne}$

Weft yarn – $4\sqrt{Ne}$

Knittability is another factor, as significant yarn breakages not only affect the knitting productivity, but also increases number of flags and flaws in knitted fabrics, thereby enhances wastages in cutting process and also may lead to breakage of costly knitting needles, sinkers and jacks. Knittability of yarn depends on its evenness, cleanliness and more so upon its winding on cones.

Winding of hosiery yarn should match with the feeding system on knitting machines. Winding must be neither loose nor hard, just right. It must be rewound to side, matching with the tension and the movement system of knitting machines. It should not be sticky.

Humidity or moisture content normally exceeding 8.5% obstructs its feeding on knitting machines and may call for drying in sun or over heating before using, particularly during monsoon. It is for this reason, that the stockability or storage life of hosiery yarn is very low of about 3 to 4 months, where after unless rewound, yarn becomes difficult to use.

3.4 SELECTION OF FABRICS

The selection of knitted garments is based on the following fabric types such as circular fabrics, flat knitted fabrics and warp knitted fabrics. In circular knitting, one machine can knit a fabric of given diameter or width only and unlike in flat and warp knitting machines, the fabric width cannot be adjusted. In circular knitted fabrics, if proper size diameter cloth is used, side seaming unless required by style of garment is not necessary. Otherwise if side seaming is to be done, corresponding change in cloth diameter is required, depending on depth of side curve and so forth. Normally one or two sizes bigger diameter cloth is required for same width if side seaming and side tapering is to be done.

Basic Machine factors such as diameter, needle gauge, number of feeders, knitting bed (Single knit/Double knit) and take down systems are to be considered while selecting fabrics to get produced in the knitting machines. Though more number of feeders increase the production, the chances of yarn breakages also increased, which necessitates good quality yarns.

Different type of fabric is suitable for different purpose knit wear and for different styles. For instance, low weight fabric is suitable for vests and singlets, but not for outer wears and briefs. Heavy weight Turkish, plush, air tex fabrics are more suitable for outerwear such as sports shirts. Hence, selection of type of knitted fabrics depends upon the ultimate end product and the climatic characteristics on which the garment is intended to be used. Type of garment is another aspect to be considered. If garment piece is knitted just as cardigans, no cutting and sewing is required. Socks, gloves, mufflers, caps are some of the products which do not require cutting and sewing. Of course linking or joining of ends is done.

The following are the different types of circular knitted fabrics:

- Sinker body fine fabrics for outer wear
- Plain fabrics (courser) for under wear fabrics
- Interlock fabrics
- 1 x 1 rib fabrics
- 2 x 2 rib fabrics
- Eyelet or perforated fabrics
- Airtex fabrics with small eye holes
- High pile fabrics
- Plush or fleecy fabrics
- Turkish or towelling fabrics
- Three thread fleecy fabrics
- Garment length fabrics
- Needle and/or cam designed fabrics
- Jacquard and wheel design fabrics

Fabric designs are achieved by manipulating colour of yarn, needles and cams, jacquard or design wheels.

4

Principal Stitches in Weft Knitting

Most of the weft knitted fabric designs are based on the combinations of four basic stitches namely, knit, purl, tuck and float or miss.

4.1 KNIT STITCH (Fig. 4.1)

When the needle is raised sufficiently high by the camming action to obtain the yarn in the hooked portion of the needle and the old loop is below the latch [i.e., the old loop is cleared], a knit stitch will be formed as the needle descends. The face back side of the knit stitch is called the purl stitch.

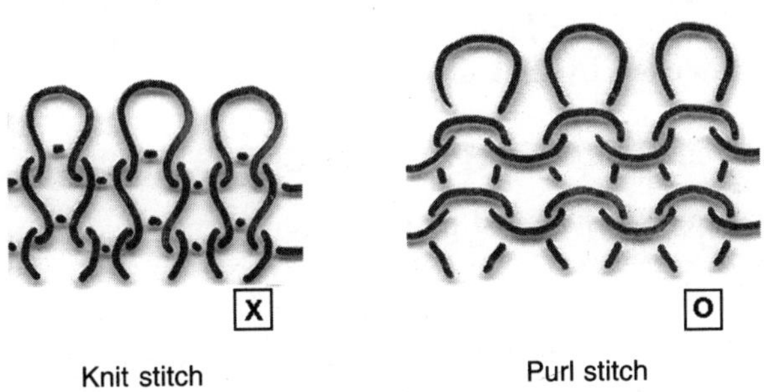

Fig. 4.1

4.2 TUCK STITCH (Fig. 4.2)

When the needle is raised by camming action to obtain yarn in the hook, but is not raised high enough to clear the previously formed loop below the latch, the needle will have two loops in the hook, and a tuck stitch will be formed when it knits at next course. A single needle may tuck at successive courses before being cleared resulting in vertical tuck which extends along a wale for a number of courses. Four course tuck is the maximum limit otherwise loops become tight and yarn breakage would occur. It is also possible to form a horizontal tuck stitch by tucking on adjoining needles.

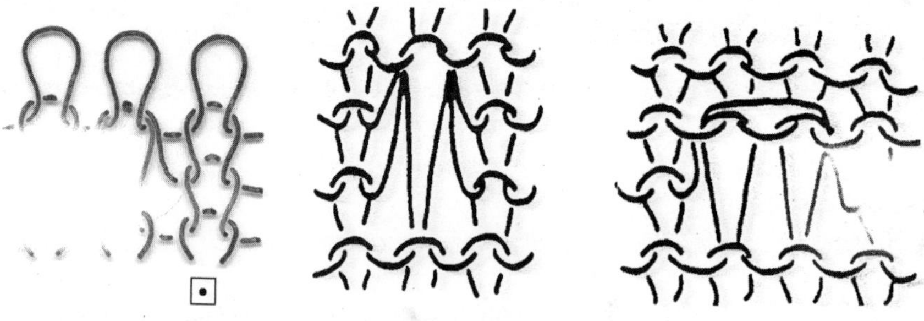

Fig. 4.2

4.3 FLOAT STITCH [MISS STITCH] (Fig. 4.3)

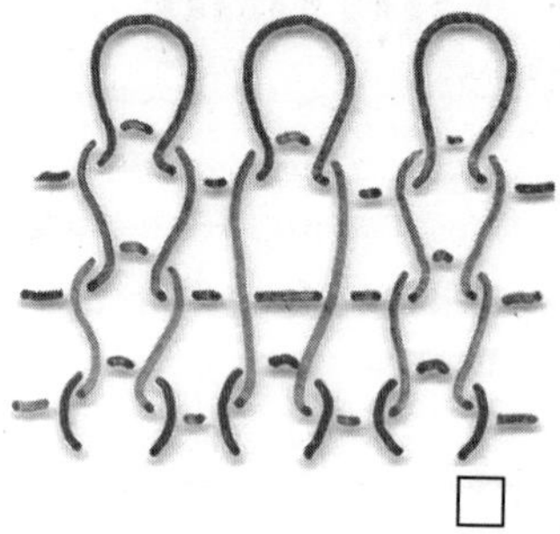

Fig. 4.3

A float stitch occurs only if the yarn is presented to the needle but is not taken by it. Here the needle is not activated upwards to receive the yarn that is presented to it and hence it will retain its old loop in the hook. Long float at back is not desirable, which causes problem of snagging. They are often checked by tuck stitches.

4.4 NEEDLE DISPLACEMENTS FOR DIFFERENT STITCH POSITIONS (Fig. 4.4)

The cams are so designed that they would give the necessary needle movement to knit, tuck, float or purl the yarn. The Figure shows the needle movements to form a knit, a float or a tuck stitch. The tuck stitch needles raises to 'feed position', but does not rise enough to 'clear position'. In float stitch, the needle does not rise even to feed position.

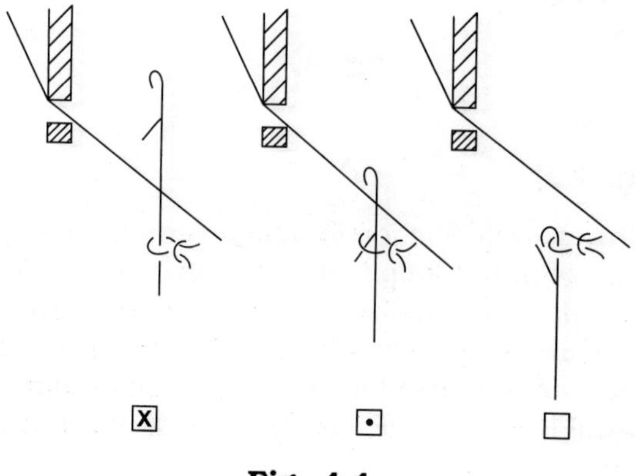

Fig. 4.4

4.5 EFFECT OF TUCK STITCHES

1. Fabric with tuck stitches is thicker than knit stitches due to accumulation of yarn in stitches at tucking places.
2. The structure with tuck stitches is wider than with knit stitches and the loop shape has a wider base at stitches.
3. Tuck stitch structure is less extensible because at every tuck stitch, the loop length is shortened.
4. Due to thicker in nature, the tuck stitched fabric is heavier in weight per unit area than the knit stitches.
5. Tuck stitched structure is more open and porous than the knit stitched fabric.

Tuck stitch is also used to get fancy effects by using coloured yarns.

4.6 EFFECT OF FLOAT STITCHES

It is used for designing, when an unwanted, coloured yarn is to be hidden completely from the surface, by allowing it to float at the back instead of tucking. By floating, uniform texture is obtained and the yarn can be saved.

1. Float stitch makes the fabric thinner than the tuck stitched one, as there is no yarn accumulation.
2. It makes the fabric narrower as there is no looped configuration and hence the whole structure is pulled to minimum width.
3. Less extensible than either knitted or tucked structures.
4. Fabric is lighter in weight due to minimum yarn used in construction.
5. Fabric is flimsy or less rigid compared to others.

5

Basic Structures and Notations in Weft Knitting

5.1 SINGLE JERSEY (Fig. 5.1)

It is a plain single knit structure with face loops on one side and back loops on the other. The structure is produced when all the needles of a single bed machine knit at each feed. This structure can be of technical face or technical back.

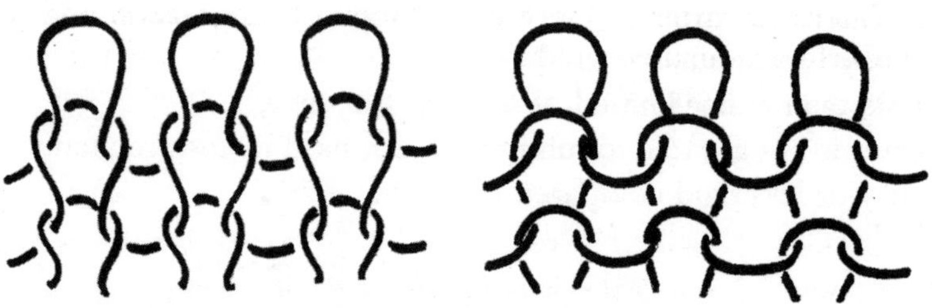

Single Jersey (Technical face) Single Jersey (Technical back)

Fig. 5.1

5.2 RIB (Fig. 5.2)

In rib structure, each courses having one face and one back loop, with succeeding courses identical to the first one, having alternating wales of face and back loops.

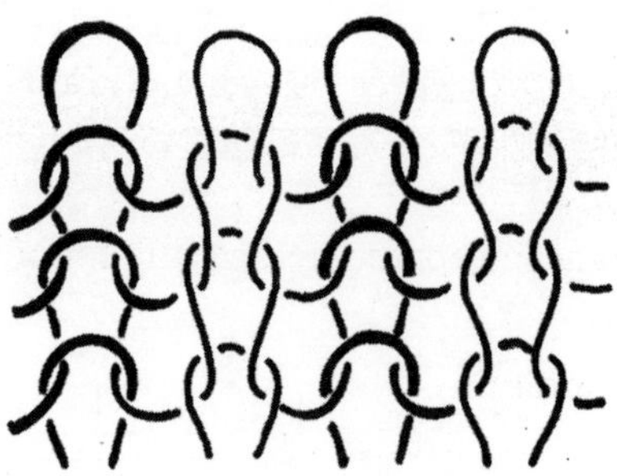

Rib (1 × 1)

Fig. 5.2

5.3 PURL (Fig. 5.3)

In purl structure, each wales having one face and one back loop, with succeeding wales identical to that first one, having alternating courses of face and back loops.

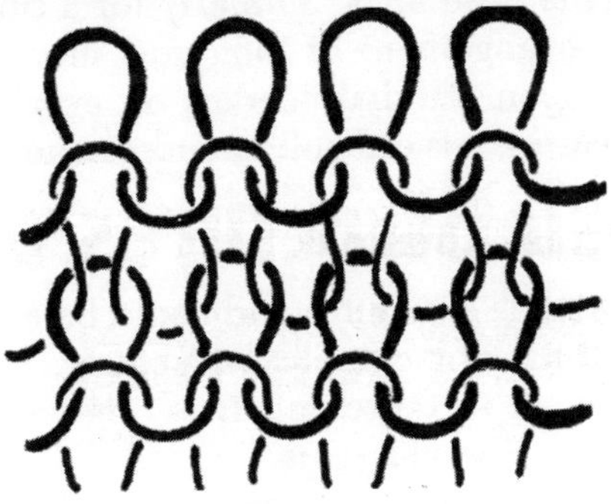

Purl (1x1)

Fig. 5.3

5.4 INTERLOCK (Fig. 5.4)

The structure essentially consists of two separate 1x1 rib fabrics intimately combined into one complete fabric. Interlock structure is made up of rib structures, which are of knitted alternately, first a course of one, then a course of the second so that the two individual structures are intermeshed in single composite structure. Unlike the other structures, a course is completed by knitting with two feeders in interlock knitting.

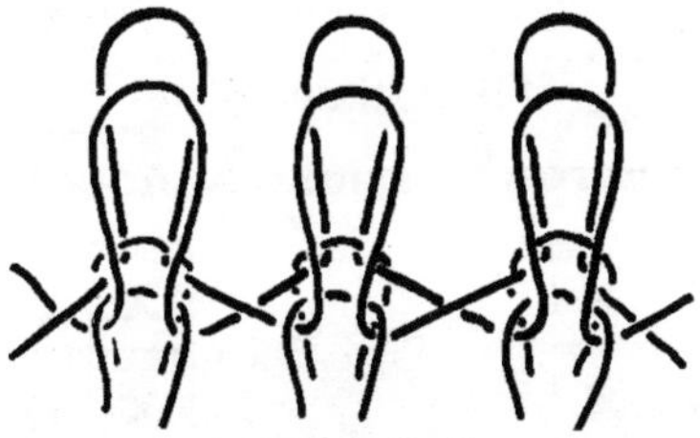

Interlock (1x1)

Fig. 5.4

5.5 NOTATIONS

The representation of weft knitted structures by using a graph paper is very difficult, because even one unit stitch is sometimes under and sometimes over the neighbouring course or wale to form a loop (e.g., tuck stitch).

The weft knitted structures can be represented by the following four methods of notations:

1. Verbal notation
2. Graphic or line diagram notation
3. Symbolic notation and
4. Diagrammatic notation

5.5.1 Verbal Notation

It describes the structure in a definition form and hence it essentially requires the technical knowledge of knitting. For example 1x1 rib structure is represented by consisting of face and back loops disposed in a alternate wale lines. Similarly for a single pique structure, it is to be described with needle layout, arrangement of long and short needles in dial and cylinder, knitting sequence that is tucking on the dial needles on every third feed etc. Unless one is familiar with machine and structures, it is difficult to understand and follow the verbal notation.

5.5.2 Graphic or Line Diagram Notation

This method is only useful for simple structures and could be easily understood by a beginner with a little knowledge of knitting. For complex structures, by means of line diagram and graphic methods, it is very difficult to represent. It is time consuming as well as confusing. Fig. 5.1 to 5.4 are the examples of line diagrams.

5.5.3 Symbolic Notation

Here the knowledge of principle stitches those used for weft knitted structures is essential.

Knit or purl stitches are ordinary loop forming stitches and are formed by needles, as they rise to feed position. A tuck stitch is formed when a needle receives new yarn, but does not "cast off" its old loop.

In float stitch, the needle does not receive the new yarn to form a loop, but instead, the yarn passes behind the previously held loops i.e., it floats on the reverse side of the fabric.

Symbolic notations for the stitches are

Knit stitch	X	Tuck stitch	•
Purl stitch	O	Float stitch	

Representation of basic structures by symbolic notations

X	X	X	X
X	X	X	X
X	X	X	X
X	X	X	X

Single Jersey

X	O	X	O
X	O	X	O
X	O	X	O
X	O	X	O

Rib (1 × 1)

X	X	X	X
O	O	O	O
X	X	X	X
O	O	O	O

Purl (1 × 1)

Feeder 1	Dial.		O		O
	Cyl.	X		X	
Feeder 2	Dial.	O		O	
	Cyl.		X		X

Inter Lock

5.5.4 Diagrammatic Notation

Whenever the face and back loops are involved in the same structure, then two sets of needles, one called the cylinder needle responsible for face loops and the other called the dial needle responsible for back loops are required. A dot represents a needle, either cylinder or dial.

A knit stitch is represented by encircling a dot, the circle facing down.

A purl stitch is represented by encircling a dot, the circle facing up.

A tuck stitch is represented as a half circle over a dot.

A float stitch is represented as line under a dot.

The diagrammatic notations for the stitches are:

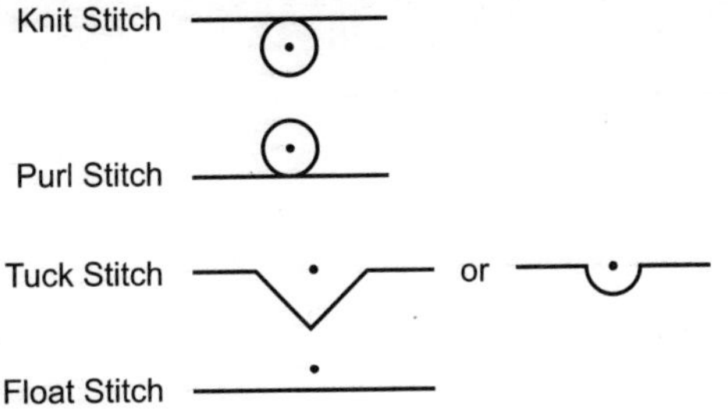

Representation of basic structures by diagrammatic notations

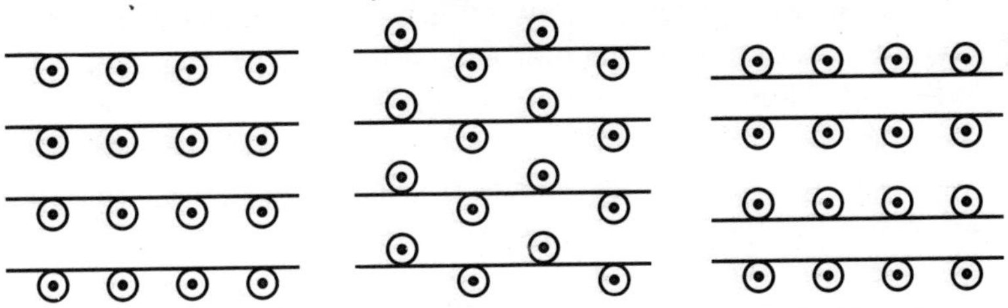

Single Jersy Rib (1 × 1) Purl (1 × 1)

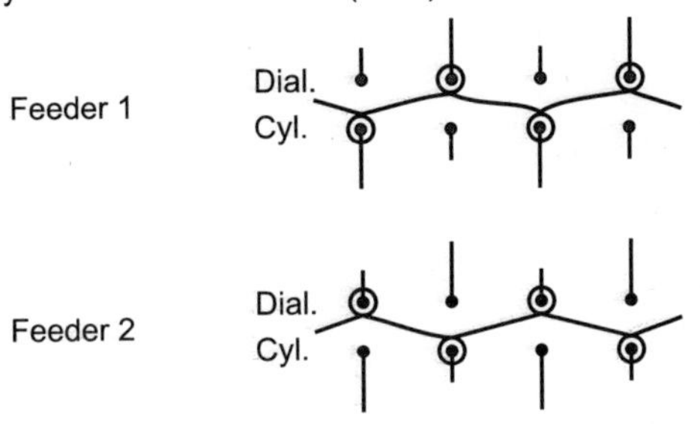

Inter Lock

6

Basic Machines and Fabrics

6.1 SINGLE JERSEY FABRICS

6.1.1 Machine Description

The single jersey fabrics are composed entirely of either face loops or back loops and are non-reversible. The machines are of one set needles with sinker top arrangement. The machines are in a variety of gauge and diameter producing underwear, outerwear plain and fancy fabrics. Machines with finer cut upto 28, diameter of 26 to 30 inches, high productive capacity are available. High productive capacity is due to each needle being capable of independent control allowing many feeders to be placed around the circumference of the machine. Machines having as many as 120 feeders are available. The number of feeds per diametrical inch and the diameter of machine decide the total number of feeders in the machine.

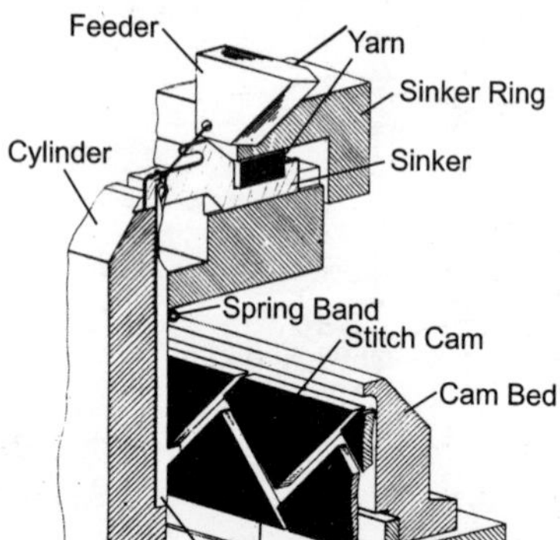

Fig. 6.1

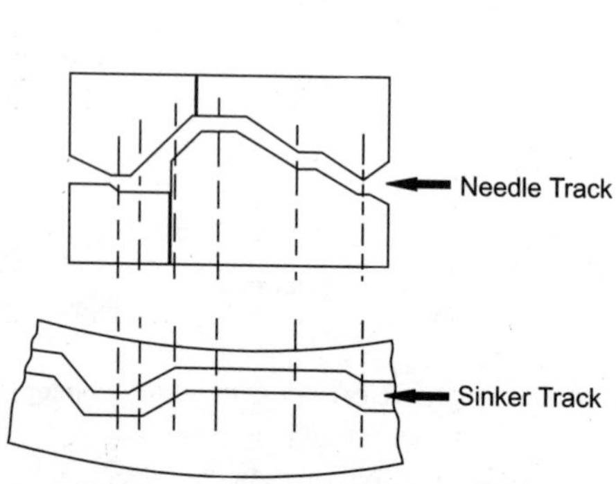

Fig. 6.2

The stitch forming elements of a single jersey circular knitting machine are shown in Fig. (6.1). The machine is functioned with single set of needles which are mounted vertically into the tricks of the cylinder and are operated up and down by the cams. The sinkers move radially in and out and assist the needles for their loop forming action. The sinker movement is by means of a sinker ring which is fixed to the cylinder. Into the track of the sinker ring, the sinker butt is engaged and operated. The sinker movement is always with reference to the needle movement through its running track as shown in Fig. 6.2.

The single jersey circular knitting machines can be of two types based on the rotation of cylinder needle bed and cam box. Most machines are of revolving needle bed type. In the revolving needle bed type, the fabric take-down rotates with the needle beds since the fabric is suspended from the needles. The cams, feeders and creel remain stationary. Of the revolving cam box type, the fabric and needle beds remain stationary, while the creel with yarn packages and guides as well as feeders rotate.

6.1.2 Fabric Characteristics

1. Appearance of face and back are different.
2. Extensibility in widthwise is approximately twice than lengthwise.
3. Curl or roll of fabrics at edges occur.
4. Unraveling of fabric course by course, from either side is possible.
5. Thickness of fabric is approximately, twice the diameter of yarn used.
6. Run (collapse of a wale) will occur if a cut or exposed loop is stressed. The direction of collapse can be either from top to bottom or vice versa.

6.1.3 Derivatives of Single Jersey

The machines described earlier are single cam track machines with only one type of needles. These machines are capable of producing plain single jersey fabrics. In order to derive various structures from the single jersey machines, certain modifications are to be carried out in the sinker body single jersey machines, with respect to their needles and cams.

Fig. 6.3

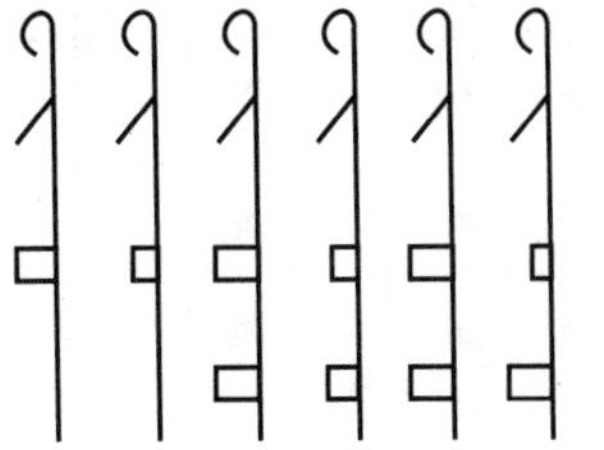

Fig. 6.4

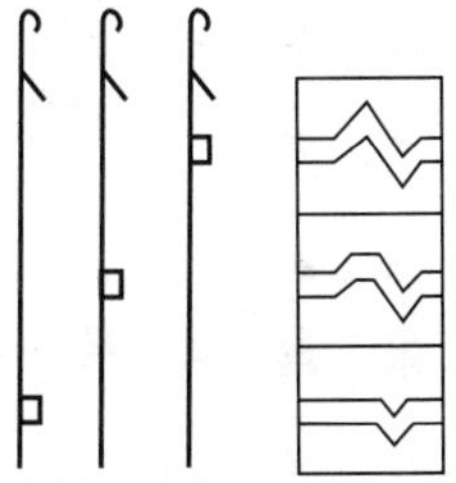

Fig. 6.5

Machines can be equipped with needles having high and low butts with needle selections by swing or split cams at the raising cam places. With such arrangements, the needles can be selected for any two stitches out of the three viz., knit, tuck and miss (Fig. 6.3).

In another type the high and low butts are fitted at different heights in the needle stem. Two cam tracks are needed to operate the needles. (Fig. 6.4).

Machines can be equipped with multicam tracks at the feeder positions and their respective needle types. Machines with three track, four track, five track cams are available depends upon the needle selection complications for different structures (Refer Chapter 8). The needles have three, four or five possible butt positions to get activated corresponding to the above multicam tracks. (Fig. 6.5).

6.1.3.1 La-Coste (Fig. 6.6): This fabric design is formed by knitting two courses on odd needles (say 1, 3, 5, 7 etc.) and tucking on even needles (say 2, 4, 6, 8 etc.). On the next two courses, the condition is reversed. The stitches are not in horizontal alignment. Both the long and short stitches are seen on the face of the fabric. A repeat is completed with two courses and hence a 24 feed machine for example produce 12 courses per revolution. Due to tuck stitches, the fabric is dimensionally stable with excellent comfort and drape. It is more suitable for sports wear.

•	×
•	×
×	•
×	•

Fig. 6.6

×	•	×	•
•	×	•	×
×	•	×	•
•	×	•	×

Fig. 6.7

	×
×	

Fig. 6.8

×	×	×	•
×	×	•	×
×	•	×	×
•	×	×	×

Fig. 6.9

×	×	×	•
×	×	×	•
×	×	×	•
×	×	•	×
×	×	•	×
×	×	•	×
×	•	×	×
×	•	×	×
×	•	×	×
•	×	×	×
•	×	×	×
•	×	×	×

Fig. 6.10

•	×	×	×
•	×	×	×
×	×	×	×
×	×	•	×
×	×	•	×
×	×	×	×

Fig. 6.11

×	×	×	×
×			
×			
×			
×			

Fig. 6.12

•	
×	×
	•
×	×

Fig. 6.13

			•
		•	
	•		
•			
×	×	×	×

Fig. 6.14

•	•	×	×
×	×	•	•

Fig. 6.15

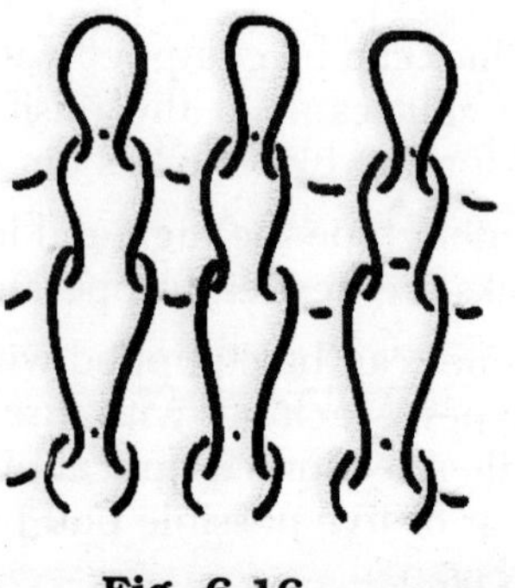

Fig. 6.16

6.1.3.2 Cross tuck (Fig. 6.7): The simplified version of lactose is cross-tuck in which knitting (with odd needles) and tucking (with even needles) takes place at every alternate course unlike every two alternate courses of lacoste. Here also the loops of the same course are not found horizontal.

6.1.3.3 Satin (Fig. 6.8): Unlike cross-tuck, here the knitting and floating takes place at every alternate courses on every alternate needles. The miss stitches impart greater stability than the cross-tuck. The structure is narrower than the cross-tuck and plain jersey. Only one course is produced per repeat and hence the production rate is reduced into half of the feeders.

6.1.3.4 Knitted twill (Fig. 6.9): These fabrics are designed with diagonal lines of tuck stitches on a plain jersey base. It is a four feed repeat produced with four cam tracks.

6.1.3.5 Knop fabrics (Honey Comb) (Fig. 6.10): These structures are characterised by the distribution of multiple tucks diagonally, or in staggered form, throughout the fabric. More of tuck stitches causes the fabric to pucker, and give a honey combed effect.

6.1.3.6 Popcorn (Fig. 6.11): The fabric is not balanced and a puckered effect is created. In a repeat, odd wales completed with 4 courses and even wales with 6 courses. The multiple tucks produce elongated and inclined stitches, which cause curved effect in the fabrics.

6.1.3.7 Jersey Blister (Fig. 6.12): More of miss stitches is the characteristic feature of this structure. The repeat is produced with five courses in wale 1 and one course in wales 2, 3, 4 and hence the fabric is not balanced. These small stitches results in a blistered and puckered effect to the fabric.

6.1.3.8. French terry (Fig. 6.13): This structure is produced by laying-in the yarns at feeder 2 and 4, which undergo brushing process to raise the surface fibres. This gives fleece effect to the fabric. However, as the laying-in frequency is less, it does not create thick fleece.

6.1.3.9 Thick fleece (Fig. 6.14): By increasing the laying-in frequency considerably the structure produce fabrics having thicker fleece. The basic is heavier, thicker and more expensive. The production rate is slower than the french terry.

6.1.3.10 Ribbed effect (Fig. 6.15): This structure is formed by incorporating more than one tuck stitches laterally. The fabric is produced by altering the needle set out to two long needles following two short needles. Ribbed effect of 3x3, 4x4, 4x2 etc., can also be produced.

6.1.3.11 Plaiting: Plaiting is accomplished by incorporating two different yarns (varied by their colour, fibre type, fineness, filament/spun, soft/hard) simultaneously into the knitted fabrics. One yarn can be displayed on the technical face side while the other on the technical back side, giving different effects on either sides. Special yarn carriers enable the yarns to pass into different needles in the same relative position.

6.1.3.12. Seer sucker effect (Fig. 6.16): The seer sucker effect is achieved by knitting one yarn with a larger stitch than the others. The extra length of fabric produced by these longer stitches causes crinkle or seer sucker effect.

6.1.3.13. Plush (loop pile): The plush fabric is obtained by simultaneously knitting two ends of yarn at each feed. One yarn forms normal stitches which appear on the technical face of the fabrics and form the ground structure. The second yarn forms a stitch with extended sinker loops. These sinker loops form the pile which appears on the reverse side of the fabric. The yarn guide for this purpose has two holes, the lower hole feed the ground yarn and the upper one feed the plush yarn. To produce plush fabrics, special sinkers of different neb heights are used. In much of the plush fabrics, textured continuous filaments are used as ground yarns and spun fibrous yarns as pile yarns. The extended loops at the back can be cut and the resultant fabric is known as Velour. If the front of this velour fabric is brushed, the fabric will have fleece on both sides producing a fairly thick fleece.

6.2 RIB FABRICS

As defined earlier, in weft knitting, alternate wales will be having face and back loops. One vertical row of wales is meshed in the opposite direction to the other vertical row of wales. Face row or loops tends to close up in one plane and so also the back row of loops in the other plane. Thus stitches of rib fabrics lie in 2 planes and hence the rib structure is also known as Double Jersey Structure. Due to their exceptional elasticity and greater weight, they find many outlets in outer wear garments production of sweaters and trimmings like cuffs, waist bands and collars.

6.2.1 Machine Description

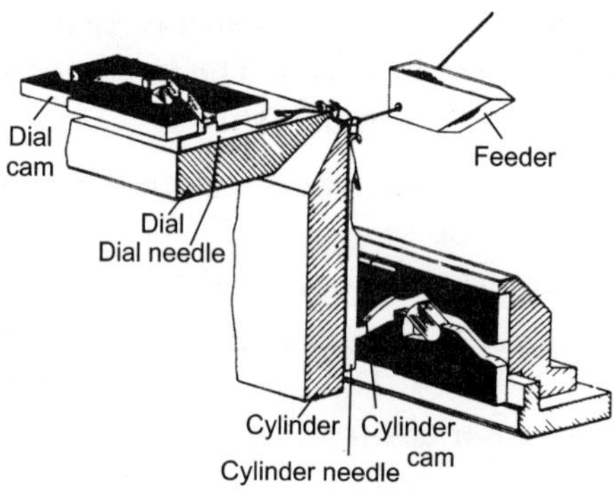

Fig. 6.17

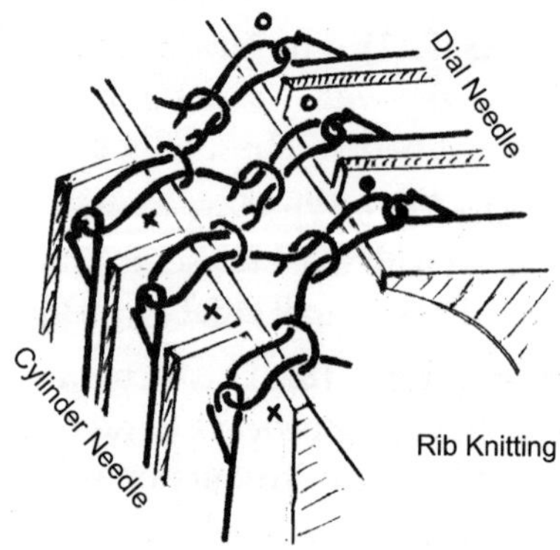

Fig. 6.18

The stitch forming elements of circular rib knitting machines are shown in Fig. 6.17 and Fig. 6.18. Circular rib knitting machines are most common with 14 to 30" dia and 10 to 12 cut. The commonly produced structures are 1×1 and 2×2. Because of two needles sets to produce the fabrics, the dial needles and cylinder needles are provided. Dial is housing for horizontally or radially moving needles and cylinder for vertical set of needles. Dial is a steel disc in which tricks are radiating from center and they should equal to the number of tricks in cylinder. Sinkers are not used in rib machines. The cylinder and dial needles are positioned alternately and the early outward movement of dial needles serves to some extent as holding down sinkers. This type of alternate positioning of cylinder and dial needles is called "rib" gaiting. Each dial and cylinder has cam tracks containing rising cams, stitch cams, swing cams. These cams occupy more space and hence rib knitting machines have less number of feeders such as 8, 12, 16, 24 etc.

Rib knitting machines are also of two types, namely stationary head and rotating head. A modern trend is towards stationary head in which cylinder is driven directly by cylinder gear ring, dial is rotated by a large gear and the take up also rotates. In rotating head the cylinder cam and dial cam will rotate and the dial cam shaft gets its motion from cylinder cam ring. Thus the relationships of dial and cylinder cam timings are properly maintained.

A suitable yarn count for a particular "cut" of the machine can be found out approximately as,

$$\text{Cotton count} = \frac{(\text{Cut})^2}{6}$$

6.2.1.1 Needle timings (Fig. 6.19): In circular rib knitting, the knock over of both the cylinder and dial needles may be simultaneous (synchronized timing) or the cylinder needles may knock over prior to the dial needles (delayed timing). The timing adjustments can be made by changing the stitch cam profile for cylinder needles.

In synchronized timing, both dial and cylinder needles move down their stitch cams, the cylinder needle first receive the new yarns and immediately the dial. They both complete their knock over action simultaneously and return to their start position to release the tension on the the loops remaining on the needles. Synchronized timing can be used on all machines and for all knitted fabric types.

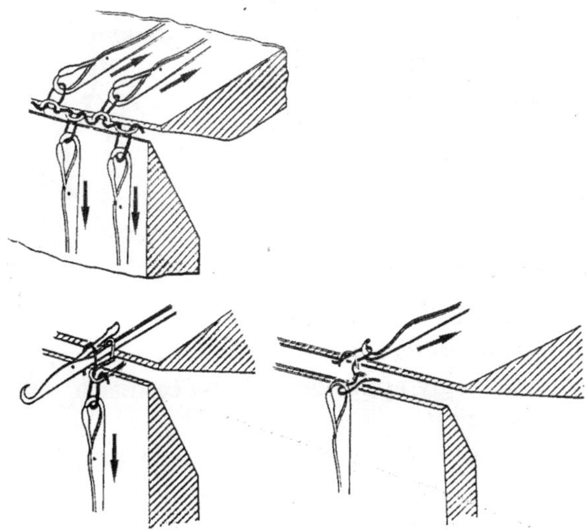

Fig. 6.19

The synchronized timing is also known as point, jacquard or two by two timings. In delayed timing, the dial knock over occurs later after the cylinder needles have drawn loops and are rising slightly to relieve the strain. The dial loops are thus little extended by robbing yarns from the cylinder loops, with less fabric strain. Knitting with dial timing results in fabrics with a more even loop appearance as compared to the synchronized timing. The delayed timing is also known as rib or interlock timing.

6.2.1.2 Operation cycle for rib knitting (Fig. 6.20): Fig. *a.* shows the position when the hooks of the cylinder and dial needles hold the old loops.

Fig. *b.* shows that the dial and cylinder needles move forward and upward respectively for the clearing heights. The old loops are on the needle stems.

At Fig. *c.* the supply yarn approaches the needles. Old loops have slided down the latches on the stem of the needles.

Fig. *d.* shows, that the cylinder needle is coming down and dial needle is retracting towards the center, positioning the feed yarn into the hooked portion of the needles of cylinder and dial.

Fig. *e.* shows that the cylinder and dial needles reach the knitting position.

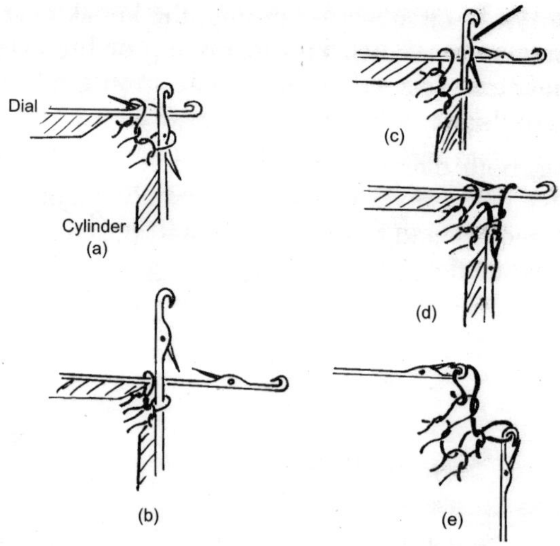

Fig. 6.20

6.2.2 Fabric Characteristics

1. The appearance of face and back are identical.
2. Fabric widthwise extensibility is approximately twice that of single jersey. Fabric lengthwise extensibility approximately twice that of single jersey.
3. Fabric does not curl at edges.
4. Fabric thickness is approximately twice that of single jersey.
5. A run will develop if the exposed loop is cut and the direction of collapse will be from the top to bottom only.

6.2.3 Derivatives of Rib Structure

Rib machines have little patterning scope and are used mainly for producing underwear in 1 × 1 rib or derivative structures. The derivatives are obtained by putting needles out of action or by introducing tuck stitches.

By making some of the inbetween needles inactive or by removing them from the machines, the following structures can be produced.

6.2.3.1 2 x 2 Rib (Fig. 6.21): This structure can be produced by removing out every third needle from the cylinder and dial in such a way that the empty slot in dial and cylinder is positioned between two cylinder and two dial needles respectively. This reversible fabric has vertical rib design effect, which is slightly wider ribbed effect than 1 × 1 ribs. There are two wales of knit stitches (cylinder loops) alternating with two wales of purl stitches (dial loops).

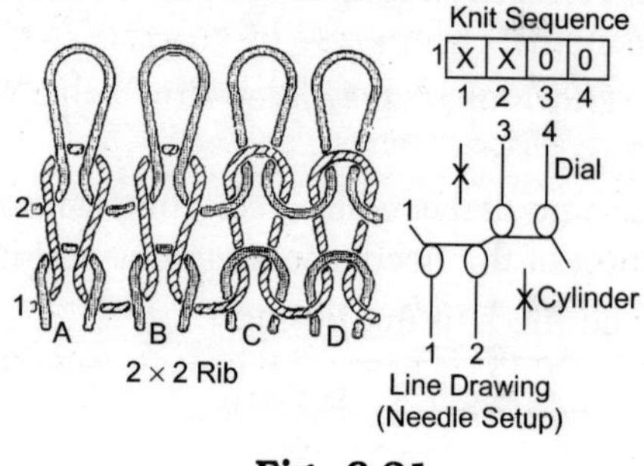

Fig. 6.21

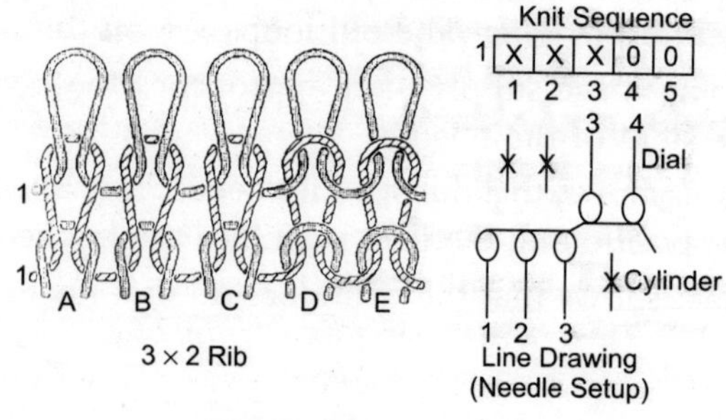

Fig. 6.22

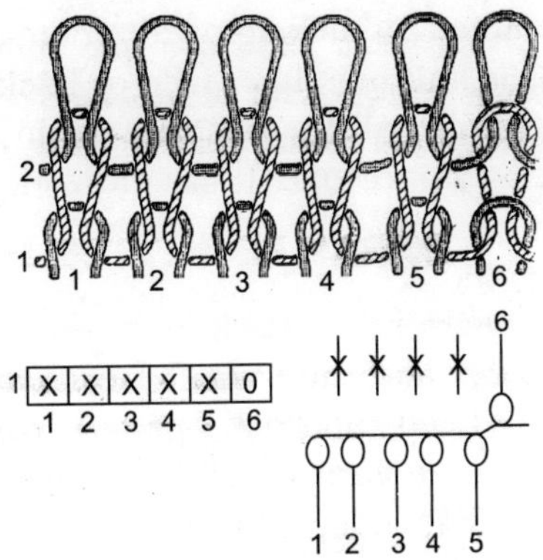

Fig. 6.23

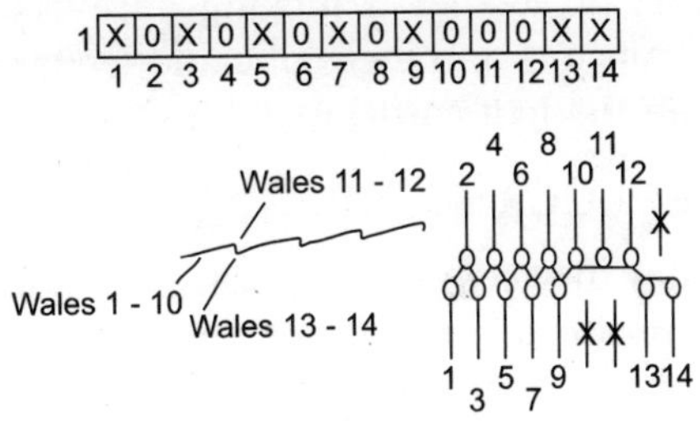

Fig. 6.24

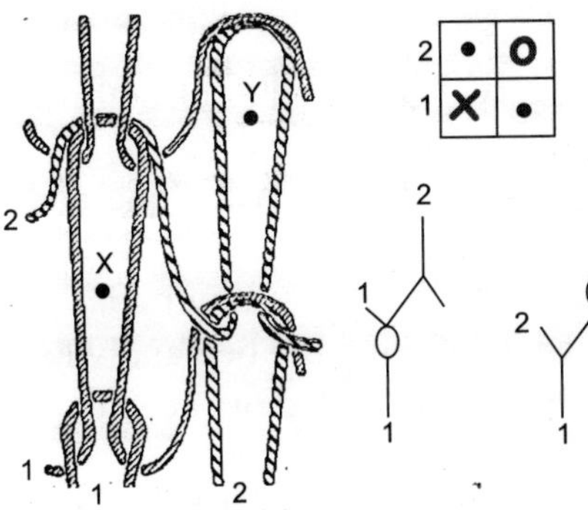

Fig. 6.25

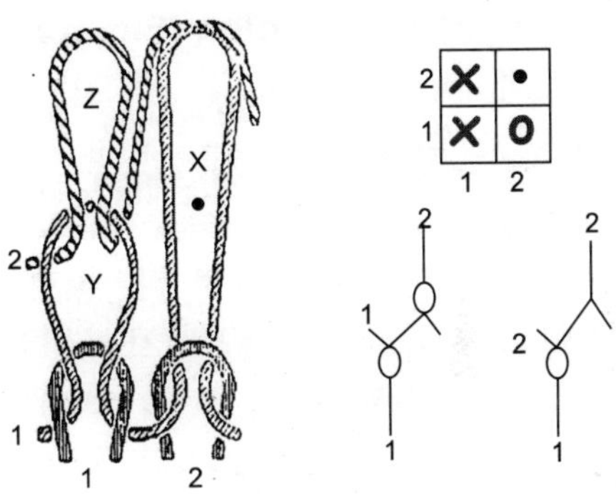

Fig. 6.26

6.2.3.2 3 x 2 Rib (Fig. 6.22): This structure consists of three wales of knit stitches alternating with two wales of purl stitches. This broad rib is non-reversible. The repeat is five wales by one course.

6.2.3.3 5 x 1 Derby rib (Fig. 6.23): There are five wales of knit stitches alternating with one wale of purl stitches. This broad rib is non reversible. The repeat is six wales by one course. This 5×1 rib has less stretch and greater shape retention than the 2×2 rib.

6.2.3.4 Pleated fabrics (Fig. 6.24): There are an infinite number of pleated fabrics that can be produced. The principle of pleated fabric production is that the needles out of action cause the fabric to curl in certain directions in specific locations with in the fabric.

In a repeat of 14 wales and one course, for example, the first ten needles of both the cylinder and dial knit, next two needles of cylinder out of action but the corresponding two dial needles knit. Next two dial needles are out of action, but the corresponding two cylinder needles knit and so on. Wherever the needles are out of action, the fabric curls in opposite direction of wales and forms pleats.

6.2.3.5 Royal rib (Half Cardigan) (Fig. 6.25): By incorporating the tuck stitches in the 1×1 rib pattern the royal rib can be produced. All the needles are in action. The structure is developed by knitting face wales on one set of needles and by tucking on alternate courses on back wales *i.e.*, feeder 1 knits one and purls one and feeder 2 knits one and tucks one. Due to tuck stitches, the fabric is wider and thicker having special design effect with elongated stitches.

6.2.3.6 Polka rib (Full Cardigan) (Fig. 6.26): This structure is obtained by tucking on alternate courses in both face and back wales i.e., when one set of needles is knitting either face or back loops, the other set of needles is tucking on every course. Here also all the needles are in action. Feeder 1, knits one and tucks one and feeder 2, tucks one and purls one. It has more elongated stitches than the half cardigan.

6.3 INTERLOCK FABRICS

Interlock has the technical face of plain fabric on both sides and the wales on each side are exactly opposite to each other and are locked together. Each interlock "pattern row" *i.e.* interlock course requires two feeder courses each with separate yarn which knits on separate alternative needles producing two half gauge 1 × 1 rib courses. Because of good drape and dimensional stability of interlock structures, they have considerable application in the field of outer wear.

6.3.1 Machine Description (Fig. 6.27)

- The cylinder and dial needles are of interlock gating i.e., The needles of both the beds should be opposite to each other.
- Both the cylinder and dial beds should have two types of needles viz., long and short needles.
- Alternating placement of long and short needles in both the beds is required.
- The long needle of one bed should face the short needle of the other bed and vice versa.
- Each bed should have two cam systems to control the long and short needles of their own.
- Minimum of two yarns are required to knit one interlock course and hence a minimum of two feeder supply.
- Diagonal and not opposite needles in each bed knit together.
- The knitting style is in such a manner that only long needles of dial and cylinder will knit with the first feeder and only short needles of dial and cylinder will knit with second feeder.

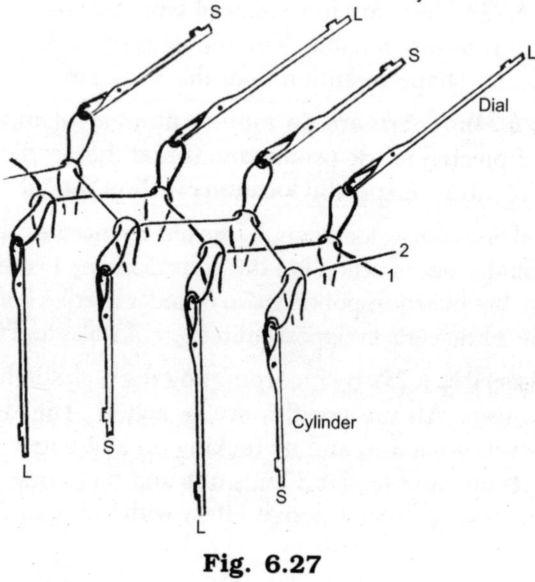

Fig. 6.27

The cut of an interlock machine may vary from 10 to 24 needles per inch. An approximate machine count is given by,

$$\text{Cotton count} = \frac{(\text{Cut})^2}{9.6}$$

For a true course of interlock structure, two feeders are needed and one feed knits only on long needles of dial and cylinder and the other feed knits only on short needles of dial and cylinder. It is noted that when the long needles in the dial and cylinder are activated by the cam tracks to knit, the short needles in the dial and cylinder are held in their rest position by the raceway and hence they are inactive. At the next feed, short needles in the dial will be caused to knit with the short needles in cylinder while long needles remain at the rest position. Thus, during the performance of one set of needles, either long or short, the other set of needle butts ride in non-knitting track.

The yarn carrier of an interlock machine should be set as close as possible to, but without touching the cylinder needles. This is an important setting. The interlock machines require great precision and a small gap variation between cylinder and dial will create horizontal lines in the fabric. Interlock machines are more susceptible to temperature changes than other circular knitting machines.

6.3.1.1 Cam system in interlock machine (Fig. 6.28):

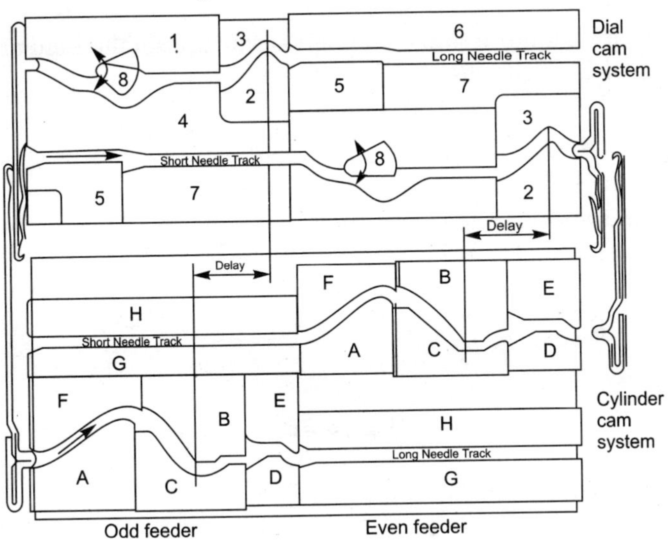

Fig. 6.28

The Fig. 6.28 shows the dial and cylinder cam arrangement to produce the basic interlock structure of one course with two feeders. The dial cam is shown with swing cams for both long and short needles to produce different stitches. The cylinder cam system consists clearing cams (A), guard cams (E,F), stitch cams (B, C) up throw cam (D) and running cam (G, H) for both long and short needles. Similarly the dial cam system consists rising cams (1), dial knock over cams (2,3), guard cams (4), running cams (6,7) and swing type clearing cams for both the long and short dial needles.

6.3.2 Fabric Characteristics

- The appearance of back and face are same.
- Widthwise and lengthwise elongations are approximately the same as single jersey.
- The fabric does not curl at the edges.
- The fabric can be unraveled from the end last knitted.
- Two yarns must be removed to unravel a complete knitted course.
- Fabric thickness is approximately twice that of single jersey.

6.3.3 Derivatives of Interlock

6.3.3.1 Colour ornamentation of interlock structures

(*i*) Vertical Coloured Stripes alternating on one wale is obtained by placing the coloured yarn feeders on the machine in the order of odd and even feeders.

(*ii*) Sharp Horizontal Lines can be obtained by feeding the machine in the order of two feeders of one colour and two feeders of other colour alternately.

(*iii*) Wide vertical stripe can be obtained by removing some short cylinder needles in one section and long dial needles in the next section.

6.3.3.2 Eight lock: (Fig. 6.29): It is an important derivative of interlock. This structure is identical to interlock except that two 2 × 2 rib structures are intermeshed instead of two 1 × 1 rib intermeshed in an interlock structure.

The eight lock structure is a more stable and tighter fabric than the regular interlock structure and has good amount of elasticity also.

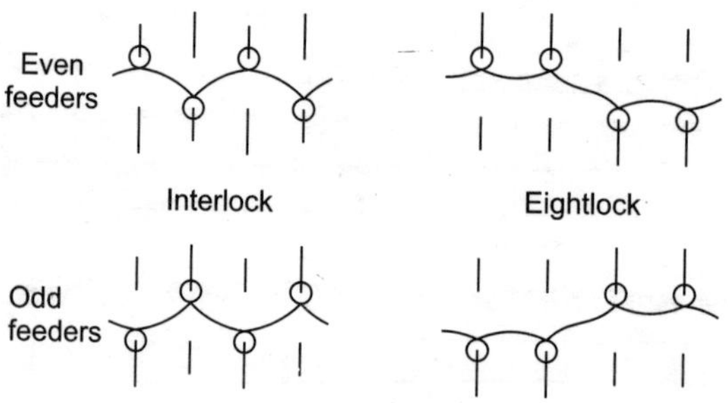

Fig. 6.29

6.4 PURL FABRICS

The purl fabrics are also referred as link-link fabrics. It is reversible, i.e., its appearance on both sides is same. In purl, the loops of one course are intermeshed in one direction and the loops of the next course intermeshed in opposite direction. i.e., the alternate courses having face and back loops.

In purl fabrics lengthwise elasticity is more than widthwise and hence they contract towards the centre in a course wise direction, by their nature of knitting. As the extensibility in lengthwise direction of a fabric is not usually a desirable feature, purl fabrics are produced to a very small extent. Due to its right fitting, the fabric can be conveniently used as garments for sports shirts, sweaters, socks etc.

The main drawback of purl fabric is that the production rate of this type of fabric is very slow compared to other structures.

6.4.1 Machine Description (Fig. 6.30 and Fig 6.31)

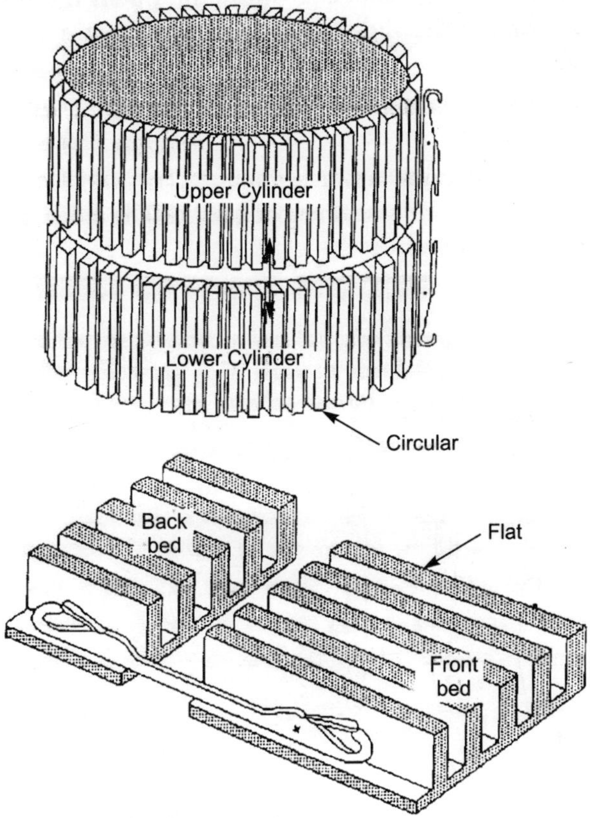

Fig. 6.30

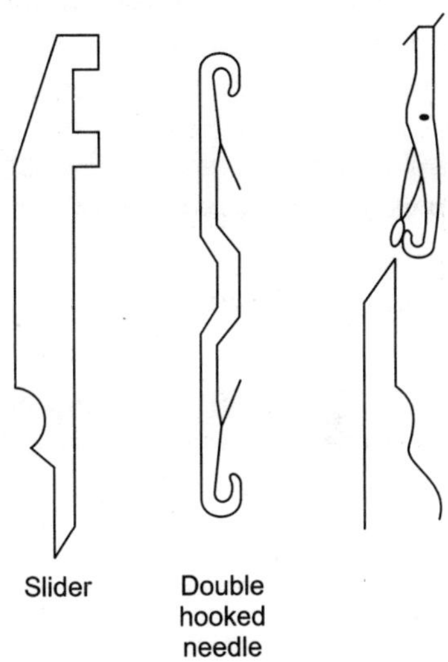

Fig. 6.31

The normal machine gauge is 14 needles per inch. In purl knitting as successive courses are drawn to opposite sides, it is necessary to draw all the stitches towards the front of machine for the first course and towards the back of the machine for the second course. For this it is necessary to have a needle with a hook and latch at each end which may pass through the loops and knit off the other end. To support the needles when they are knitting, it is necessary to have 2 beds in the same plane with needle slots or tricks opposite to each other in direct alignment.

The needles have no butts and are activated by sliders, small sheet metal pieces, which fit in the slot of the needle beds to control the double hooked needle. The butt of the slider fits with the cam track and notch fits over the hook of the needle. There are two sliders for each needle, one in each bed on opposite side to control the movement of needles for knitting and transferring.

The purl knitting machines are basically of flat and circular types. The flat is having two horizontal beds for needle movement and a central gap for fabric formation. The circular type has two cylinders, one above the other and is thus referred as super imposed cylinder machine.

6.4.1.1 Needle transfer principle (Fig. 6.32): The Fig. 6.32 represents two cylinder circular machines in which the tricks of the lower cylinder (I) and upper cylinder (II) are aligned. The double hooked needles (*b*) are controlled by the sliders (*a*) and (*a′*) and are transferred from one cylinder to the other.

In A, the slider (*a′*) pushes the needle (*b*) towards the slider (*a*) whose tip opens the needle latch. The needle latches are opened by the tips of the sliders, for which the latches of these needles are slightly longer than the normal.

In B, the slider (*a′*) continues its descent, controlling the movement of the needle. The other head of the hook lifts the slider (*a*) slightly out from its cylinder (I) trick.

In C, the needle is controlled by both the sliders (*a*) and (*a′*).

In D, a cam (*e*) causes the slider (*a′*) to tilt, thus release the needle (*b*). The needle now continues its descent under the action of the slider (*a*).

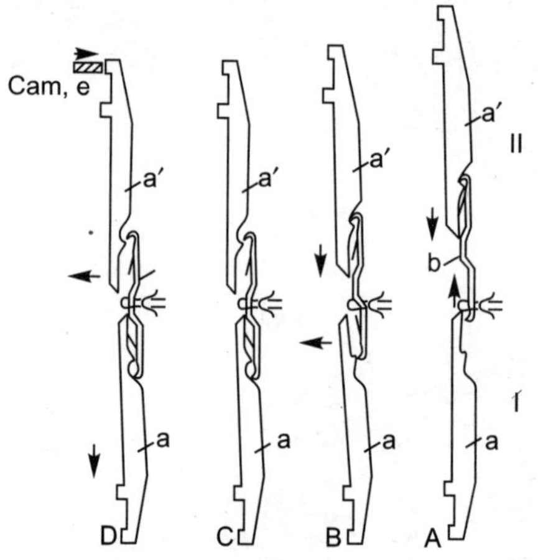

Fig. 6.32

6.4.1.2 Knitting cycle (Fig. 6.33)

The Fig. 6.33 shows the knitting cycle for a two cylinder circular bed purl knitting machine.

At A, after the loop formation (old loop) is completed by the lower hook part of the needle with the upper bed, the slider (*a′*) pushes the needle to the lower bed to produce the next course of loops.

At B, the needle further descends to the lower bed. The lower hook part of the needle comes in contact with slider (*a*), its latch gets opened and is now under the control of slider (*a*).

At C, the cam (*e*) causes the slider (*a′*) to tilt, thus release the needle. The needle is now taken down by the slider (*a*). The upper hook part of the needle catches a new yarn (NY) that is supplied.

At D, the needle further descends with the lower bed, thereby knitting the new yarn through the old loop.

At E, the upper hook part of the needle completes its loop formation with the lower bed. The slide (*a*) pushes the needle to the upper bed to produce the next course of loops.

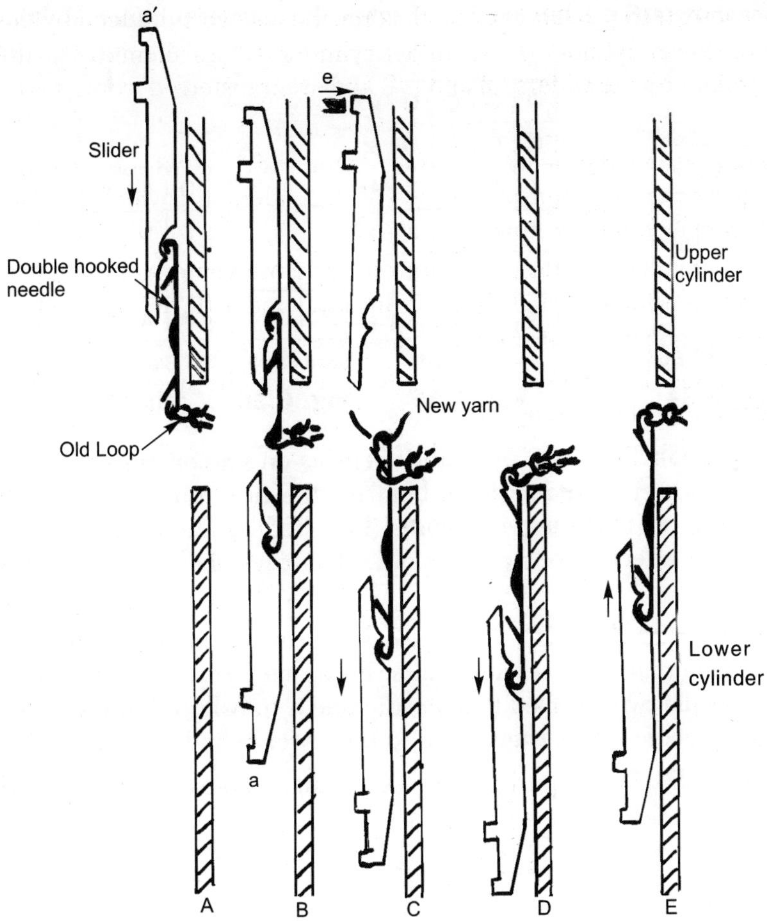

Fig. 6.33 *Purl Knitting Cycle*

6.4.2 Fabric Characteristics

- Fabric has the same appearance in face and back.
- Extension in all directions and lengthwise extension is more.
- The fabric does not curl at the edges.
- The fabric is two or three times thicker than single jersey.
- The fabric will run in the wale direction starting from either end.
- The fabric may be unraveled course by course starting from either end.

6.4.3 Derivatives of Purl Structure

The greatest advantage of purl fabrics is that a three dimensional design effect can be created without the use of tuck or miss stitches.

6.4.3.1 2 x 2 Purl (Fig. 6.34): The 2×2 purl repeats on four courses and one wale. All the needles remain in the upper cylinder (back bed) for the first two feeds produce purl stitches. They are then transferred to the lower cylinder (front bed) for the next two feeds to produce knit stitches.

The courses produced by purl stitches are raised, while the courses produced by the knit stitches are recessed.

X	X	X	X	4
X	X	X	X	3
O	O	O	O	2
O	O	O	O	1

2 x 2 Purl

Fig. 6.34

O	O	O	O	6
O	O	O	O	5
X	X	X	X	4
X	X	X	X	3
X	X	X	X	2
X	X	X	X	1

4 x 2 Purl

Fig. 6.35

O	X
X	O

Seed Stitch

Fig. 6.36

O	O	X	X
O	O	X	X
X	X	O	O
X	X	O	O

Basket Purl

Fig. 6.37

6.4.3.2 4 x 2 Purl (Fig. 6.35): The 4 × 2 purl fabric repeats on six courses and one wale. All the needles remain in the lower cylinder (front bed) for the first four feeds, and produce knit stitches. They are transferred to the upper cylinder (back bed) for the next two feeds, to produce purl stitches. The design effect is four courses of knit stitches alternating with two courses of purl stitches.

6.4.3.3 Seed stitch or Moss stitch (Fig. 6.36): The repeat is two courses by two wales. It consists of face and reverses loops in alternate courses and wales. Feed one forms a knit stitch on needle one and a purl stitch on needle two. At feed two, needle one is transferred to the upper cylinder to form a purl stitch and needle two is transferred to the lower cylinder to form a knit stitch.

6.4.3.4 Basket Purl (Fig. 6.37): Basket purl consists of rectangular areas of all face or all back loops which alternate with each other.

7

Double Knit Structures

7.1 INTRODUCTION

Double knit is the general terminology that is applicable for the fabrics produced with two sets of needles viz., the cylinder needles and dial needles in circular knitting machines.

The following characteristics such as :

(*i*) Knitting on two types of needles viz, long and short

(*ii*) Producing on machines of 16 cut or finer

(*iii*) Incorporating tuck or float stitches in addition to knit stitches are applicable to double knit structures. The fabrics are more suitable for women's outer wear rather men's suitings due to their crease resistant and poor crease retention. These fabrics are generally knitted on 30 inch diameter machines with rib/interlock gating. The machine gauge ranges from 18 to 28. The double knit designs can be produced from non-jacquard and jacquard machines. In addition to the stitch combination, artistic designs can be obtained with color combinations.

The machines are also referred as eight system machines. They are gated for either rib on interlock. Transition from rib to interlock or vice versa is made by displacing the dial half a trick spacing with respect to the cylinder with the help of adjusting screws. The cams of dial and cylinder have raising cams, which may have any of the three positions viz., knit, tuck and miss. The raising cams may be either pivoting or bolt cam type. Both the stitch cams on the cylinder may be fixed together and likewise the stitch cams on the dial.

The eight system machines differ from the interlock machines with respect to the following (Fig. 7.1):

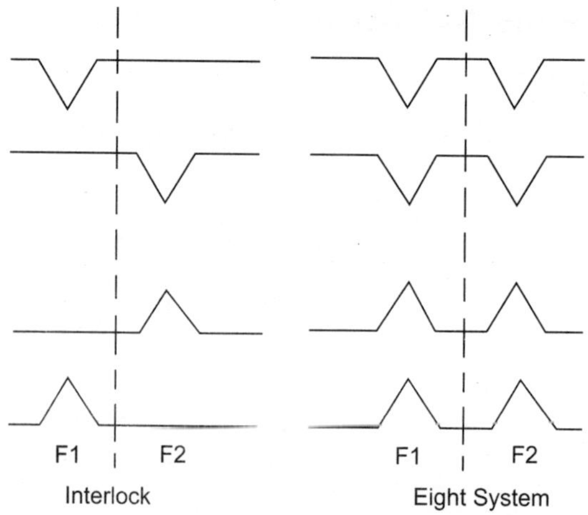

Fig. 7.1

- The eight system machines are gating convertible, i.e., form rib gating to interlocking gating and vice-versa.
- In interlock machines, at any feed it is only possible to work with one type of needle i.e., at feeder 1 with long needles and feeder 2 with short needles.

  Where as on eight system machines, (*a*) with rib gating position, it is possible to activate all the needles at the same feed and (*b*) with interlock gating position, only alternate needles may be cleared at any feed.
- Increased patterning possibilities due to the presence of raising cam in each cam track.

Plain double knits, ripple double knits and flat rib jacquard double knits are the classes of double knits. The plain knits have no visual pattern, but they are introduced by stitch variations. Ripple double knits are generally formed with an excess of face or back loops. They are also called as blister or relief double knits. Flat rib jacquard uses two or more than two colours with a stripe on birds eye backing.

Some popular double knit structures are discussed in the following, with their thread diagrams and needle layout such as rib or interlock gating with long and short needles in dial and cylinder. The structures are represented by their diagrammatic notations. The following points are observed while representing the structures.

(*i*) The rows of the dots are arranged in pair to represent the needle position.

(*ii*) The dots arranged in zig zag manner represent rib gating and dots arranged directly opposite each other represent interlock gating.

(*iii*) To represent top row dial needles and bottom row cylinder needles, the letters D and C are used respectively.

(*iv*) Diagrammatic stitch notations such as continuous line encircling a dot for knit stitch, line passing round the dot without encircling for tuck stitch and straight line without touching the dot for float stitch are used.

(*v*) First pair of rows indicates dial and cylinder needles with long and short needles.

(*vi*) Similar assumption is to be made for the subsequent rows of needles.

7.2 STRUCTURES OF RIB GATING

7.2.1 Milano Rib (Fig. 7.2)

This fabric has good elasticity. It has horizontal striped appearance. The repeat completes on three feeds.

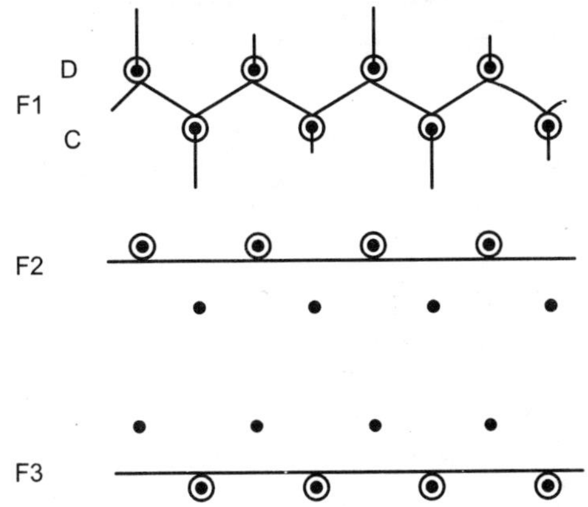

Fig. 7.2

Stitch Pattern

Feed 1 : knit on all dial and cylinder needles.

Feed 2 : knit on all dial needles (only back loops).

Feed 3 : knit only on all cylinder needles (only face loops)

7.2.2 Double Pique

Two varieties of double pique are French pique and Swiss pique. Both these are produced on a 4-feed cycle. Both the structures are almost similar and there is a little difference in knitting cycle. Both are heavy, stable structures and they do not curl. These structures are also termed as drap (or sheet fabrics) because of the poor width wise extensibility.

7.2.2.1 French pique (Fig. 7.3):

Stitch Pattern

Feed 1 : Knit on all (short and long) needles in cylinder and only on short needles in dial.

Feed 2 : Knit only on long needles in dial. (no knitting on cylinder needle).

Feed 3 : Knit on all (short and long) needles in only on long dial needles.

Feed 4 : Knit only on all short needles in dial (no knitting on cylinder needles).

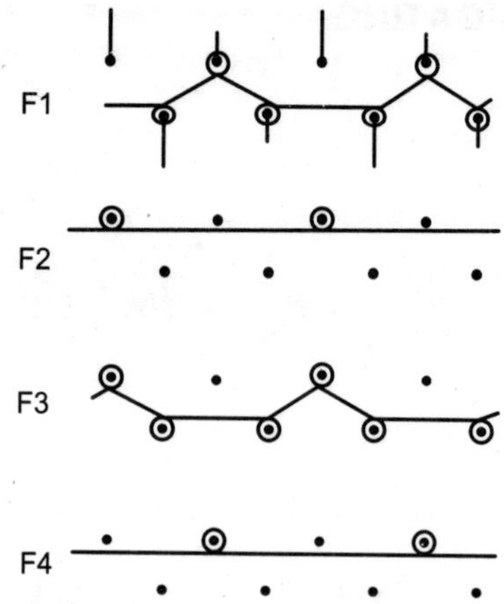

Fig. 7.3

7.2.2.2 Swiss pique (Fig. 7.4): This fabric is produced with smaller and finer loops than French double pique.

Stitch Pattern

Feed 1 : Knit on all long and short needles in cylinder and only on long needles in dial.

Feed 2 : Knit on all long dial needles only.

Feed 3 : Knit on all long and short needles in cylinder and only on short needles in dial.

Feed 4 : Knit only on short needles in dial.

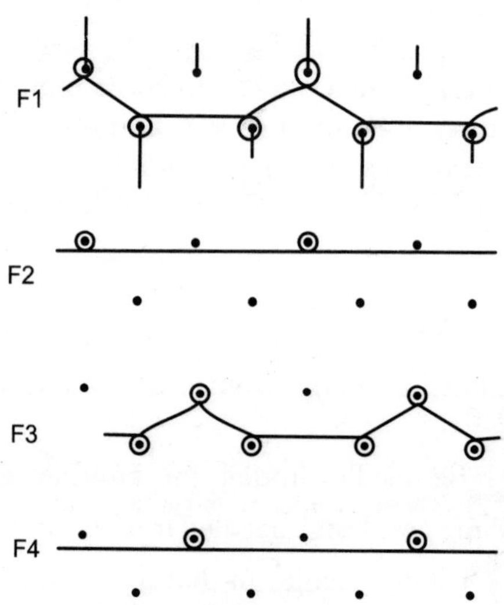

Fig. 7.4

7.2.3 Pique Poplin (Fig. 7.5)

It is a six feed structure with knit and float stitch combination. At feeds 2,3,5 and 6, there is no knitting with cylinder needles.

Stitch Pattern

Feed 1 : Knit all needles of cylinder and only short needles of dial
Feed 2 : Knit only long needles of dial
Feed 3 : Knit only short needles of dial
Feed 4 : Knit all needles of cylinder and only long needle of dial
Feed 5 : Knit only short needles of dial
Feed 6 : Knit only loop needles of dial

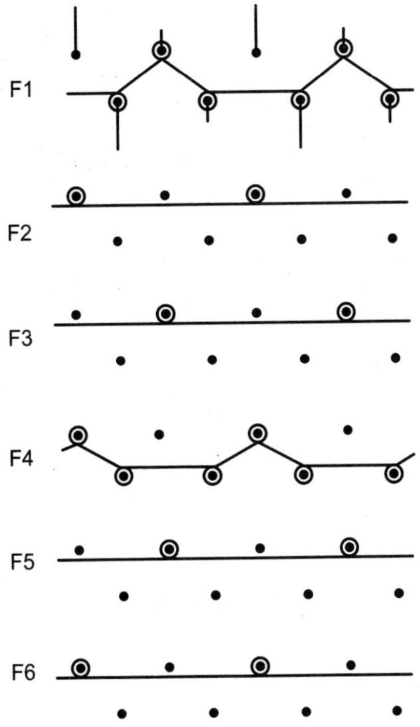

Fig. 7.5

7.2.4 Evermonte (Fig. 7.6)

In this eight feed repeat structure, four courses (1,2,5 and 6) use every other needle on both needle beds. Course 1 and 2 tuck on dial and courses 5 and 6 tuck on cylinder. These tuck stitches cause pearl effect to the fabric.

Stitch Pattern

Feed 1 : Knit all short needles of cylinder and tuck all long needles of dial
Feed 2 : Knit all long cylinder needles and tuck all short dial needles
Feed 3 : Knit all dial needles only

Feed 4 : Knit all cylinder needles only
Feed 5 : Tuck all long cylinder needles and knit all short dial needles
Feed 6 : Tuck all short cylinder needles and knit all long dial needles.
Feed 7 : Knit all cylinder needles only
Feed 8 : Knit dial needles only.

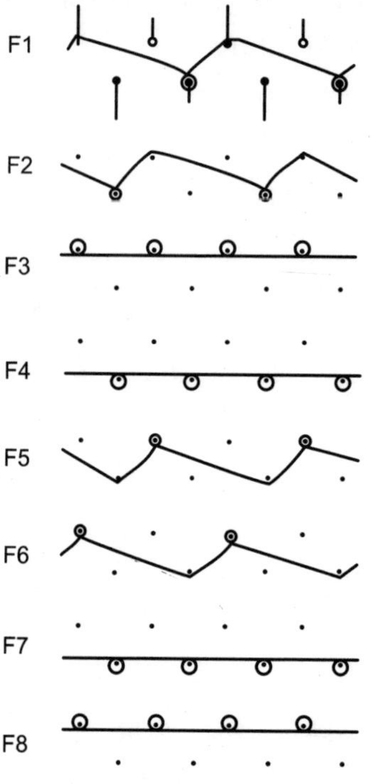

Fig. 7.6

7.3 STRUCTURES OF INTERLOCK GATING

7.3.1 Single Pique (Fig. 7.7)

In this interlock structure, the standard interlock knitting cycle is modified by tucking instead of knitting at every third feed and sixth feed dial needles. It is a six feed repeat or multiples of six. The knit loops at fabric face and tuck loops at fabric back produce a pique effect.

Stitch Pattern

Feed 1 : Knit on all long needles of dial and cylinder
Feed 2 : Knit on all short needles of dial and cylinder
Feed 3 : Tuck on all long dial and knit on all long cylinder needles.
Feed 4 : Knit on all short needles of dial and cylinder.
Feed 5 : Knit on all long needles of dial and cylinder.
Feed 6 : Tuck on all short dial and knit on all short cylinder needles

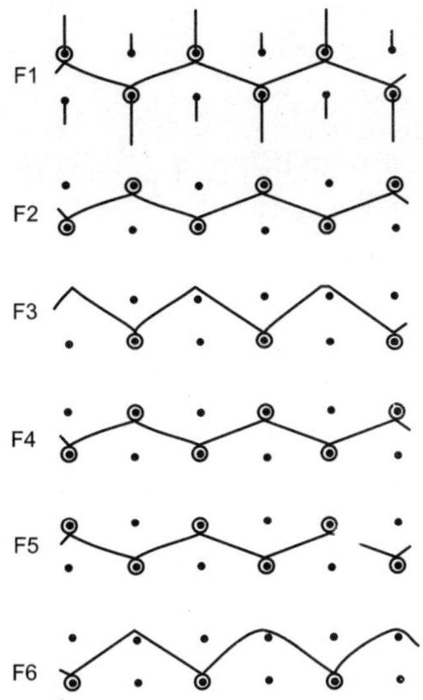

Fig. 7.7

7.3.2 Punto-Di-Roma (Fig. 7.8)

This float stitch interlock structure has a horizontal striped characteristics feature. It needs four feeds to complete two full courses. At feeders one and two a normal interlock knitting occurs while at feeder three only dial needles knit and at feeder four only cylinder needles knit.

Stitch Pattern

Feed 1 : Knit on long needles of dial and cylinder.

Feed 2 : Knit on short needles of dial and cylinder.

Feed 3 : Knit on all long and short dial needles only.

Feed 4 : Knit on all long and short cylinder needles only.

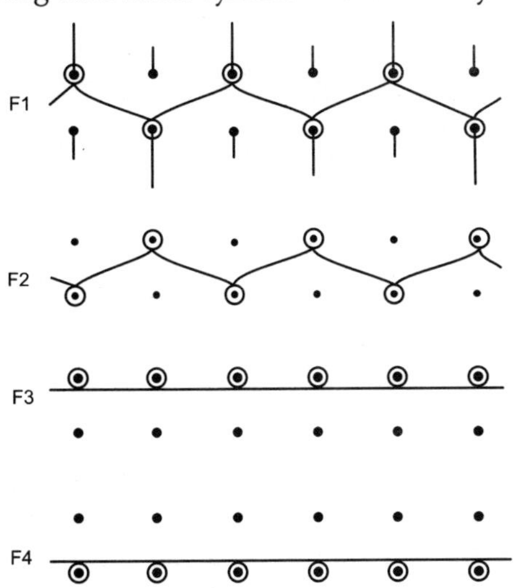

Fig. 7.8

7.3.3 Ottoman Rib (Fig. 7.9)

This type of fabric is obtained with knit and tuck as well as knit and float combination. They have ripple-like appearance on the face side. Tucking takes place for four successive feeds in an eight feed cycle. The tucking effect bunches and pushes the fabric upwards to form a roll like appearance.

Stitch Pattern

Feed 1 : Knit on all long needles of dial and cylinder.

Feed 2 : Knit on all short needles of dial and cylinder.

Feed 3 : Tuck on all long dial needles and knit on all long cylinder needles.

Feed 4 : Tuck on all short needles and knit on all long cylinder needles.

Feed 5 : Tuck on all long dial needles and knit on all short cylinder needles.

Feed 6 : Tuck on all short dial needles and knit on all short cylinder needles.

Feed 7 : Knit on all long needles of dial and cylinder.

Feed 8 : Knit on all short needles of dial and cylinder.

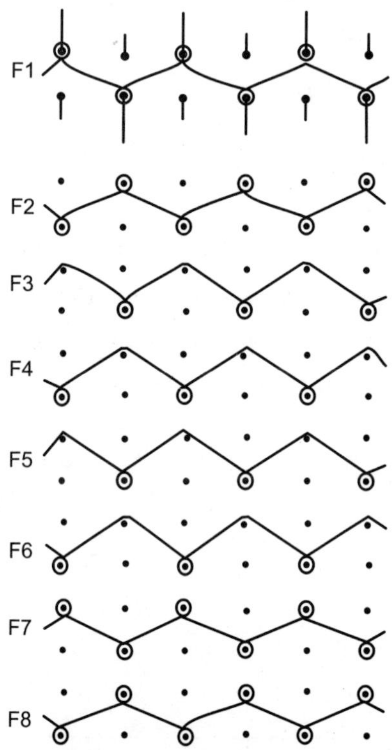

Fig. 7.9

7.3.4 Bourrelet (Fig. 7.10)

This structure can be made by floating and knitting the yarn. In a six feed structure, dial needles remain idle for 4 successive feeds and only cylinder needles knit. Two feeds form an interlock knitting.

Stitch Pattern

Feed 1 : Knit on all long needles of dial and cylinder.

Feed 2 : Knit on all short needles of dial and cylinder.

Feed 3 : Knit on all long cylinder needle only.

Feed 4 : Knit on all short cylinder needles only.

Feed 5 : Knit on all long cylinder needles only.

Feed 6 : Knit on all short cylinder needles only .

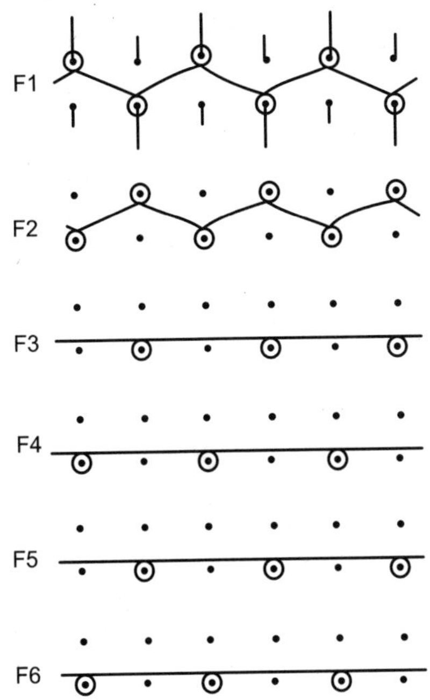

Fig. 7.10

7.3.5 Texi-Pique (Fig. 7.11)

In this interlock structure, working on all the active needles at feeders three and six takes place. This fabric is heavier and more resistant to deformation than the single pique.

Stitch Pattern

Feed 1 : Knit on all long needles of dial and cylinder.

Feed 2 : Knit on all short needles of dial and cylinder.

Feed 3 : Tuck on all long needles of dial and cylinder.

Feed 4 : Knit on all short needles of dial and cylinder.

Feed 5 : Knit on all long needles of dial and cylinder.

Feed 6 : Knit on all short needles of dial and cylinder.

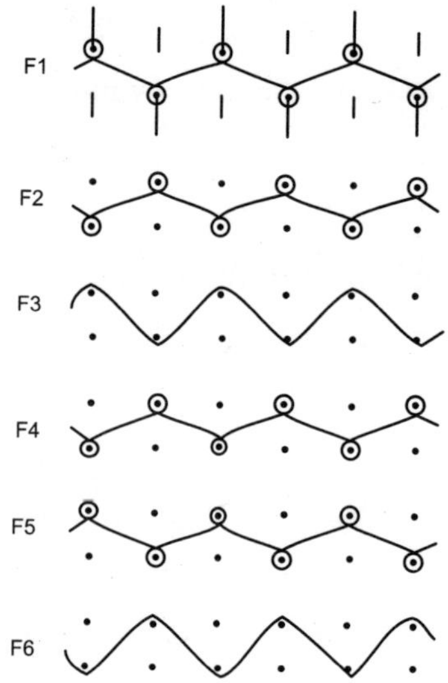

Fig. 7.11

7.3.6 Pin-tuck (Fig. 7.12)

This structure has the repeat on 6 Feeders. Feeders 1 and 4 produce regular interlock whereas, feeders 2,3,5,6 produce tuck interlock.

Stitch Pattern

Feed 1 : Knit on all long needles of dial and cylinder.
Feed 2 : Tuck on all short dial needles, knit on all short cylinder needles.
Feed 3 : Tuck on all long dial needles, and knit on all short cylinder needles.
Feed 4 : Knit on all short needles of dial and cylinder.
Feed 5 : Tuck on all long dial and knit on all long cylinder needles.
Feed 6 : Tuck on all short dial and knit on all short cylinder needles.

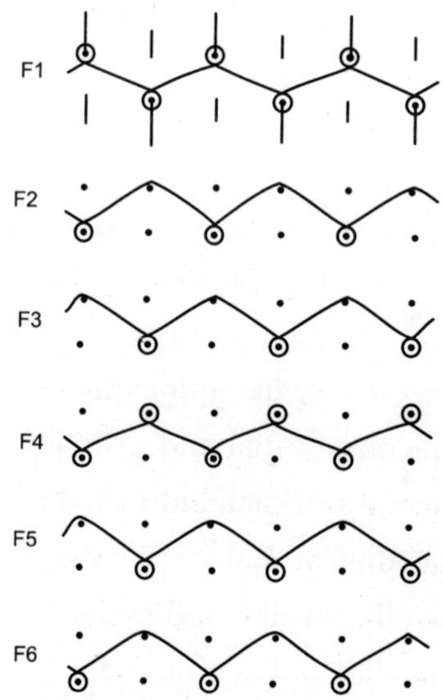

Fig. 7.12

7.3.7 Super Roma (Fig. 7.13)

This eight feed structure is formed with knit and tuck stitch combination. In this eight feed repeat, first four feeders make regular interlock. The stitches formed by the next four courses form in relief on the fabric face without the tuck stitches.

Stitch Pattern

Feed 1 : Knit all the short needles of dial and cylinder.

Feed 2 : Knit all the long needles of dial and cylinder.

Feed 3 : Knit all the short needles of dial and cylinder.

Feed 4 : Knit on all short needles of dial and cylinder.

Feed 5 : Knit all the short needles of cylinder and tuck all the short needles of dial.

Feed 6 : Knit all the long needles of cylinder and tuck all the long needles of dial.

Feed 7 : Knit all the short needles of cylinder and tuck all the short needles of dial.

Feed 8 : Knit all the short needles of cylinder and tuck all the short needles of dial.

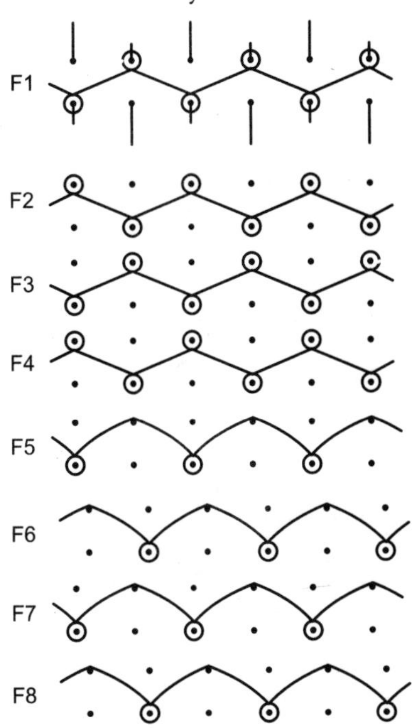

Fig. 7.13

7.4 BLISTER FABRICS

7.4.1 Single blister (3 Miss Blister)(Fig. 7.14)

These are surface textured fabrics with raised effect on its face side. It requires two coloured yarns, one for ground yarns and another for raised surface effect or blister effect. The ground yarn will knit only on selected dial and cylinder needles and blister yarns only with the selected cylinder needles. The ground will knit with odd dial needles and also with cylinder needles which are not going to knit with blister yarns. In a four course cycle, each dial needle knit only once, but every cylinder needle knits at least two times, and hence there are more stitches on face side of fabric, than its back side and hence it is called blister fabric.

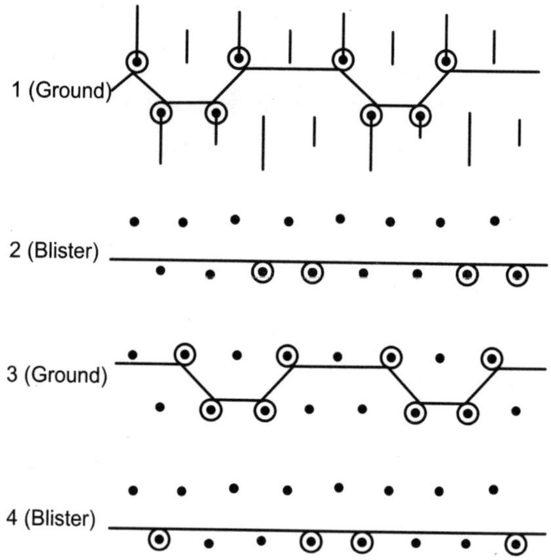

Fig. 7.14

The fabric is also named as three-miss blister fabric, as every needle (dial) is held out action for 3 courses after having formed a new loop either on long or on short dial needle. The raised effect is obtained by knitting more frequently with cylinder needles and missing dial needles.

Stitch Pattern

Feed 1 (Ground yarn) : Knit on selected long and short cylinder needles and long dial needles.

Feed 2 (Blister yarn) : Knit on all other long and short needles in cylinder. No knitting with dial needles.

Feed 3 (Ground yarn) : Knit on all short dial needles and on selected short and long cylinder needles on which the blister yarn will not knit on the next course.

Feed 4 (Blister yarn) : Knit on selected short and long cylinder needles. No knitting on dial needles.

7.4.2 Double Blister (5 Miss blister) (Fig. 7.15)

Here there are 2 blister courses to one ground course and two blister yarns knit in the same manner. After knitting a new loop each dial needle is held out of action for five courses. Hence it is five-miss blister. Each dial needle knits only once but the cylinder needle knits either 2, 3 or 4 times in a 6 feed cycle. At feed 1 all long dial needles form stitches and at feed four all short dial needles form stitches. One or more colours for two blister courses and one colour, for ground can be used.

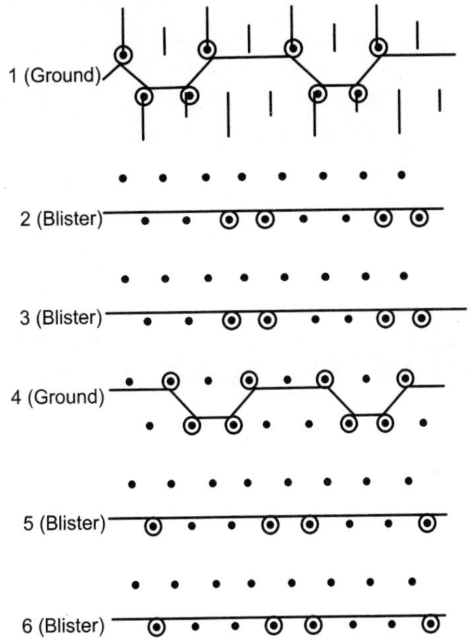

Fig. 7.15

Stitch Pattern

Feed 1(Ground)	:	Knit all long dial needles and selected short and long cylinder needles.
Feed 2(Blister)	:	Knit on remaining long and short cylinder needles. No knitting on dial system.
Feed 3 (Blister)	:	Same knitting as feed 2.
Feed 4 (Ground)	:	Knit all short dial needles and alternate short and long cylinder needles.
Feed 5 (Blister)	:	Knit on remaining long and short needles in cylinder. No knitting on dial needles.
Feed 6 (Blister)	:	Knitting is the same as feed 5.

8

Patterning in Weft Knitting

8.1 INTRODUCTION

In circular weft knitting machines, fabric structural designing or patterning is another interesting section of the industry. Patterning is principally effected by the selection of needles at different feeders. Although a number of methods are available to select the needles for patterning, multicam tracks and pattern wheels are the most popular means of mechanical needle selection, on multifeed circular machines. The jacquard designs produced on circular machines, with either one or two needles beds, generally have restricted length and width dimensions. Machine builders attempted to employ the pattern wheel on circular rib jacquard machines because of its compactness. For a given design of knitted fabrics, the principles of selection of right type of needles and the right choice of cams for multicam track single jersey machines are discussed in this chapter. Jacquard designs of different stitch variations and multicolour combinations are worked out with selected examples.

8.2 SINGLE JERSEY PATTERNING WITH MULTICAM TRACKS

At the place of the rising cams, inorder to lift the needles of different types having different butt positions, for the required stitch variations, multicam tracks are employed. Machines of 3 cam tracks, 4 cam tracks and 5 cam tracks are common for the production of jacquard designs. The possible fabric structure variations are numerous and they permit individual selection of needles for knit, tuck and miss stitches and thereby increase the machine potential.

In the following examples, while selecting needles according to the stitches at different feeders, the following points are to be observed:

(*a*) Similar type of needles are to be observed vertically i.e., needle trick wise. That is any particular trick should accommodate only one type of needle i.e., long or short or medium.

(*b*) While knitting with a feeder, course wise same needle cannot produce two different stitches i.e., any particular type of needle should knit only one type of stitch coursewise.

In the following examples, pattern graphs in the form of point paper representation for different knitted fabric structures or designs are given. In order to produce them on a multifeed circular single jersey machines with knit, tuck and miss stitch combinations, needle set out (selection of needle types) and cam set-up (selection of respective cam track for the needles) are to be worked out. To start with, the point paper design which is in symbolic notation is to be converted into diagrammatic notation (i.e., yarn path diagram of each feeders.)

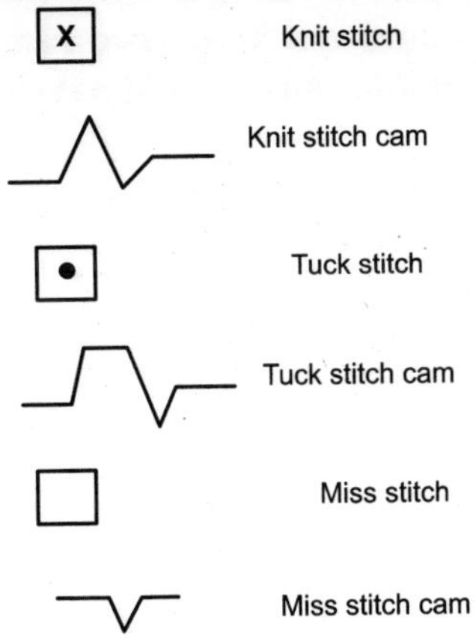

The following symbols are used for their respective terms in the examples

C1, C2, C3 etc ...– Courses
F1, F2, F3 etc ... – Feeders
L – Long Needle
S – Short Needle
M – Medium Needle
M1, M2, M3 ... – Medium1, Medium2, Medium3 needles and so on.

The needle types L, S, M etc., are named based on the butt positions in the needles. Instead, the needle types may also be referred as A,B,C,D etc.

8.2.1 Plain Stitch with Single Feeder (Fig. 8.1)

1. The design consists, all of knit stitches at every feeders.
2. Convert the design into yarn path diagram in the form of diagrammatic notations and select the needles.
3. Since all the stitches are knits, a single needle type is sufficient and hence the single cam track at each feeder.

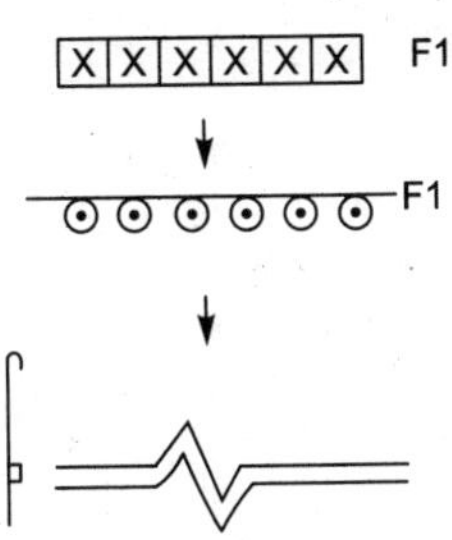

Fig. 8.1 *Plain Stitch with Single Feeder*

8.2.2 Knit and Miss Stitches Having Two Feeder Supply (Fig. 8.2)

1. As the design consists knit and miss stitches at every feeders, single needle type is not sufficient and two different needles are required.

2. From the yarn path diagram, needle selection can be made.
3. By observing the conditions while selecting needles wale wise and course wise, it is found that two needle types are required and hence two cam tracks at each feeder as shown in the Figure. At each feeder position, cams should have two tracks of knit and miss cams to produce the design.

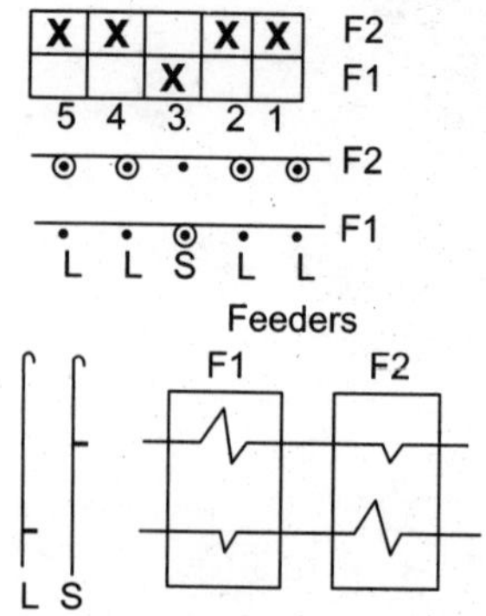

Fig. 8.2 *Knit and Miss Stitches with Two Feeder Supply*

8.2.3 Knit, Tuck and Miss Stitches - Three Feeder Supply (Fig. 8.3)

1. This twill effect is a repeat of three stitch combination at all the 3 feeders.
2. Three needle types are found to be required based on the yarn path diagram.
3. The cam set up is in such a manner at each feeder position, three track cams having knit tuck and miss possibilities are to be incorporated. Therefore in order to produce the above design the machine should have 3 types of needles and 3 track cams at each feeder.

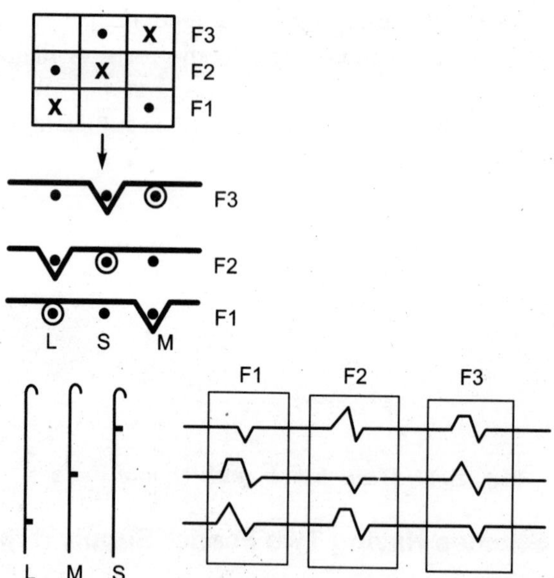

Fig. 8.3 *Knit, Tuch and Miss Stitches with Three Feeder Supply*

8.2.4 Twill Effect with Four Feeder Supply (Fig. 8.4)

1. This twill effect is a four feeder repeat with three stitch combinations.
2. Based on yarn path diagram, the needle selection is made.
3. The cam set-up is in such a manner, at every feeder position, four track cams having all stitch possibilities are incorporated. This design can be produced with the knitting machines working with 4 needle types and 4 track cams accordingly at each feeder positions to select the needles.

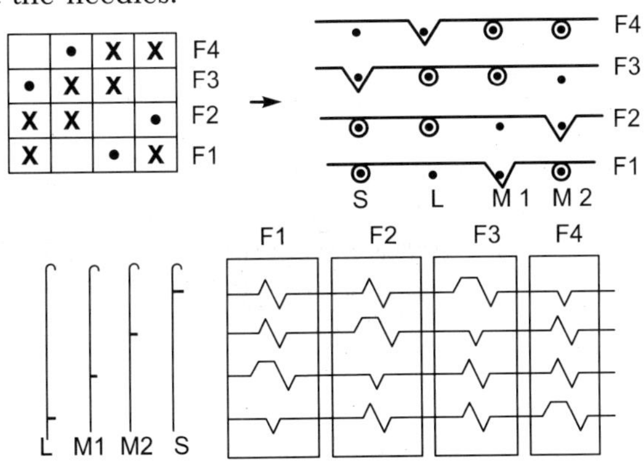

Fig. 8.4 *Twill Effect with four feeder supply*

8.3 COLOURED PATTERNING IN WEFT KNITTING

While going for colour patterns, the feeder requirement is increased according to the number of colours. While working with one colour at a course, wherever the colour is to be incorporated, at those places they are treated as knit stitches and if there is no colour yarn, there they are miss stitches. The following examples give an idea about the selection of needle types, cam set up so that the design can be produced in the knitting machine accordingly.

8.3.1 Two Colour Patterns (Fig. 8.5)

From the given structure, one repeat is selected, which is consisting of two coloured yarns. The number of feeders to produce such repeat is worked out, considering one feeder supply one coloured yarn. Course 1 is completed with one colour where as course 2 is completed with two colours supplied by feeder 2 and feeder 3. The colour which is knitted first by the first needle is to be given first preference while selecting the feeders. From the yarn path diagram, it can be noted that two needle types and two cam tracks are required to produce the design. Out of the 3 feeders, red yarn is knitted with feeders 1 and 3 and black yarn with feeder 2. Other than the knitting points, at all other places, the needles are inoperative (i.e. 2nd needle for F2 and first and third needles for F3) and the loops are not formed.

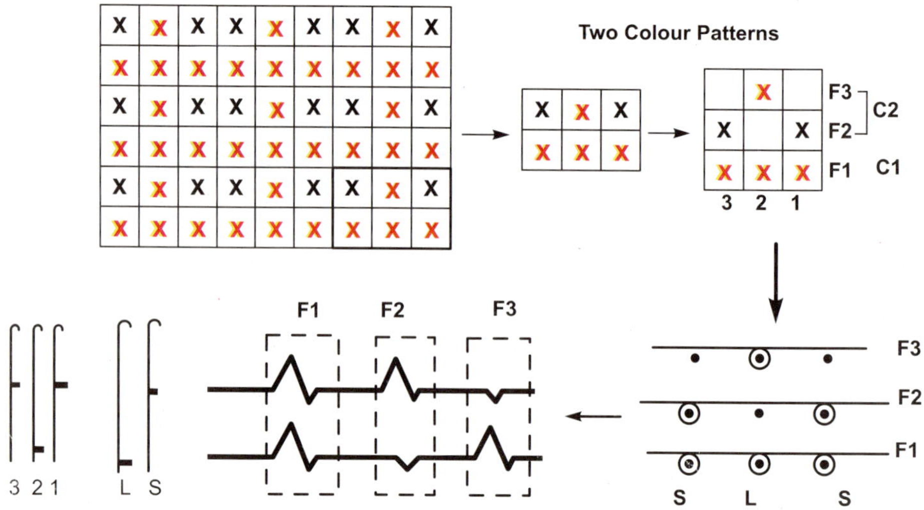

Fig. 8.5

Therefore in the production of the above design with a knitting machine having a total of 12 feeders, 3 feeders are required to produce 2 courses, a total of 8 courses can be produced per revolution of the cylinder.

8.3.2 Three Colour Patterns (Fig. 8.6)

The repeat of a three colour design is given. The first step is to expand the design so that for this 2 course repeat, the number of feeders required is to be found out. 6 feeders are required, in such a manner, the feeders 1 and 4 supply red yarn, feeders 2 and 5 supply black yarn and feeders 3 and 6 supply blue yarn. From the yarn path diagram of six feeders, four types of needles and four track cams at each feeder are decided. The needles are mounted in the tricks in such a manner that the first five places are short, Medium 1, Medium 2, short and long needles respectively and so on. Therefore in the production of the above design in a knitting machine having twelve feeders, four courses are produced at each revolution of the machine.

At feeder 1, short needle starts knitting with red yarn at needle places 1 and 4

At feeder 2, medium 1 and long needle knit with black yarn at needle places 2 and 5

At feeder 3, medium 2 long needles knit with blue yarn at needle places 3

At feeder 4, medium 2 and long needle knit with red yarn at needle places 3 and 5

At feeder 5, short needle knit with black yarn at places 1 and 4

At feeder 6, medium 1 needle knit with blue yarn at places 2 and so on.

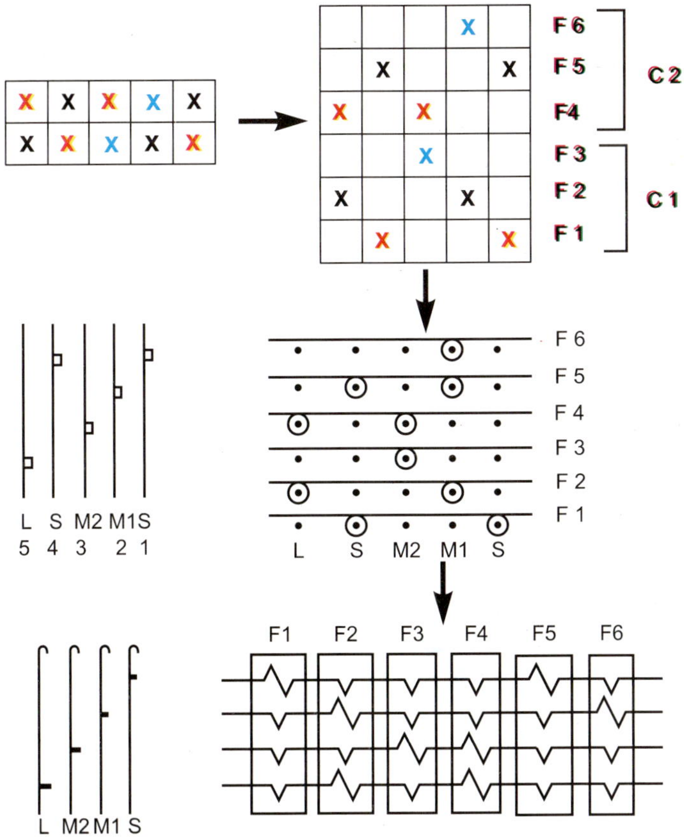

Fig. 8.6

8.4 STRIPED PATTERNS

8.4.1 Horizontal Stripe Patterns (Fig. 8.7)

Multiple feed single jersey machines with large number of feeds are usually employed for such patterns. Coloured yarns are supplied alternatively in uniform manner. For machines fitted with stripers, each feed may be supplied at will, and the selection is controlled by a predetermined pattern programme.

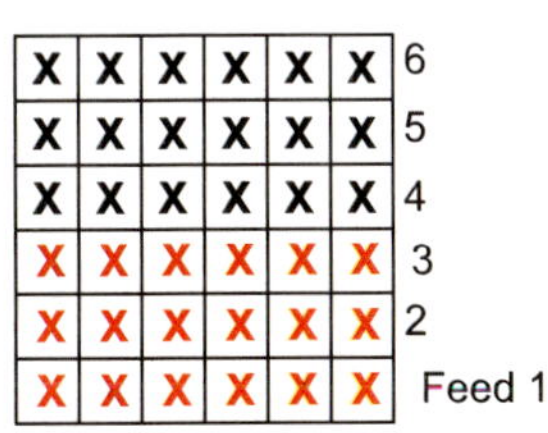

Fig. 8.7

As each course is knitted fully with single coloured yarns, single needle type and single cam tracks at each feeder positions are sufficient. To start with, red yarns knits with feeders 1, 2 and

3 and green yarn with feeders 4, 5 and 6. The next three feeders supplied with red yarn and so on. To produce this design, the number of feeders employed should be multiples of six.

8.4.2 Vertical Stripe Patterns (Fig. 8.8)

The design consists wale wise vertical stripes of black and red yarn. The repeat consists of first two wales having red yarn, next two with black, fifth red, sixth black, seventh red and eighth black. As each course consists 2 colours, two feeders are required. From the yarn path diagram, it can be noted that two types of needles and hence two track cams at each feeders are required. Half the number of courses to the available number of feeders is produced. Red yarn knits with odd feeders (1, 3, 5, 7 etc.,) with needles 1,2,5,7 and so on. Green yarn knits with even feeders (2, 4, 6) with needles 3,4,6,8 and so on. By arranging the needles in the order as shown in the diagram and 2 track cams at feeder positions in the machine, the above design can be produced. To produce vertical coloured stripes with more than two colours, machines with more number of cam tracks are required.

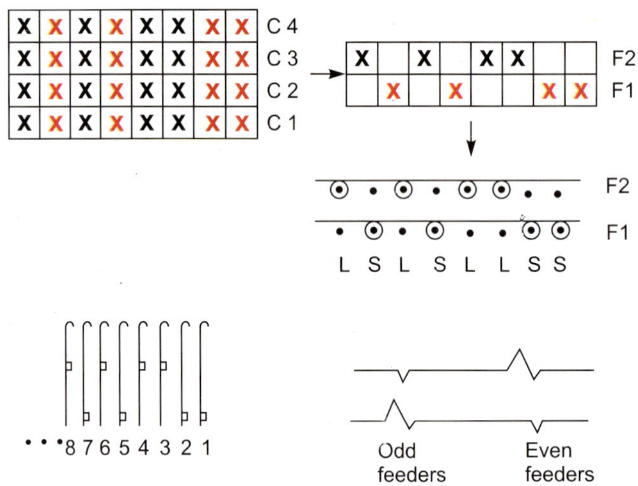

Fig. 8.8

8.5 SQUARE PATTERNS (Fig. 8.9)

This squared design consists of red and black yarns knitted with four wales and four courses repeat. Each course consists of two colours and a full repeat is completed with four courses. Eight feeders per repeat are required and the number of feeders in the machine is multiples of eight. From the yarn path diagram, it can be noted that two needle types and two cam tracks at each feeders are required to produce the fabric.

Red yarn is supplied with feeders 1 and 3 for the first two courses and with feeders 6 and 8 for the next two courses in order to being knitted by both short and long needles. Similarly the black yarns which are supplied with feeders 2 and 4 for the first two courses are originally knitted by long needles and with feeders 5 and 7 are being knitted by short by needles for the next two courses. For every four feeders, the colour supply pattern is reversed.

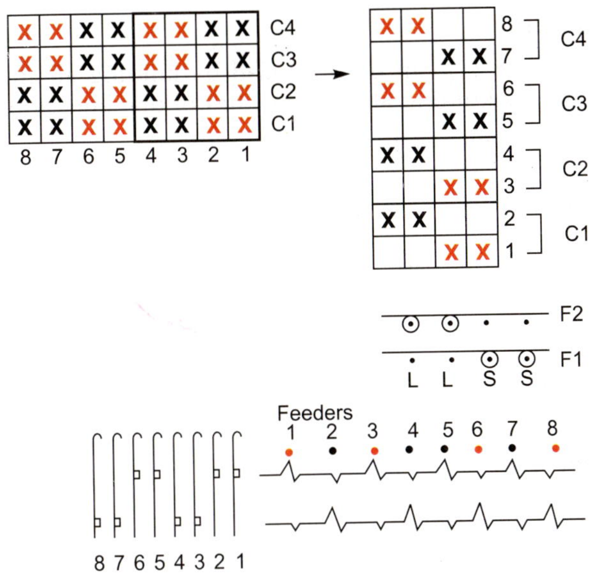

Fig. 8.9

8.6 DOUBLE JERSEY PATTERNS

The double jersey machines in use today may be gated either for rib or interlock which are named as eight system machines. The transition from one position to the other may be made easily by displacing the dial half a trick spacing with respect to the cylinder. The two needle setouts viz., rib gating and interlock gating are commonly employed. Both the dial and cylinder have two cam tracks and at their rising cam places they have the possibilities of knit, tuck and miss positions. This possibility may be effected by either pivoting or bolt cams. In interlock knitting one needle type work with one feeder i.e., long needles with feeder 1 and short needles with feeder 2. Where as in eight system machines, in rib position, all the needles can be activated at the same feed and in interlock position, only alternate needles may be cleared at any feed. The following symbols (Fig. 8.10) can be used for the selection of cam and needles in the machine.

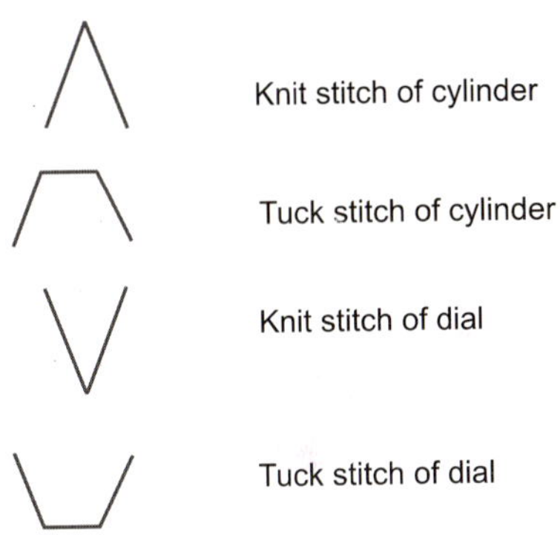

Fig. 8.10

8.6.1 Interlock Structure

The interlock structure (Fig. 8.11) can be produced by knitting all short needles of dial and cylinder with feeder 1 and by knitting all long needles of dial and cylinder with feeder 2.

The cam set up for the interlock is given in (Fig. 8.12)

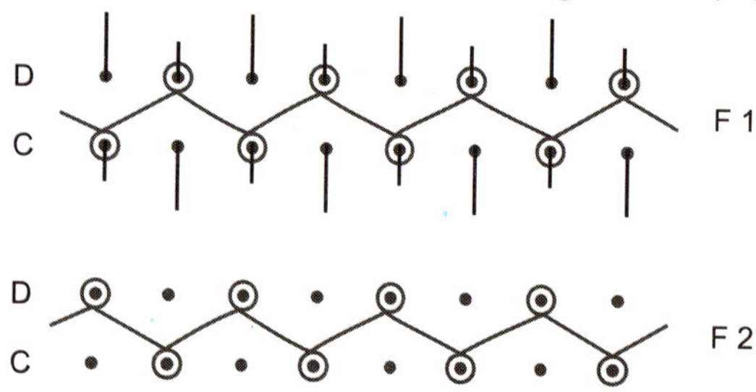

Fig. 8.11

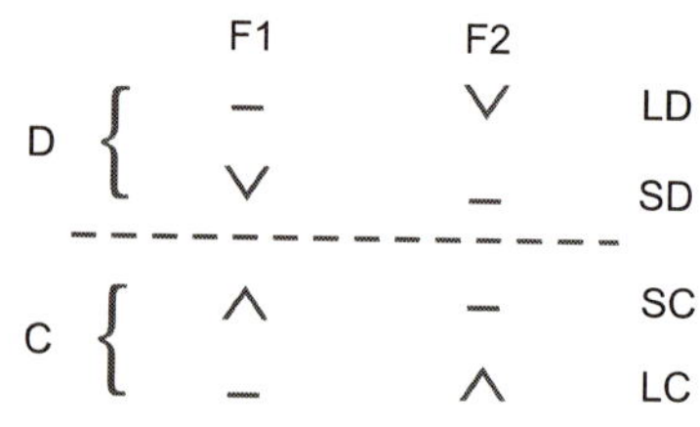

Fig. 8.12

For a machine of 22 inch diameter, to run with 34 feeders, the cam setting order is such that all the odd feeders (*i.e.*, 1,3,5,7.......) act identical to that of feeder 1 so that knitting takes place on all the short needles of dial and cylinder. Similarly for all the even feeders (*i.e.*, 2,4,6,7..........) Knitting takes place on only long needles of dial and cylinder.

In the above figures,

D - Dial Bed
C - Cylinder Bed
LD - Long Dial Needles
SD - Short Dial Needles
LC - Long Cylinder Needles
SC - Short Cylinder Needles

8.6.2 Interlock Pique

The repeat of the structure is as given below (Fig. 8.13) with 6 feeders

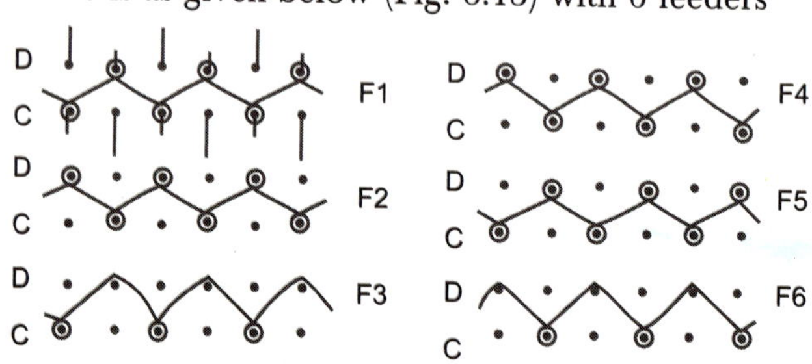

Fig. 8.13

Except at feeders 3 and 6 having tuck stitches, at all other feeders the regular interlock type knitting takes place. The cam set-up for this structure is as given below for a 34 feeder machine.

In the cam settings shown below, the 6 feeder repeat completed only for 30 feeders. The other four feeders out of the 34 are left as missed cams.

	F1	F2	F3	F4	F5	F6	
D	–	∨	—	∨	—	∪	LD
	∨	–	∪	–	∨	—	SD
C	∧	–	∧	–	∧	—	SC
	–	∧	—	∧	—	∧	LC

Feeders

7	8	9	10	11	12
13	14	15	16	17	18
19	20	21	22	23	24
25	26	27	28	29	30

As per the above cam set up diagram, with
Feeders 1, 7, 13, 19, 25 short needles knit on dial and cylinder
Feeders 2, 8, 14, 20, 26 long needles knit on dial and cylinder
Feeders 3, 9, 15, 21, 27 short dial needles tuck and long cylinder needles knit
Feeders 4, 10, 16, 22, 28 long needles knit on dial and cylinder
Feeders 5, 11, 17, 23, 29 short needles knit on dial and cylinder
Feeders 6, 12, 18, 24, 30 long dial needles tuck and short cylinder needles knit.

9

Needle Selection Techniques in Circular Knitting Machines

9.1 INTRODUCTION

The patterning capabilities have been considerably improved in the past few years and designing of fabrics has become an important branch in knitted fabric development. Surface effect with various stitch combination and figure ornamentation with colour combinations are the two general ways of accomplishing designs in weft knitting. For patterning in weft knitting, the selection of needles is very important which can be achieved by:

(*i*) Direct needle control

(*ii*) Indirect needle control and electronic control.

The control systems, which enable free selection and guidance of more than 4 needles into appropriate working positions, are denoted as jacquard controls. The following are the various important mechanisms, through which the selection of needles is mainly achieved:

1. Swing cam
2. Split cam
3. Multi cam track
4. Pattern wheel
5. Battery selector
6. Pattern drum
7. Mono film
8. Revolving cam boxes
9. Electronically controlled jacquard

In the case of direct needle control, the control is carried out by cam parts or pattern wheels, which directly act on pattern/knitting butts. In the case of indirect needle control, one or several control sinkers/pins are arranged after the needle in the same groove of the needle carrier. Pattern drums battery selectors/sliders are such examples. Needle selection through film tapes, control magnets etc., comes under electronic control category.

9.2 SWING CAMS (Fig. 9.1)

They are also called as wing cams. Here the patterning can be made by selecting the needles to knit, tuck or float positions depending upon the tracks for the needle butts provided by the cam mechanism. For machines of single jersey type, single swing cam track can be provided, whereas for rib or interlock types, as many as four tracks can be provided for four types of needles used. These cams are placed at raising cam positions.

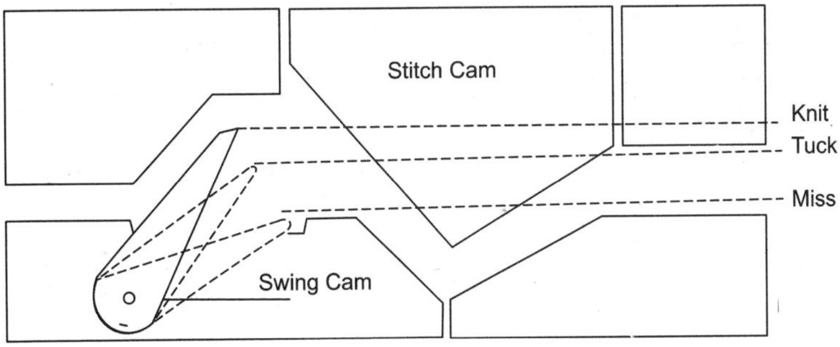

Fig. 9.1

9.3 SPLIT CAMS (Fig. 9.2)

When a knitting machine is fitted with high and low butt needles, the cams must be of different depths to act selectively according to the cam shape (knit, tuck or miss) and the size of the butts. The split cams can be used for 6 possible positions such as only knit, only tuck, only miss, knit and tuck, knit and miss and tuck and miss. Accordingly the cams will be of different depth and thickness.

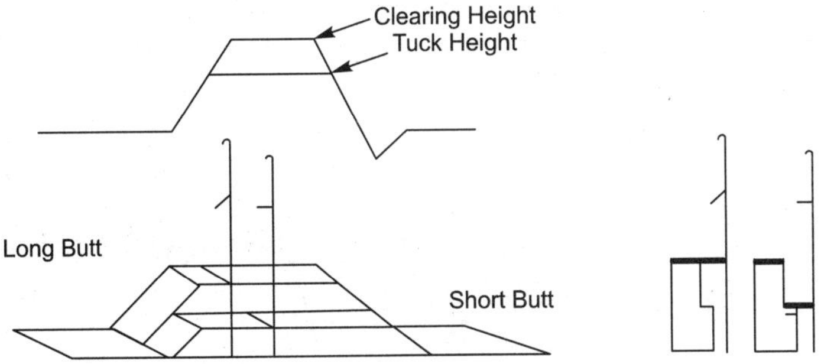

Fig. 9.2

9.4 MULTI CAM TRACKS (Fig. 9.3)

In multi cam track machines, needle butts may be positioned in one of between two and five tracks which at every feed position have fixed but exchangeable knitting, tucking or missing cams. Accordingly the needles also have different butt positions. In some machines, a common top butt is controlled by a stitch camtrack whereas in high speed machines the exchangeable cams also incorporate the stitch and guard cam shape.

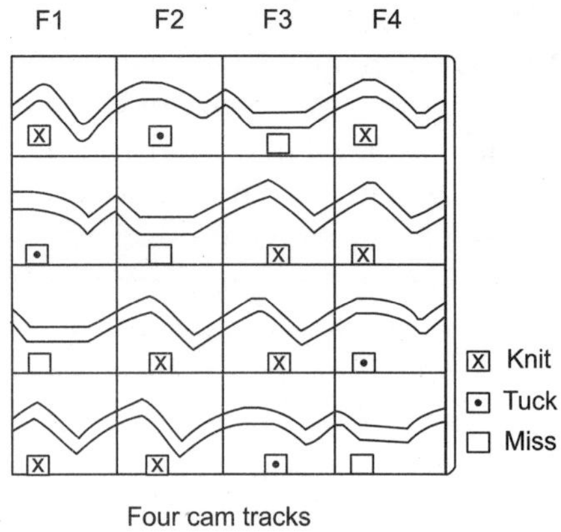

Fig. 9.3

9.5 PATTERN WHEELS (Fig. 9.4)

The pattern wheel is the most popular technique of mechanical jacquard selection on multifeed circular machines. In pattern wheel, its periphery is cut at the same gauge as the knitting machine and it revolves at the same linear speed as the machine. The pattern wheels are fixed at an inclination with their slots opposing the cylinder tricks and mounted one wheel per feed. High and low bits are put in the tricks of the pattern wheel which allows the needles to be selected either to knit or to tuck and also to remain out of action. Thus, three needle positions may be obtained. These pattern wheels generally comprise two steel discs of different diameters whose periphery is pre-segmented to form bits. If the number of pattern wheel tricks can be divided exactly into the number of needles in the machine, the patterns can be arranged without displacement, otherwise displacement occurs. In practice, all the machines equipped with pattern wheels have displacement inherent in the repeat units.

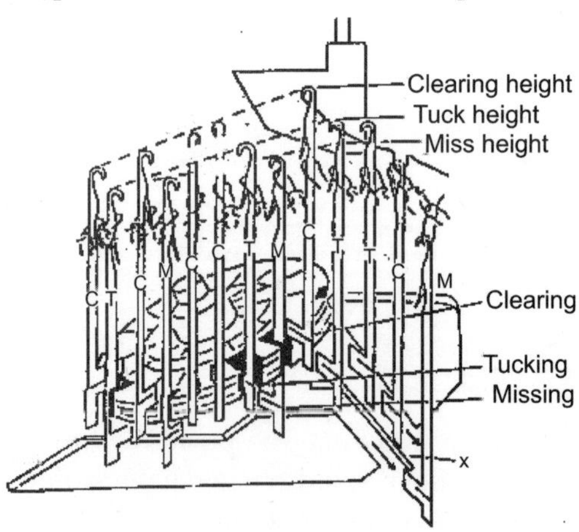

Fig. 9.4

9.6 BATTERY SELECTOR (Fig. 9.5)

This selection device is simple in its principle. On each cylinder trick, just under the needles, pressers are placed. This presser which may be depressed into its trick, has two control butts and a selection butt. Usually, the butts on the Jacks are at a number of different heights so that they can be arranged in a diagonal or chevron formation along a straight needle bed or around a needle cylinder. In a circular machine, a presser raising cam, similar to a knitting cam, is provided to raise the needle into a cam track that will cause it to knit or tuck. As the cylinder rotates, the presser passes in front of a selection mechanism (battery selector-battery of 13 selectors, in general) which in turn either deflects or misses the presser to select or unselect the needles to form stitches. The selector elements at each feeder are arranged one above the other at heights corresponding to the position of the butts on successive pressers presented to the selector by the rotation of the needle cylinder. The number of different heights used determines the width of the pattern and the pattern depth is determined by the number of feeders.

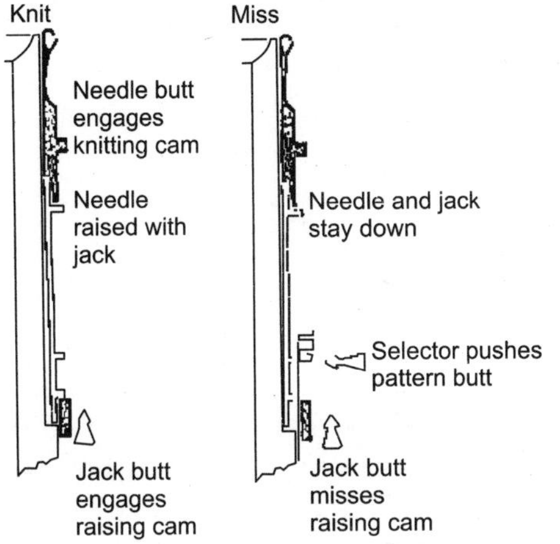

Fig. 9.5

9.7 PATTERN DRUM (Fig. 9.6)

Machines fitted with drum selection devices can produce patterns of large dimensions, without displacements. It also increases the height of the pattern area. In this technique, a pattern drum is filled with an array of pegs that push or pivot a stack of independent selector elements. Each column of peg positions on the drum provides the selectors with instructions for one course, and once this is completed, the drum is racked to present a new column of pegs and spaces with instructions for the next course. The pattern drums are of two types:

(*i*) Multi step mechanism in which selectors inserted in the grooves of the pattern drum operate on the jacks, and

(*ii*) Peg drum machines, in which metal pegs are inserted in the holes of the pattern drum, which directly operate on the jacks.

The latter is simple and effective and the drum can be racked clockwise or anti-clockwise or can be kept stationary in a predetermined manner. The jacks in the cylinder have butts at a height corresponding with each horizontal row of pegs in the drum, and are arranged in echelon or chevron fashion, similar to the battery selector, depends upon the pushing or missing of the jack by the peg in the drum, the needle may miss or knit. It is increasingly difficult to accommodate the drums around the periphery of the machine and trends to finer gauges and steeper cams further restricts the space available.

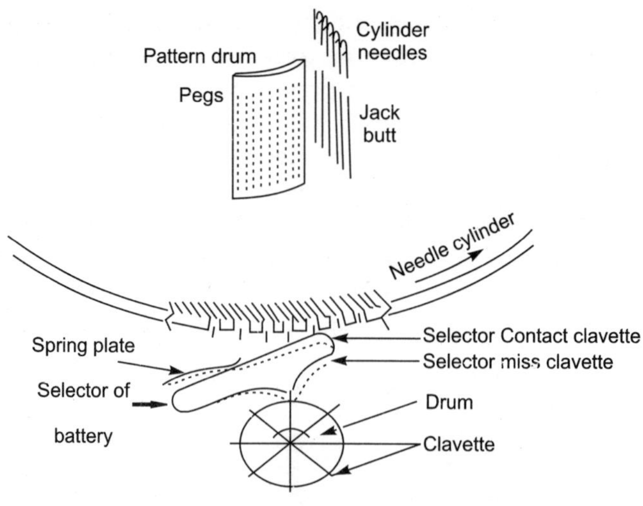

Fig. 9.6

9.8 MONO-FILM (Fig. 9.7)

In mono-film system, each feed is furnished with a horizontal selection wheel. A mono-film passes around the assembly of wheels and transmits a selection to each of them, the selections are arranged at each machine revolution. It is a modification of a pattern wheel mechanism. The wheel contains special types of jacks and the tape/film acts on the pointed portion of the jack. If there is a hole in the tape, the jack remains undisturbed and consequently the jack butt engages the cam and therefore the needle is actuated to be raised to knit. If a solid, unpunched part of the tape faces the pointed projection of the jack, it moves the jack, and its lower portion is acted by a cam, which raises it. As a result the presser is away from the raising cam and hence the needle makes miss stitches. The length of the punched tape determines the depth of patterning area.

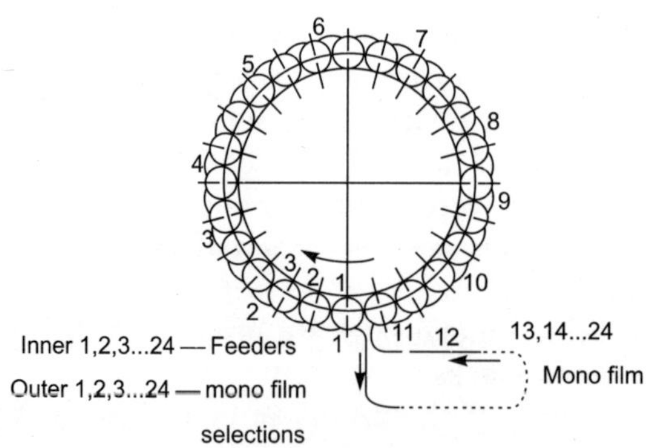

Fig 9.7

9.9 REVOLVING CAM BOXES (Fig. 9.8)

This method of needle selection is distinctly different which produce pattern areas without displacement. When exploited to the full, a single design may encompass the entire periphery of the fabric. These machines are equipped with a mechanical means of selecting each needle individually. They usually have stationary cylinders and revolving cam boxes. There are twice as many selection drums as there are feeds on the machine and these drums are positioned around the machine. A perforated film is placed on each drum. Each track on the film contains punched and intact areas and controls one needle.

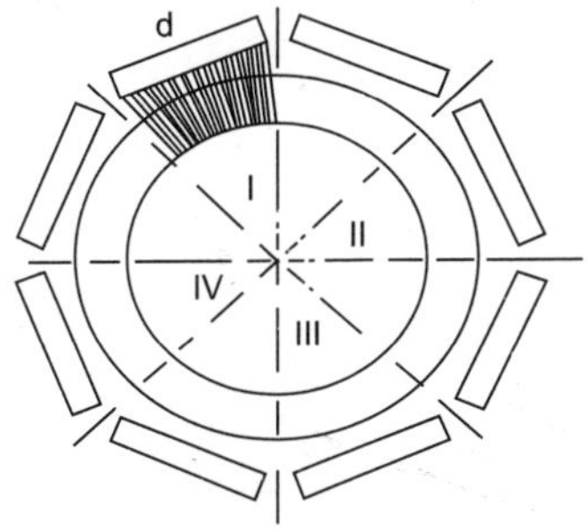

I,II,III,IV - Four feeds

d - Eight drums

Fig. 9.8

9.10 ELECTRONIC SELECTION DEVICE (Fig. 9.9)

This technique enables very large pattern areas and facilitates rapid sampling. Each feed is fitted with an electromagnet which pre-selects pressers or springs; this selection requires only a minimum amount of energy. The slight displacement is then mechanically amplified by the intermediary clavettes, jacks, etc., in order to introduce the needles into the cam track of the raising cams. A small magnet is polarised by a solenoid at the same frequency that the needles pass by the feed, and according to the information given by the control unit thus making the needle selection.

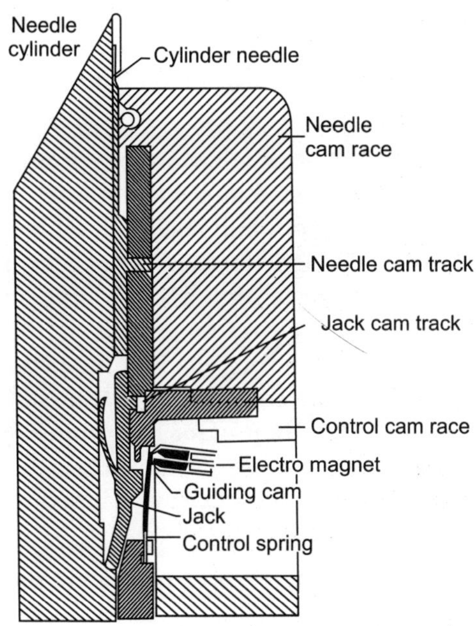

Fig 9.9

10

Weft Knit Fabric Geometry

10.1 DIMENSIONAL STABILITY

The yarns after knitting, which were originally straight and desire to return to the straight state were prevented from doing so by the frictional forces acting over the regions of loop interlacing. Thus an apparent equilibrium of loop shape may be purely due to instantaneous equilibrium between restorative and frictional forces and not because the loop has attained the state of minimum energy level. In practice, the hosiery yarns are not perfectly elastic and the interlocking points are of high friction in addition, the yarns are often temporarily distorted by the throw of the needle during knitting. The knitting strains, therefore are dissipated only gradually on standing, as the fabric tends towards its state of minimum energy. Thus this gradual relaxation mechanism causes instability in the dimensional behaviour of knitted fabrics.

The dimensional stability of knitted fabrics is an important area of the knitting industry. Fabric shrinkage is the ultimate problem if the dimensional stability of the knitted fabrics is not properly taken care. There are various factors influencing the dimensional stability as well as the shrinkage of the knitted fabrics. Though the factors such as fibre characteristics, stitch length, machine gauge, yarn twist, knitting tension and washing and drying methods causes dimensional variations, the first factor mostly responsible is the relaxation of internal stress, since these have been imposed on the yarn during the knitting processes. The second factor is the swelling of the yarn when the fabrics are subjected to wet treatments. The configuration of the loops change due to yarn swelling. It is important to examine the exact geometrical loop change upon swelling and further to associate the fabric shrinkage with loop geometry. So, it can be concluded that the geometry of knitted fabrics are based on two fundamental factors such as:

(*i*) The different dimensionally stable states (*i.e.*, relaxation states) to which the knitted fabrics are imposed, and

(*ii*) The loop length *i.e.*, the length of yarn in the knitted loop (stitch length).

10.2 FABRIC RELAXATION STATES

Changes in dimensions after knitting can create major problems in garments and fabrics especially those produced from hydrophilic fibres such as wool and cotton. As synthetic thermoplastic fibres can be heat set, their shape or dimensions are least affected. The changes of dimensions with wool fibres are magnified by felting shrinkage, however with shrink (felting) – resist finish in wool yarns, the problem can be much reduced. During knitting, the loop structure is subjected to tension from sources such as yarn feeding and fabric take down mechanism. In order to avoid further dimensional changes as the fabric is taken out of the machine, the knitted fabric should attain a stable state of equilibrium. A number of stable states are suggested by various research workers by which relaxation conditions are possible so that the yarn and structure can create high internal restrictive forces and thus inhibit recovery.

The three important dimensionally stable states are:

(*i*) Dry relaxed state
(*ii*) Wet relaxed state
(*iii*) Finished relaxed state

10.2.1 Dry Relaxed State

The fabric has been taken off the knitting machine and allowed to lie freely for a sufficient length of time. Eventually the fabric attains a dimensionally stable condition, called the dry relaxed state. In this state, the natural configuration of yarn is almost straight so that when it is unroved, the yarn takes up an approximately straight form. A plain fabric knitted from worsted yarn will recover from a 60–80 % extension in length to its natural length after 48 hours if allowed to relax freely in the dry state, whereas a cotton fabric will retain permanently 10–20 % of the extension in length. In dry state, therefore plain knit wool fabrics may be expected to return to their strain free conditions more freely than the cotton fabrics.

10.2.2 Wet Relaxed State

If the fabric is soaked in water and allowed to dry flat, the wet relaxed state is attained, again a dimensionally stable condition. The equilibrium is reached after static relaxation in water and subsequent drying. In this state, the natural configuration of yarn is not straight but is set into a form approximating the loop shape in the fabric. Wet relaxation is carried out in water at 30°C containing 0.1 % wetting agent, allowed to lay for 24 hours, hydro extracted and dried naturally for atleast three days.

10.2.3 Finished Relaxed State

In order to reach this stable condition, the fabric is subjected to agitation in water or steam, and a denser fabric results. The fully relaxed condition is obtained by subjecting the samples into gentle agitation at 80° C for 2 hours, tumbles dried at 80° C for 2 hours tumble dried at 80° C for 2 hours in a domestic top loading washing machine, and finally conditioned in the standard atmosphere for atleast 24 hours. A satisfactory relaxation technique applied during the finishing of cotton fabric in continuous length form is the compacting or compressive shrinkage technique. The fabric is passed between two sets of roller nips with the feed rollers turning at a faster rate than the withdrawal rollers so that the courses are pushed towards each other and the fabric is positively encouraged to shrink in length. This technique can create difficulties with interlock fabric which tends to buckle outwards three-dimensionally to produce ripples on the surface known technically as "orange peel".

10.3 DIMENSIONAL CHARACTERISTICS

A Plain knitted structure is shown in Fig. 10.1. The lines of loops across the fabric are known as courses and the lines down the fabric as wales. One loop is referred to as a

W = Wale spacing

C = Course spacing

AB = Stitch length, l in mm (*i.e.* loop length)

S = Number of stitches per square cm, stitch density

w = Wales/cm = 1/W

c = Courses/cm = 1/C

The dimension of plain knitted fabric is related as below:

(*i*) Courses per unit length is inversely proportional to stitch length

i.e., $c \propto 1/l$ (or) $C = Kc/l$

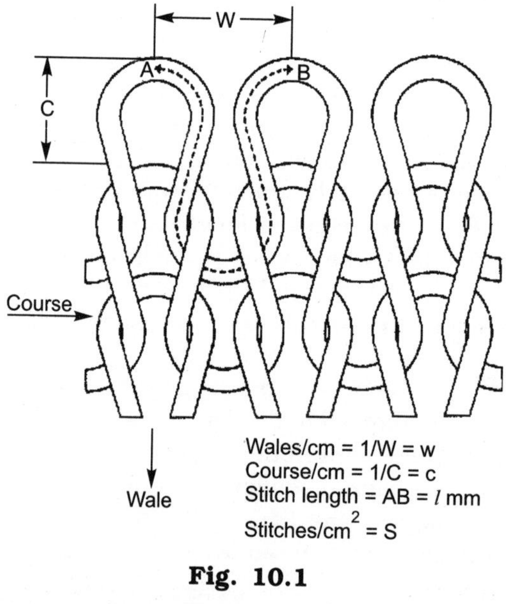

Fig. 10.1

(*ii*) Wales per unit width is inversely proportional to stitch length

i.e., $w \propto 1/l$ (or) $w = Kw/l$

(*iii*) Stitches or loops per unit area is inversely proportional to the stitch length squared

i.e., $S \propto 1/l^2$ (or) $S = Ks/l^2$

(*iv*) Loop shape factor;

R = Courses per unit length/Wales per unit width

$= c/w = Kc/Kw$

Where Kc, Kw, Ks are the geometrical constants for the plain knitted fabrics at different stable states, as given in Tables 10.1 and 10.2.

Table 10.1 Constants for Geometry of plain knitted worsted fabrics at different stable states (Imperial units)

Parameters / *Fabric state*	*Ks*	*Kc*	*Kw*	*Kc/ Kw*
Dry–relaxed	19.0	5.0	3.8	1.31
Wet–relaxed	21.6	5.3	4.1	1.29
Finished–relaxed	23.6	5.6	4.22	1.32

In Imperial units, courses and wales per unit length and stitch length are in inches:

Table 10.2 Constants for Geometry of plain knitted worsted fabrics at different stable states (SI units)

Parameters / *Fabric state*	*Ks*	*Kc*	*Kw*	*Kc/ Kw*
Dry–relaxed	1900	50	38	1.31
Wet–relaxed	2160	53	41	1.29
Finished–relaxed	2360	56	42.2	1.32

In SI units, courses and wales per unit length are in centimeter and stitch length in millimeter. This causes the values of Ks to be multiplied by 100 and those of Kc and kw by 10. Of course, the ratio Kc/Kw is not affected.

The constants are derived by using similar length units for courses per unit length, wales per unit length and stitch length. By plotting the values of S against those of $1/l^2$ and finding its slope, gives the value of Ks. Similarly the slopes obtained with c against $1/l$ gives Kc and w against $1/l$ gives kw.

10.3.1 Areal Density

By knowing the stitch density (loops per sq.m) loop length (mm) and yarn linear density, the areal density of the knitted fabric can be calculated as,

$$\text{Areal density} = S \times l \times \text{tex} \times 10^{-6} \text{ grams/m}^2 \qquad \text{...(10.1)}$$

$$(1 \text{ Km} = 10^{-6} \text{ mm})$$

If S is in stitches/cm^2, then

$$\text{Areal Density} = \frac{S \times 10^4 \times l \times \text{tex}}{10^6} = \frac{Sl \text{ tex}}{100}$$

Also, $S = Ks / l^2$

Therefore,

$$\text{Areal Density} = \frac{Ks}{l^2} \times \frac{l \times \text{tex}}{100}$$

$$= \frac{Ks \times \text{tex}}{100\, l} \text{ g/m}^2 \qquad \text{...(10.2)}$$

Example 10.1: *A wet relaxed plain-knitted worsted fabric is made from 55 tex yarn of stitch length 5 mm. Estimate the fabric areal density.*

From Table 10.2 the value of Ks for wet relaxed fabric is 2160.

The stitch density, $S = Ks/l^2 = 2160/5 \times 5 = 86.4$ stitches/cm^2

By relating equation, (10.1)

Areal density $= 86.4 \times 5 \times 55 \times 10^{-6} = 237.6$ g/m^2

Also, by relating equation (10. 2)

Areal density $= \dfrac{2160 \times 55}{100 \times 5} = 237.6$ g/m^2

10.3.2 Courses/cm and wales/cm

The Tables of geometrical constants gives the ratio of courses/cm and wales/cm. If the product of these two is also known, it is possible to estimate wales/cm and courses/cm.

Example 10.2: *For a wet relaxed worsted plain knitted fabric, the stitch density is 86.4. Estimate the number of courses/cm and wales/cm.*

From Table 10.2

$$\frac{Kc}{Kw} = \frac{c}{w} = 1.29$$

$$c = 1.29\ w$$

Also, we have, $S = cw = 86.4$

By substituting for c, $1.29\ w \times w = 86.4$

$$w = \sqrt{\frac{86.4}{1.29}} = 8.184$$

Since, $c = 1.29\ w$

$$c = 1.29 \times 8.184 = 10.557$$

Courses/cm $= 10.557$ and

Wales/cm $= 8.184$

10.3.3 Fabric width

If "*n*" is the number of needles knitting on the machine, then

Fabric width $= \dfrac{\text{Number of needles}}{\text{Wales/cm}}$

We know that, wales per cm is given by

$$w = kw/l$$

Therefore,

Fabric width $= \dfrac{n}{Kw/l} = \dfrac{nl}{kw}$

but nl is the course length, L

Therefore, Fabric width $= \dfrac{L}{kw}$ ...(10.3)

10.3.4 Fabric cover

Similar to woven fabric, the cover of the knitted fabrics is also important for the study of their tightness.

The total area covered by the yarn per cm^2 of fabric is given by

$$\text{Fractional cover} = \frac{S \times l \times d}{100} \text{cm}^2$$

Where d is the yarn diameter in mm

As we know, $S = \dfrac{Ks}{l^2}$

$$\text{Fractional cover} = \frac{Ks \times l \times d}{l^2 \times 100} = \frac{Ks \times d}{100 l} \qquad \text{... (10.4)}$$

By expressing the yarn diameter in terms of its linear density in tex and by deleting various constants and by considering the fact that the yarn diameter is proportional to the square root of linear density, a new expression, cover factor, known as Tightness factor for knitted fabrics is obtained.

$$\text{Tightness factor,} \quad K = \frac{\sqrt{\text{tex}}}{l} \qquad \text{... (10.5)}$$

10.4 SPIRALITY

Spirality is a dimensional distortion in circular plain knitted fabrics. The lengthwise rows of stitches, called wales or needle lines, should occupy a truly vertical line in the fabric and should always be at right angles to the cross wise courses of stitches. This perpendicularity of wales to the courses is frequently, not the case and many times the wales may skew to the right or left forming an angle, which appears in the form of a twilled surface. This geometrical defect has been termed spirality of knitting in circular fabrics. The following Fig. 10.2 shows the fabrics with normal loop position and with spirality having wale skewness.

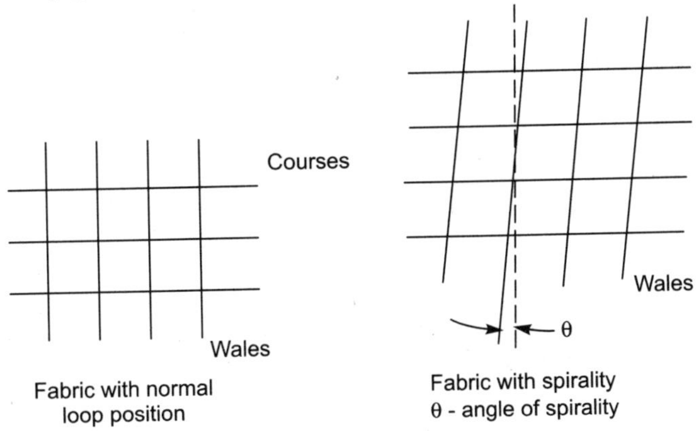

Fig. 10.2

Spirality has definite influence on both the functional and aesthetic performance of knitted fabrics and their garments. Displacement or shifting of seams during the garment make-up, mismatched patterns due to wale skewness, sewing difficulties etc., are some important practical difficulties due to spirality. As the dimensional properties of the fabrics are affected by spirality, it is very essential to minimise or eliminate it altogether. This spirality problem is often corrected in finishing treatments by imposing distortion to fabrics so that the wales straighten out and subsequently set in new form. Though the setting by finishing treatments are normally achieved by using resins, heat, steam, mercerisation etc., it is not permanent and after repeated washings, the wale skewness takes place.

10.4.1 Causes of spirality

The following are some predominant causes of spirality in knitted fabrics.

- Yarn twist multiple is the principle cause of spirality and it is directly proportional
- Residual torque in the yarn or the twist liveliness
- Number of feeders–though higher feeder numbers increases production, spirality also increased.
- Different spinning technologies such as ring, rotor, airjet etc., also influence spirality. The physical properties of these yarns, their geometrical characteristics, their basic fibre properties (*i.e.*, modules, fineness, cross section etc.,) and blends are the causes.
- Variation in knitting tension, yarn frictional properties, yarn/metal coefficient of friction, yarn lubrication, number of contact points in the knitting zone (*i.e.*, needles and sinkers) also influence spirality.
- Washing – wet treatments increases fabric relaxation and also increases spirality.
- Direction of machine rotation has little influence on spirality. Slight inclination of loops occurs in the direction of machine rotation. Multifeed machines rotating clockwise produce spirality to the left and machines rotating anticlock wise produce spirality to the right.

10.4.2 Theory of spirality

The spirality occurred in knitted fabrics is shown in Fig.10.3. The fabric is assumed to be knitted with Z twist yarn on a multifeed circular machine, revolving clockwise.

Let F = total number of feeders, c = courses per unit length

n = total number of needles, w = wales per unit length

Let DD^1 = position of a wale when total spirality occurs

BB^1 = position of a wale when total spirality occurs due to number of feeders

XX^1 = position of a course when total spirality occur

XA = position of a course when spirality occurs due to number of feeders

$X^1A = F/C$ = displacement between two consecutive courses knitted by the same feed

$$\overline{XX^1} = \frac{n}{w} = \text{Open width of the fabric}$$

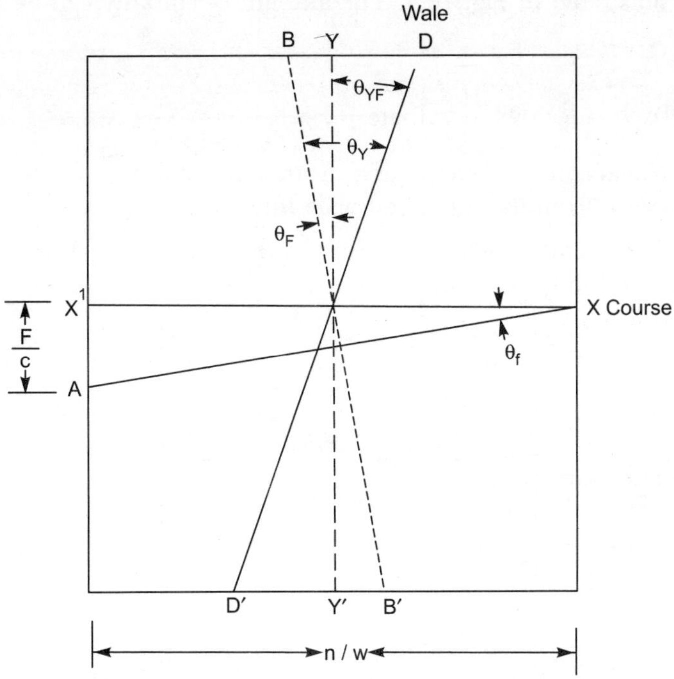

Fig. 10.3

Let θ_Y = angle of spirality due to yarn

θ_F = angle of spirality due to number of feeders

θ_{YF} = total spirality

Then, $$\tan \theta_f = \frac{\text{Course displacement}}{\text{Fabric width}} = \frac{\frac{F}{c}}{\frac{n}{w}} = \frac{F}{c} \cdot \frac{w}{n}$$

From the geometrical constants, as we know

$$w = \frac{kw}{l},\ c = \frac{kc}{l} \text{ and loop shape factor, } R = \frac{Kc}{Kw} = \frac{c}{w}$$

Now, $$\tan \theta_f = \frac{F\omega}{cn} = \frac{F}{nR} \text{ and, } \therefore\ \theta_f = \tan^{-1}\left(\frac{F}{nR}\right) \quad \text{... (10.6)}$$

The above relationship shows that the angle of spirality depends on

(*i*) Number of feeders of the machine,

(*ii*) Shape of the loop in a particular state of relaxation, and

(*iii*) Number of active needles in the machine which depends on machine gauge and diameter.

10.4.3 Measurement of Spirality

The fabric sample is laid freely over a platform without any wrinkles and crimp. The spirality can be analysed using a mini drafter/L–Square scale and the angle of deviation is calculated.

The test method is illustrated in Fig. 10.4. The amount of spirality can be obtained using the following formula.

$$\text{Degree of spirality,} \qquad \theta = \tan^{-1}\left(\frac{d}{L}\right) \qquad \text{... (10.7)}$$

Where d is the displacement of the course from a normal line to the wales of the fabric measured at a distance L from the identified wale line.

The spirality measurements can be taken for the three states of fabric samples such as dry relaxed, wet relaxed and fully relaxed.

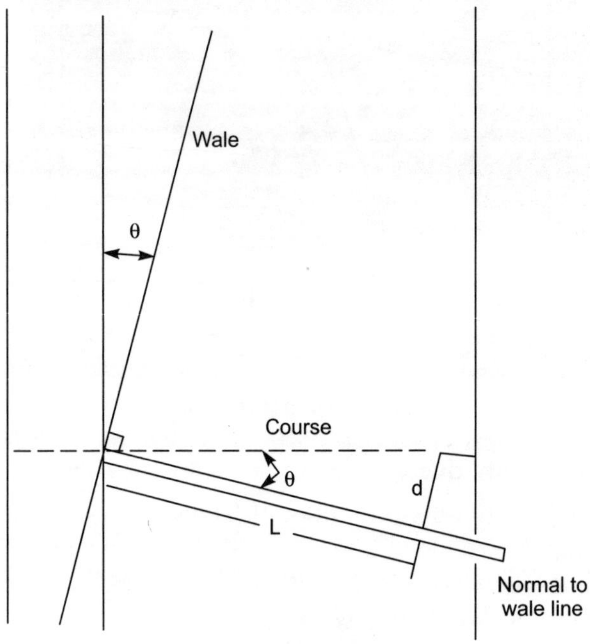

Fig. 10.4

10.4.4 Reduction of Spirality

As the spirality is caused by several aspects such as yarn, machine and fabric finishing treatments as discussed earlier, in the same way it can be reduced or eliminated by the following techniques.

(*i*) With machines that rotate clockwise, the wales will be inclined to left, giving an S-skew spirality and with machines of counter, clockwise rotation to the right giving Z-spirality. The Z-twist yarns make wales to go to the right, giving Z-skew and S-twist yarns make the wales to go to the left, giving S-skew to the fabric. Therefore, to minimise spirality, the Z-twist yarns should be knitted on clockwise rotating machines and S-twist yarns on counter clockwise rotating machine.

(*ii*) As derived by equation 10.6 the angle of spirality increases with more number of feeders. However, with the knitting of yarns that give no spirality (e.g., flat yarns) and running them on clockwise rotating machines, the total spirality would be to the left. Under such circumstances, use of large number of yarn feeders is beneficial rather than detrimental with respect to spirality.

(*iii*) Fabrics knitted with more tightness factor (*i.e.*, tighter fabric) exhibit less loop movement and yarn movement. Such fabrics give less spirality compared to looser fabrics.

(*iv*) High twist multiple yarns have high snarling tendency and thereby increases spirality. Finer yarns and high twist multiples resulted in higher yarn twist liveliness. Therefore, lively yarns have to be knitted as tightly as possible on a machine of the finest possible gauge.

(*v*) As per equation (10.6), use of more number of needles *i.e.*, finer gauge machines reduce spirality as they increase fabric tightness factor as well as take care of yarn liveliness.

(*vi*) Torque relief or twist setting in yarn reduces spirality. As untreated yarns have highest snarling tendency and exhibit high spirality, the following yarn treatments may reduce spirality:

(*a*) Autoclave treatment of steam and vacuum considerably reduce yarn snarling and hence angle of spirality. Dry-relaxed state exhibits greater reduction of spirality than the fully relaxed states. Higher number of treatments (*i.e.*, number of cycles) reduce snarling and thus spirality.

(*b*) Sizing treatment set the yarn, reduces the fibre movement during bending, reduce snarling and thus reduce spirality.

(*c*) Small percentage of low melt polyester fibres can be blended with cotton and heat treated. Low melt polyester will bind to cotton fibres, thus reduce fibre movement, results in low snarling and less spirality.

(*d*) Yarn dyeing is a promising way to reduce spirality. During dyeing, a wet hot treatment set the yarn twist, reduce snarling thereby the spirality. Spirality increases with fully relaxed states.

(*vii*) By plying two identical single yarns and twisting them in a direction opposite to the spinners twist, twist stable yarns can be produced. They are more stable, if they are package dyed. Such yarns produce zero spirality in both dry and fully relaxed states. This plying is a very effective way to avoid completely the spirality.

(*viii*) Plating is an effective way to produce spirality free fabric. By plating one yarn of Z-twist and other yarn of S-twist in knitting, zero spirality fabric can be produced both in dry and fully relaxed states.

(*ix*) In 3-thread fleece fabrics, the relationship between the binder and ground yarn is as similar as plaiting 2 yarns. By using two yarns of different twist distribution as binder and ground yarns, spirality can be reduced considerably.

(*x*) Use of 50/50 Polyester/Cotton blended yarns exhibit lower snarling tendency than 100 % cotton yarns and give less spirality. Auto clave treatment effectively sets the polyester fibres in the blend.

(*xi*) The yarns spun with different spinning technology produce different angle of spirality. The angle of spirality of friction, ring, rotor and airjet spun yarns are of the decreasing order as studied by De Araujo'1et. al. This trend applied for both dry relaxed and fully relaxed states. The lowest twist multiples and minimum snarling tendency of both rotor and airjet yarns lead to lowest angle of spirality in both dry and fully relaxed states.

(*xii*) Interestingly, this spiral effect can be used for the production of fancy fabrics. For example by knitting four threads of normal twist and four of reverse twist yarns consequently on a eight feeder machine, a fabric closely resembling the familiar shadow stripe fabric of the warp loom can be made. Also by knitting alternate ends of normal and reverse twist yarns, quite a successful imitation of crepe fabric is possible.

References

1. De Araujo M.D., Smith G.W., "Spirality of Knitted Fabrics", Part II, TRJ, June 1989, P-350.
2. Davis. W and Edwards C.H., "Spirality in Knitted Fabrics", JTI, April 1934, T122.

11

Knitting Dynamics

The knitting elements are the heart of the knitting machines and particularly the design of needles and cam play an important role in dynamic knitting conditions. The knitting motions of circular knitting machines are simple, but the dimensions, quality and performance of knit fabrics vary widely if there are only slight differences in the simple motions of the cams and needles. The make up of cams, the shape and motions of the needles, the make up of the needle bed and yarn guide, the relative positions between each elements and the timing of their motions are important factors in designing and running the machines.

11.1 NEEDLE DYNAMICS

The shape of the needle hook and latch and the stroke of the needle play an important bearing on the look and quality of knitted fabrics. The movement length of the needles i.e., the stroke of the needles depends on the length of the latches. If the hooks are the same in size, the hook with a longer latch is likely to receive a smaller increment ratio of stress, resulting in a knit fabric of good look. The longer the stroke, the smaller the number of yarns fed and the lower the productivity, resulting in excessive centrifugal force being loaded on the latch and in more shock on the hook.

A latch short in length makes a knit fabric poor in outward, reduces the space between the traps as shown in Fig. 11.1 and causes yarn to be gripped between the latch and the cheek, resulting in knitting defects.

With longer stroke, the impact load on the needle hook also increases. To overcome this, it is suggested that the cheek can be made high. Kiso[1] et al have conducted experiments to find out the relationship between cheek height and the stroke of the latch motion. For the two cheek heights of 0.040 inch and 0.030 inch, the strokes of the latch motion are 0.430 and 0.325 respectively and they have concluded that the shock load is reduced with the cheek of 0.04 inch high. It is also suggested that reducing the cheek height by 0.01 inch reduces the stroke of the latch motion by 0.105 inch, thus resulting in higher productivity.

Also the yarn tension as well as tightness of the old stitches during knock over impose load on the needle hook. The tension imposed on the needle hook is assumed to exceed 50 grams as studied by Heinrich[2].

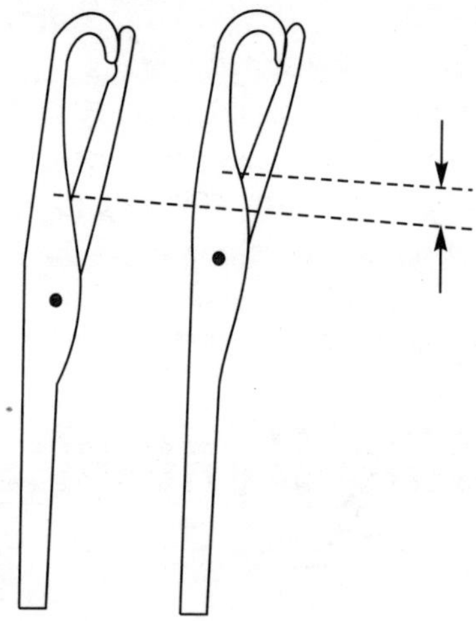

Fig. 11.1 *Clearing between traps*

Hosada et al[1] calculated the strength of the hook of the latch at the knock over and the lowest position of the needle.

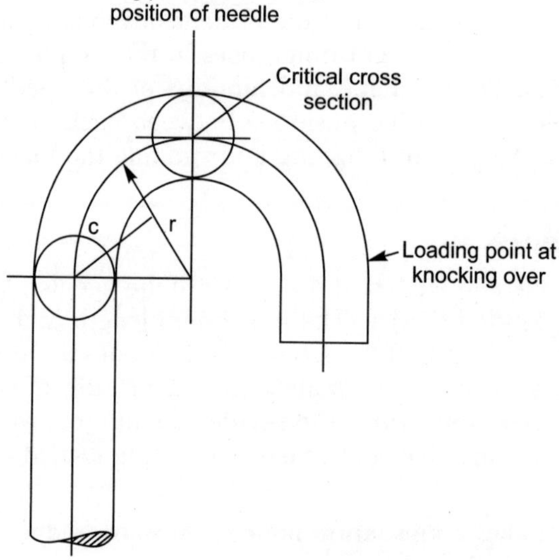

Fig. 11.2 *Load on hook*

He found that the critical cross section and its stress in both the cases are same. Henshaw[3] studied that the needle reaction forces are influenced by a wide range of variables, including cam setting, yarn input tension, machine speed, needle finish, takedown force and the use of positive yarn feed systems. Traces of needle reaction force throughout the knitting cycle shows that with linear cams, the maximum forces are obtained during the knitting of the new loop. He also found the importance of surface finish of the needle in reducing both the overall level of the force and their coefficient of variation.

Munden[4] has given the following expression for calculating the forces acting on the needle in linear cam systems, using the following symbols (Fig. 11.3).

θ = cam angle
P = Force of yarn on the needle
R = Force of needle butt on the cam face
S = Force on the needle caused by side pressure due to machine rotation
μR = Frictional force appropriate to R
μS = Frictional force of the needle against the cylinder walls

$$R = \frac{P}{(1-\mu^2)\cos\theta - 2\mu\sin\theta} \quad \text{... (11.1)}$$

$$S = \frac{P(\mu\cos\theta + \sin\theta)}{(1-\mu^2)\cos\theta - 2\mu\sin\theta} \quad \text{... (11.2)}$$

With μ changed to 0.1 – 0.3, the forces S and R become infinite by θ = 78° – 59° and the machine stops. For this reason, Munden said that θ had better be less than 55° and close to 45°.

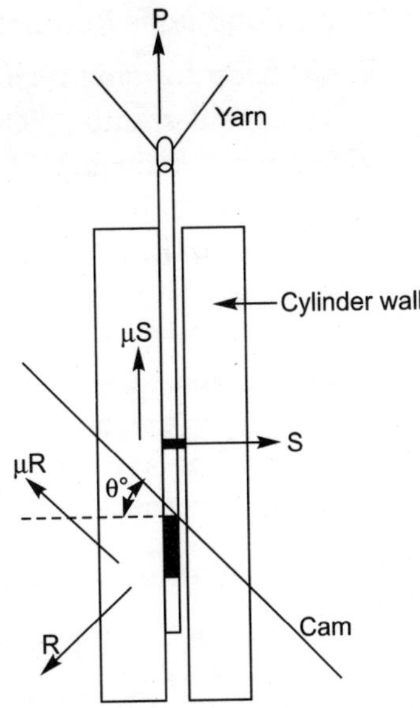

Fig.11.3 *Forces acting on the needle*

11.2 CAM DYNAMICS

The knitting cam controls the motion of the needles throughout the knitting cycle as well as the production rate of the machine. The pressure to increase the rate of production of good quality fabric has initiated attempts to change the design of the cams.

There are two ways of increasing the rate of production from a machine. The first is to increase the number of yarn feeders around the machine. This increases spirality and also the cam angles must be steepened to accommodate additional feeders. Above a certain angle of steep, this results in high incidence of needle butt failures when the machine is started up. The alternative is to raise the rotational speed of the machine. For this to succeed greater attention must be paid to the cam dynamics. For this, the conventional cams consisting entirely of straight lines have been modified with cam profiles consisting of straight lines and circular arcs. However, these produce a step change in acceleration at each transition point. So, any further attempt to modify the cam profile should hold the needle acceleration to reasonably low values. The introduction of non-linear cam shapes, with simple design changes effectively controls the knitting dynamics.

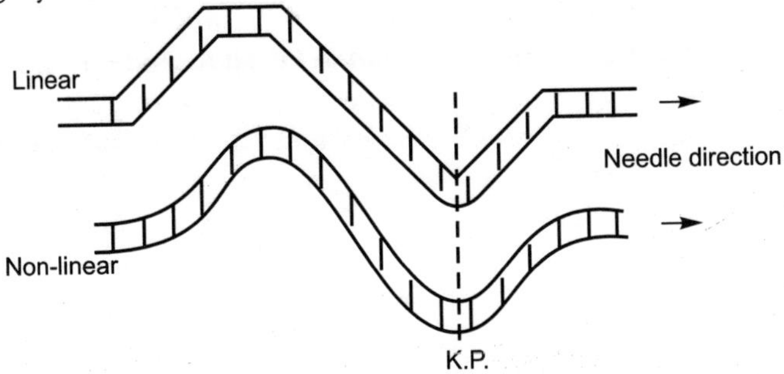

Fig.11.4 *Standard linear and high speed non-linear cams*

Any attempt in changing the design of cam profiles for high speed knitting should satisfy the following conditions for improving quality and productivity.

- The loop forming region should be kept short in order to reduce the friction between.
- Yarns and knitting elements, the maximum yarn tension thereby being kept low.
- The maximum angle should not be so high that the needle butts break when the machine is started up.
- The cams should allow for variation in the stitch draw so that variations between one knitting section and the next can be compensated for.
- There should be no rapid changes in the displacement, velocity and acceleration of the needles since these cause impulsive forces.
- Preferably the new cam design should not be significantly longer than the existing straight-line and circular-arc cams so that maximum of feeders can be accommodated around the machine circumference.

11.2.1 Conventional Linear Cams

Figure 11.5 shows a simple commercial cam system in which the cam pieces are essentially of linear design with cam angles at 45°. The linear cams have the advantage of allowing the greatest number of feeders to be fitted on to the machine circumference as the number of feeders on a machine is dependent on the circumferential length of a cam system. On the other hand, the severe drawback is that the needle butts change direction very abruptly in moving from one linear cam to another, they have infinite acceleration at the ends of the linear cams and they therefore break. However, in practice this does not normally happen, because the needles have a natural resilience, which enables them to absorb the shock. Nevertheless it has been suggested that if the cam angles were increased[4,5] from say 45° to 60° or if the linear speed of the needles were increased[6] much above the normal running speed, then the forces on the needle butts would be excessive and the needle breakage could occur.

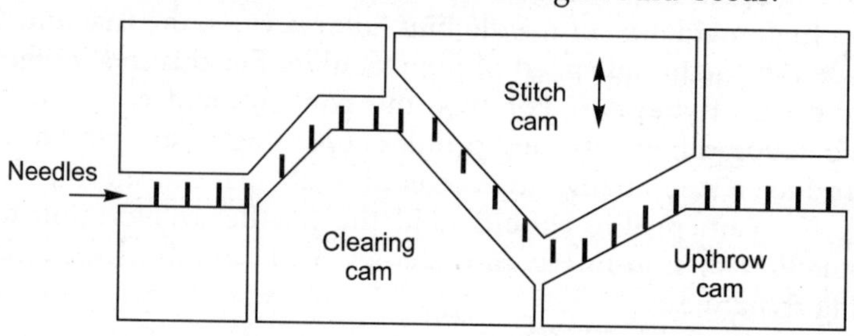

Fig.11.5 *Simple commercial cam system*

Fig. 11.6 represents the true behavior of the relative motion between knitting elements and yarn when the loop is being formed.

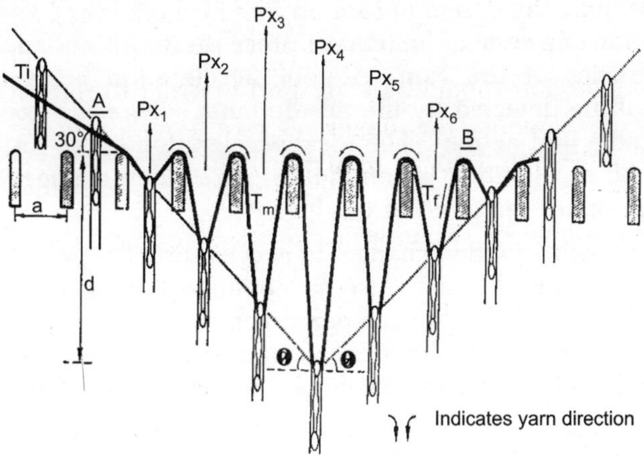

Fig. 11.6 *Schematic representation of the knitting zone of a typical 'pointed' weft-knitting cam system*

The phenomenon robbing-back[7] builds up yarn tension rapidly, but is dessipated at the same rate before completion of the loop. This results in a series of forces acting vertically on the needles as shown in Fig. 11.6, which depend entirely on the developed yarn knitting tensions. The maximum yarn tension in this zone depends largely on two factors viz., the coefficient of yarn metal friction (μm) and the number of contacts between the yarn and knitting elements. The effect of μm on knitting performance is very important. For two identical yarns, one waxed and the other unwaxed, when they are knitted, the developed knitting tension are critically different and they differ by a factor of approximately six[7] and obviously the resultant fabric faults with unwaxed yarns are more, which necessitates the importance of yarn lubrications.

The other way to reduce the maximum yarn tension at the knitting zone is by reducing the number of yarn knitting element contacts, which is possible by increasing the cam angle θ. Having a linear cam of say $\theta = 60°$, the developed knitting tensions are drastically reduced[4]. Whether the yarns are waxed or unwaxed, the developed knitting tensions are very much less than for the conventional 45°cam. Better knitting performance results, in terms of efficiency of production and fewer fabric faults occur. Also by increasing θ by 15° (*i.e.*, from 45° to 60°), many more feeders can be accommodated to the circumferential length of the machine.

Also it is advocated that the cam angles as steep as 60° may not necessarily be commercially realistic because of needle failure due to the forces of impact on such steep cams. Yet otherwise they appear to offer considerable advantages. Besides the advantage of reduced knitting tensions, the total work required to rotate the machine at a constant speed, given in terms of the total torque i.e., Ty+Tn is also lesser with steeper cams of 60°.

Tn – Torque to move the needles alone through the cams

Ty – Torque required to manoeuver the needles through the cam against the resistance of the knitting tensions

11.2.2 High speed Non-Linear Cams

The design of a high speed knitting cam system should be a mathematically designed non-linear cam which would provide complete control over the needle-butt acceleration. While designing high speed non-linear cam, it is not sufficient merely to join straight lines with smooth curves, or even to use circular arc curves, since these will not appreciably improve the acceleration characteristics of the cam. A poor acceleration curve will cause dynamic disturbances, which will be detected by the needle butts, whereas a good acceleration curve will give a smooth needle butt action[8]. The ideal acceleration curve for a non-linear knitting cam is one in which the acceleration is continuous (no sudden changes in acceleration) with maximum acceleration values that are not too high.

In commercial cams, due to sudden changes in acceleration the needles bounce and tend to wear and even cause grooves in the cam. Also the continuous bouncing can weaker the needle butt and eventually breaks off. Other needle parts such as the latch, latch pivot and the hooks also undergo continuous wear due to sudden acceleration changes. The abrupt opening and closing of the latch leads to bent latches and hooks and latch pivot fatigue.

Black and Munden[8] conducted exhaustive studies in order to determine a mathematical design to give continuous acceleration characteristics for the knitting cam system. They examined basic curves such as simple harmonic, double harmonic, parabolic and cycloidal and showed that these curves had unsuitable acceleration characteristics when applied to an unsymmetrical knitting action. Although a combination of the basic curves could be used to give the desired acceleration characteristics, the mathematics involved is much complicated. A method of solution by using polynomial equation is more appropriate for unsymmetrical cam design, since this method is easier, more direct and more accurate than the method of combining basic curves.

A displacement curve of the type shown in Fig.11.7 may be defined by a polynomial equation of the form,

$$y = c_o + c_1 x + c_2 x^2 + c_3 x^3 + \ldots + c_n x^n \quad \ldots (11.3)$$

where x = horizontal displacement

y = vertical displacement

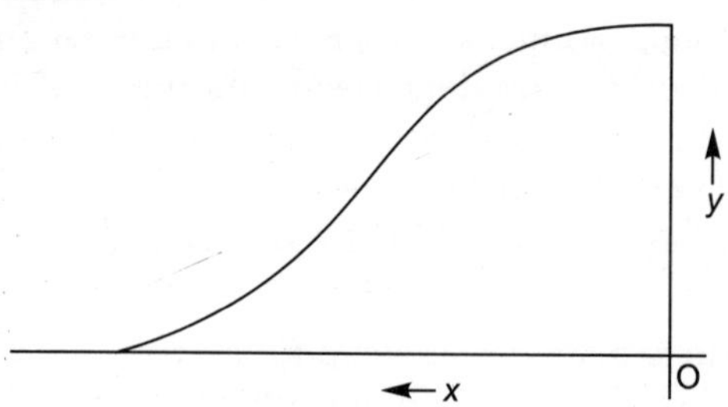

Fig.11.7

The vertical velocity $\left(\frac{dy}{dt}\right)$, acceleration $\left(\frac{d^2y}{dt^2}\right)$ and pulse $\left(\frac{d^3y}{dt^3}\right)$ of the needle are given by,

$$\text{Vertical velocity} = \frac{dy}{dt} = \frac{dy}{dx} \cdot \frac{dx}{dt} = v. \frac{dy}{dx} \quad \ldots (11.4)$$

Where v is the constant horizontal velocity of the needle.

$$\text{Vertical acceleration} = \frac{d^2y}{dt^2} = v^2 \frac{d^2y}{dx^2} \quad \ldots (11.5)$$

and, Vertical pulse $= \dfrac{d^3y}{dt^3} = v^3\dfrac{d^3y}{dx^3}$ ... (11.6)

The characteristics of the polynomial curve depend upon the boundary conditions chosen for control.

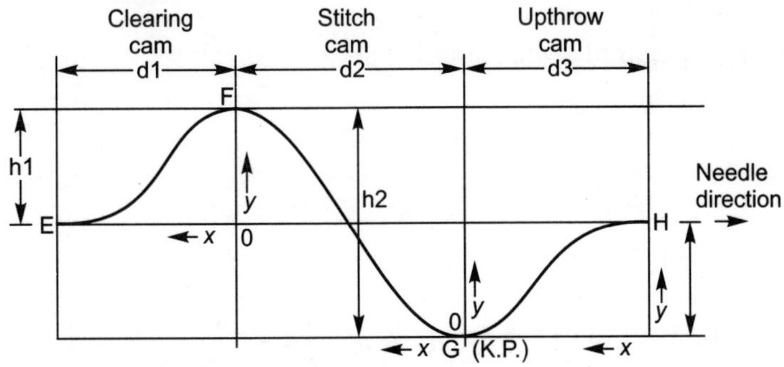

Fig.11.8 *Displacement curve of a non-linear knitting cam*

The polynomial equation (11.3) is used for the solution of symmetrical knitting cam. In the above non-linear displacement curve (Fig. 11.8) for the knitting action, the boundary conditions are at E and F for the clearing cam, F and G for the stitch cam, and G and H for the up throw cam. The principal dimensions h and d of each part of the cam are taken into account. Equations for the non-linear displacement curves are designed to give continuous displacement, velocity and acceleration throughout the knitting cam at all boundary conditions. It is essential that the boundary conditions of displacement, velocity acceleration and pulse agreed at the positions F and G, to ensure that the above curve characteristics were continuous throughout the knitting action.

Oldham et al[9] have investigated the suitability of three non-linear profiles and evaluated according to the various knitting and dynamic requirements. Three types of cam profiles viz., polynomial cams, simple harmonic cams and composite-sine-curve cams were considered for their suitability of high speed knitting.

(*i*) *Polynomial cams*: The equation for the polynomial profile as advocated by Black[10] has the form

$$y = \sum_{n=0}^{6} c_n x^n \quad \text{... (11.7)}$$

The designer is free to prescribe seven boundary conditions corresponding to seven constants.

The maximum slope of the cam profile at zero velocity is found as

Maximum slope $= \left(\dfrac{26}{15}\right)\cdot\left(\dfrac{h}{l}\right)$ ... (11.8)

where h is the cam height and l is the length within which the movement takes place. A degree six polynomial has five turning points and pairs of these may fall within the region of interest for the curve to match the prescribed conditions. Consequently, the designer may have to supply several sets of conditions to achieve a suitable profile.

(*ii*) *Simple Harmonic Cams* : Knapton and Lau[11] preferred simple harmonic cam profile for knitting plain single jersey fabrics at high speed. The profile is

$$y = a \sin\left(\frac{\pi}{b}x\right) \text{ in the range } -\frac{b}{2} \leq x \leq \frac{b}{2} \quad \text{... (11.9)}$$

where a and b are constants.

Having prescribed cam height h and length l as

$$h = 2a \quad \text{... (11.10)}$$

$$l = b \quad \text{... (11.11)}$$

the maximum slope occurs when x is zero, is given by

$$\text{maximum slope} = \left(\frac{\pi}{2}\right) \cdot \frac{h}{l} \quad \text{... (11.12)}$$

Both the simple-harmonic and polynomial cams have a finite acceleration at each end of the segment instead of the infinite acceleration obtained with linear cams.

(*iii*) *Composite sine-curve cams* : Composite sine-curve cams can be formed by joining sine segments and straight lines or dissimilar sine curves. The profiles for a typical double sine curve as shown in Fig. 11.9 is given by

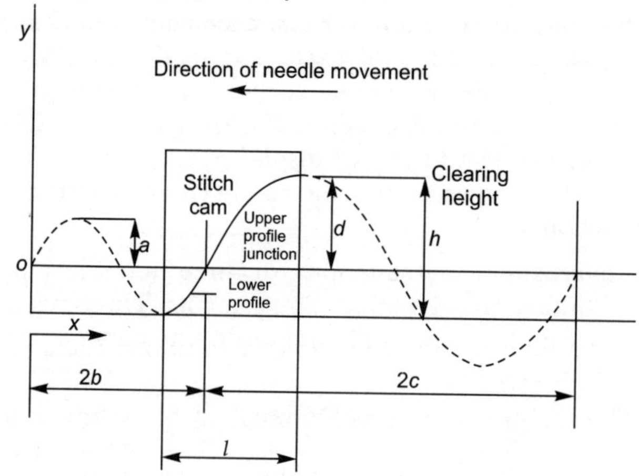

Fig.11.9 *Double-sine cam profile*

$$y_{\text{lower}} = a \sin\left[\frac{\pi x}{b}\right] \text{ for } \left[\frac{3b}{2}\right] \leq x \leq 2\,b \quad \text{... (11.13)}$$

$$y_{\text{upper}} = d \sin\left[\frac{\pi}{c}(x - 2b)\right] \text{ for } 2b \leq x \leq 2b + \frac{c}{2} \quad \text{... (11.14)}$$

Where a and d are the amplitudes. Positioning the intersection of the two curves on the x axis ensures continuity of both displacement and acceleration.

If h and l are prescribed as

$$h = a + d \quad \text{... (11.15)}$$

and $$l = \frac{b+c}{2} \qquad \text{... (11.16)}$$

Again, as for the simple harmonic cam,

$$\text{Maximum slope} = \left(\frac{\pi}{2}\right)\cdot\left(\frac{h}{l}\right) \qquad \text{... (11.17)}$$

The maximum slope does not depend on the ratio of the amplitudes *a* and *d*. The amplitude and period of the lower sine curve are critical because it is this part of the cam profile that controls the needle through the knitting zone. Generally, it is desirable that all the needles should maintain cam contact until the very bottom of the stitch cam is reached.

It has been concluded that the composite sine curve cam uses equations that are as simple as those of the simple harmonic cam, but are more flexible, since they have an extra degree of freedom. Both have a direct relationship to the needle characteristics. On the other hand, the polynomial equations are more complex and difficult to solve, and the relationship to the needle characteristics is correspondingly less direct.

11.3 HIGH SPEED CAMS WITH ADJUSTABLE STITCH CAMS

The high speed knitting cams, in which a separate adjustable stitch cam could be fitted, might be required for the present day commercial weft knitting machines. The general shape of such a cam is shown in Fig.11.10.

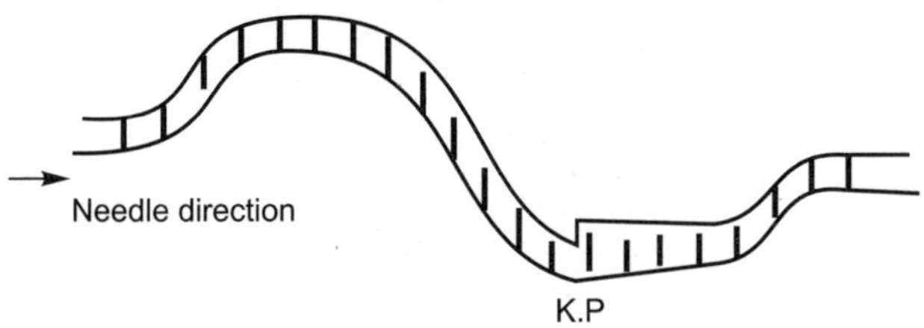

Fig.11.10 *General shape of the high speed straight-and-curved cam*

By using low angled ramps at position K and L it would be possible to fit in adjustable stitch cam, since the needle butts would meet the same slow start of the ramp when the stitch cam was set to different vertical heights. It was considered that, if the angles of ramp were reasonably low (say 10–15°) and the rest of the cam designed with mathematically curved position to give positive needle butt control, then it should be possible to run the machine well above the commercial speed. The portion after the knitting point (KP) was cut away to allow the tension in the yarn, which builds up during the knitting operation, to be relieved instantaneously after a new loop is formed. For the mathematical design of this cam shape, a cycloidal curve was used for the clearing cam, a double harmonic curve for the stitch cam and a polynomial curve for the up throw cam. It is a straight and curved cam system and the cam was calculated to give a maximum cam angle of 55° for each rise and fall of the displacement motion.

References

1. Michio Okamoto, Journal of Text. Machinery Society of Japan, Vol.16, No.1, 1970
2. Heinrich R, Ray.fib.synth. 19, 2 133, 1963
3. Henshaw D.E., TRJ, 1968, 38, 592
4. Munden D.L, JTI, 1960, 51, P712
5. Knapton JJF and Munden D.C., TRJ, 1966, 36, 1089
6. Grosberg P in Textile Engineering Processes, Butterworths, London, 1959 P 227
7. (Knapton JJF, T11, Nov, 1971
8. Black D.H. and Mudan D.L , JTI Part I, 1970, 61, 313
9. Oldham K., Burns N.D. and Simms G.J., Non linear cams for weft knitting machines, JTI, 1985, No.1
10. Black D.H. and Mundon D.L, JTI, part I, 1970, 61, 313
11. Knapton JJF and T.W.Y.lau. JTI, 1978, 69, 161

12

Quality Control in Circular Weft Knitting

The term quality refers to obtaining the required specification in a product. Quality control refers to control over the product's quality throughout the process of manufacturing. Before imparting quality control over a process, the areas and the methodology should be identified. The areas refers to the portion where the quality has to be checked and the methodology refers to the nature of quality control program. To obtain a quality end product, control over the quality of raw material, machine and process is essential.

In any manufacturing process, the quality is controlled by three ways such as preventive measures, fault identification and rectification by knowing the causes and on-line quality control.

12.1 PREVENTIVE MEASURES

Requirements on good yarn quality, well maintained knitting machine and good production condition for knitting are to be considered under preventive measures.

12.1.1 Yarn Quality Requirements

Yarn quality parameters such as evenness, breaking strength, elongation, twist, moisture content, yarn winding, yarn lubrication, yarn hairiness are to be considered for quality raw material feed to knitting. (Refer section 1.6 and 3.3)

12.1.2 Machine Quality Requirements

Long lasting and trouble free quality functioning of the knitting machine could be possible by proper maintenance care and lubrication. Proper horizontal installation of the machine, tension free yarn feeding, flawless yarn guides and needles, exact centering of needle bed towards one another, proper fabric take-off and proper lubrication are the basic quality needs of a knitting machine.

Following maintenance schedule may be carried out:

Checking the lubrication unit for sufficient oil and compressed air for specified pressure are the daily checks before switching on the machine.

Cleaning all the yarn passages and yarn path guides elements are the weekly checks to be carried out.

As a monthly maintenance routine, flush needles and cams. If needle beds and cams are heavily fouled, remove the cams and air-blast both the cam and needle beds. Half a litre of needle oil heated to 70° C is poured into the oiling points of the cams and onto the needle heads. Remove cams once more and wipe oil and dirt.

Once in every three months, remove all cam segments, needles and sinkers, thoroughly clean them with compressed air or paraffin. Liberally oil all components and replace.

Once in every three years replace the counter memory battery. Machine bed should be cleaned with petroleum, Hand wheel, take up and cylinder gear should receive few drops of light machine oil everyday.

12.1.3 Knitting Production Conditions

(*i*) *Suitable yarn count:* Selection of suitable yarn count should be based on:

(*a*) Machine gauge

$$\text{Yarn Tex} = \left(\frac{100}{G}\right)^2 .$$

(*b*) Machine types which are having vari d needle strength hook sizes and dial and cylinder distances

(*c*) Knitted structures which are produced with from one feeder (Plain, rib etc.,) to 3 or 4 feeders (blister and multicolor jacquards). More number of feeders necessitates the use of finer counts.

(*ii*) *Machine setting* : Optimum setting is based on yarn type and knitted structure.

Following factors are to be considered during setting:

(*a*) Balanced yarn tension prior and after to feeder

(*b*) Lower fabric take up tension

(*c*) Proper needle timings of dial and cylinder needles in order to obtain loose or tight structures.

(*iii*) *Yarn storage* : For the yarns to have sufficient moisture for processing, they should be stored at 20°c and 65% RH. Storage under extreme temperatures must be avoided. Higher temperature leads to paraffin migration and lower temperature leads to water condensation.

(*iv*) *Air conditioning* : Air conditioning of knitting plant prevents yarn dry-up, reduce yarn breaks (holes) and improves the surface structure of fabric. The recommended conditioning is 55% + 10% RH and 25°C + 3°C temperature.

12.2 ON-LINE QUALITY CONTROL

Preventive measures and faults identification and rectification are the conventional practices of quality control. The modern and effective approach in on-line quality control through computer aided tools. In such cases many sensors and/or transducers act as a monitoring unit for faults and deviation from set values. This gives message to the central computer which will interpret the results and send the control signals to the controllers which are mostly electronically controlled mechanical devices. And these controllers either control the process directly or stop the process temporarily. In Modern machines, the sensors and controllers are paced as,

- Yarn monitoring sensor.
- Automatic lubrication monitoring and controlling unit.
- Fabric monitoring CCD cameras.
- Speed control servo motors.
- Production monitoring and displaying unit.
- Needles monitoring and electronic needle control system.

12.3 DEFECTS IN KNITTED FABRICS, THEIR DEFINITION AND CAUSES

SI. No.	*Defects*	*Definition*	*Causes*
1.	Holes or cracks	No difference between holes and cracks. During loop formation, already broken yarn forms holes of different sizes	Bad needle, take down mechanism too tight, high tension on yarn, bad yarn needle too tight in their slots, dial height too low or too high, badly tied knots, improper stitch setting
2.	Drop stitches/cloth fall out	Yarn not properly laid in the needle hook	Take down mechanism too loose, defective needles, too loose yarn tension not sufficient, wrong needle timing set, needle tricks closed
3.	Vertical Lines	Longitudinal gaps in fabric – irregular space between adjacem-ent wales – broken appearance in fabric	Defective needle, dirt in needle slots, needle too loose or two tight in the tricks, needles not enough lubricated
4.	Tuck or Double loops	Badly or non knitted loops, showing thick places or small beads in fabric – a shadow appearance	Take down mechanism too loose, improper stitch cam setting, needle move too freely in their slots, dial height too low.
5.	Barre or horizontal stripes	Uneven courses traverse horizo-ntally and repeat regularly or irregularly	Bad yarn, uneven tension, yarn slippage in positive feed, improper stitch cam setting.
6.	Bunching–ups/Piles ups	Fabric appear as bands	Fabric too tight, takedown mechanism too low, too much blister in fabric.
7.	Colour fly/tinges	Consists single fibers, bunches of fibers or yarn pieces of varying colours	Congested machine installation, producing colour jacquards along with single colour fabrics, proce-ssing a large number of colours
8.	Bow	Courses or rows of stitches curved instead of straight	Dial tilted, fabric not level in takedwon rollers, nip pressure not constant from end to end, fabric spreader off centre or tilted

Contd.

9.	Skew	Courses or rows of stitches straight but not at perpendicular to the fabric edges.	Mainly associated with take down rollers and spreader. At stenter, piece ends not being joined course to course.
10.	Needle trace	A properly knitted wale which, for a variety of reason stands out prominently from those adjacent to it on either side.	A new needle which is tight in its trick, a worn trick a bout needle, tight latch.
11.	Centre line	A line down the centre of the fabric.	Fading of fabric along folded edge due to exposure to sun-light.
12.	Soiled yarn	Dirty streak or streaks along a course or courses.	Contamination of cones due to miss handling, knot tying with dirty hands.
13.	Rust mark	Reddy stain on fabric usually associated with a hole in fabric.	Rusty needle, rust in tricks
14.	Slub	Short bulky length of yarn, identified by broken end of theread.	Broken filaments being pushed back along the thread line excessive tails of knitter–knot.
15.	Loop distortion	Distorted stitches lead to a very unsettled fabric appearance. The fabric appearance is skittery.	Bad and bent needles, bent trick walls, uneven yarn tension, needle timing set wrong, yarn carriers set wrong

Though the above faults are due to various sources such as yarn, machine and processing imbalances, some differences in yarn properties may directly be related to some common faults as given below:

(*i*) Low strength weak yarn leads to holes and cracks.
(*ii*) Very uneven yarn produces cloudy fabric.
(*iii*) In sufficiently waxed yarn leads to drop stitches and cloth fall out.
(*iv*) Bad dyed yarn produced horizontal stripes.
(*v*) Too hairy yarn leads to diffused stitch appearance and fluff build up.
(*vi*) Yarn with long period unevenness produce fabric with stripes.

12.4 QUALITY TESTS FOR WEFT KNITTED FABRICS

To assure quality production, it is necessary to have on the spot inspection of the cloth as it comes off the machine as well as post inspection of the fabric. The usual method of examination is to feed the fabric over a lighted frame of a table at a normal speed of 10–12 meters per minute for single colour.

Following are the tests that the finished knitted fabrics normally undergo:

1. **Fabric yield:** Piece weight gives a correct idea about the quality of the fabric. Sample ourses are marked for required number of revolution and the fabric is cut and weighed to determine the yield and width. However machines with revolution counters enable to get accurate number of courses to be knitted per piece.

2. **Fabric appearance:** The fabric is inspected for appearances to determine its acceptability from quality view point AAMA (American Apparel Manufacturers Association). Four point grading system is widely used. The maximum points per 100 yd^2 should be less than 40. The size of defects and their respective points are considered for evaluation.
3. **Fabric pilling:** 12 × 12 cms fabric sewn on 15 cms long tube of rubber is placed inside the cork-lined pill box, rotated for a given number of revolutions. The pills on the fabric surface are compared to standard photographs and graded.
4. **Fabric extension:** In an extension tester, the sample piece is applied between two jaws and a uniform weight is applied to the fabric and the percentage extension is measured.
5. **Air permeability:** The apparatus measures the rate of airflow through the fabric. The sample is clamped between two circular discs and air is drawn through the fabric by a suction pump, the area of the fabric being specified.
6. **Mass per unit area:** Fabric weight in terms of grams per square meter (GSM) can be obtained either by using template or by calculating through yarn linear density, stitch density and loop length. Also refer section (14.3)
7. **Fabric width:** Width shall be measured with an accurate tape after laying circular knitted fabric flat on the table without tension or elongation (Ref: ASTM 3887–80). Width determination is based on one of the following methods:
   (*a*) Width between gummed edges of gummed fabrics.
   (*b*) Width between tenter frame pin marks when pin marks remain in shipped fabrics.
   (*c*) Overall width of circular knit fabric when (*a*) and (*b*) not exist. For a width stated as 60/62 inches for example, the lower figure governs cuttable width.
8. **Length:** Length shall be measured with any surface contact device (Trumeter or equivalent) that is calibrated regularly. The device contacts the back or a smooth surface of the fabric. (Ref. ASTM D1910 – 64 hand method) Actual yardage of each piece shall be accurate to within plus or minus 2 % when measured by the above method.
9. **Visible courses and wales (WPI & CPI):** Measured using counting glass, after laying fabric in flat horizontal
10. **Loop length and course length** (Refer section 14.3)
11. **Yarn tex** (Refer section 14.3)
12. **Tightness factor** (Refer section 14.3)
13. **Reference relaxation:** A fabric is relaxed by being subjected to one cycle of washing and tumble drying followed by four cycles of rinsing and tumble drying.
14. **Dimensional changes:** A fabric is subjected to a specified washing procedure, dried under the appropriate conditions and any changes in dimensions are determined.

$$\% \text{DC} = 100 \times (\text{B} - \text{A})/\text{A}$$

    Where DC = Dimension changes
    A = Original dimension
    B = Dimension after laundering
15. **Angle of spirality:** The angle of spirality is measured before and after a specified relaxation process and the change obtained (Refer section 10.4.3).
16. **Abrasion resistance:** Evaluation is made on the basis of weight loss of the specimen. Laboratory type abrasion testing instruments are used.

17. **Bursting strength:** It is the strength of a fabric against multi directional flow of pressure. Knitted fabrics cannot be tested easily in strip form and hence this is the preferred method. A bursting type of force is applied perpendicular to the surface of the fabric. The test can be conducted on hydraulic pressure principle.
18. **Visual assessment of barre:** Barre intensity is rated by visual comparison with the uniformity reference scale. (Ref. AATCC 178 – 1994)
19. **Fabric bow:** It is a fabric condition resulting when knitted courses are displaced from a line perpendicular to the selvages and form one or more arcs across the width of the fabric as in Fig. 12.1 The measurement is as follows:

A straight edge is placed across the fabric between the points at which a marked knitted course meets the two edges. The greatest distance between the straight edge and the marked course is measured parallel to the edges *i.e.*, Distance D in Fig. 12.1.

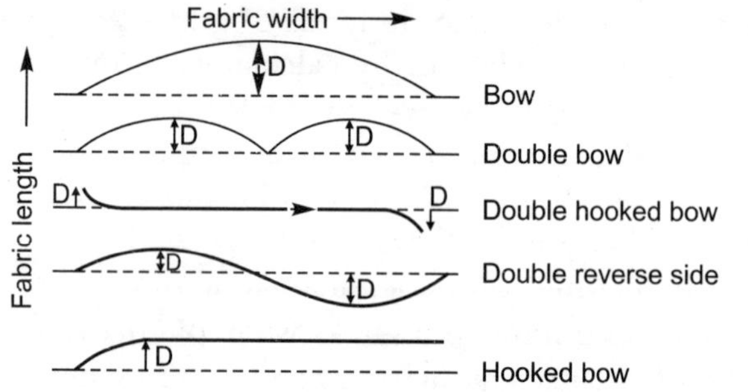

Fig. 12.1 *Typical bow conditions*

20. **Fabric skewness:** Skewness or bias is a fabric condition resulting when knitting courses are angularly displaced from a line perpendicular to the edge of side of the fabric Fig. 12.2.

Skewness is measured in three places spaced as widely as possible along the length of the fabric (1yard). If possible make no measurements closer to the ends of the roll or piece of fabric than 1 m. Draw a line perpendicular to the selvage across the fabric from a point C where the marked course meets one selvage, meeting the other selvage at point B. Measure the distance between points A and B or D and B as shown in Fig. 12.2. Record the three or more skewness or bias as a percentage of the fabric width.

Skewness (bias) % = Distance AB or DB × 100 / Width BC

(Reference; ASTM D 3882 – 85)

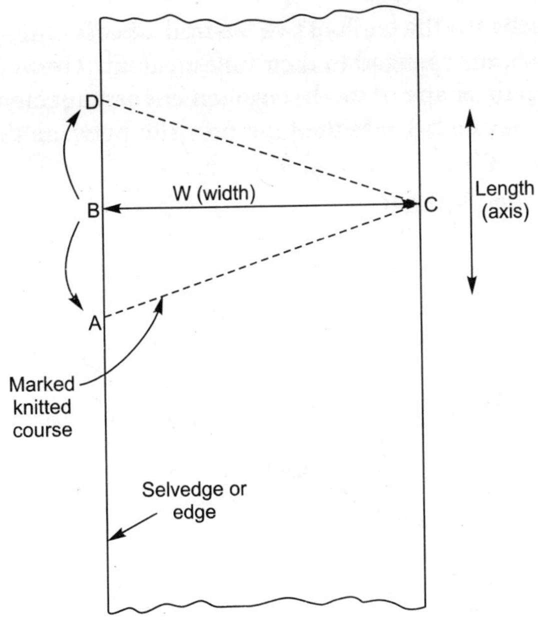

Fig. 12.2 *Typical skewness (Bias) condition*

12.5 YARN DELIVERY DEVICES IN CIRCULAR KNITTING

The most fundamental factor that influences the knitted fabric quality is the quantity of yarn that knitted into loops in a course. It is very essential to feed exact amount of yarn instead of allowing the needles to take yarn according to the knitting conditions, so that the needles can consume fixed amount of yarn and thus the stitch length will remain constant. The devices which feed the required quantity of yarn to the knitting zone at different feeder positions are basically of two types viz., yarn metering devices and yarn storage devices.

12.5.1 Yarn Metering Devices

These devices work on the principle of constant yarn consumption per unit time per feeder. The devices consist of an element which meters the amount of yarn to be delivered into the knitting zone. There are two types viz., toothed delivery wheel and tape delivery device.

12.5.1.1 Toothed delivery wheel (Fig.12.3) :

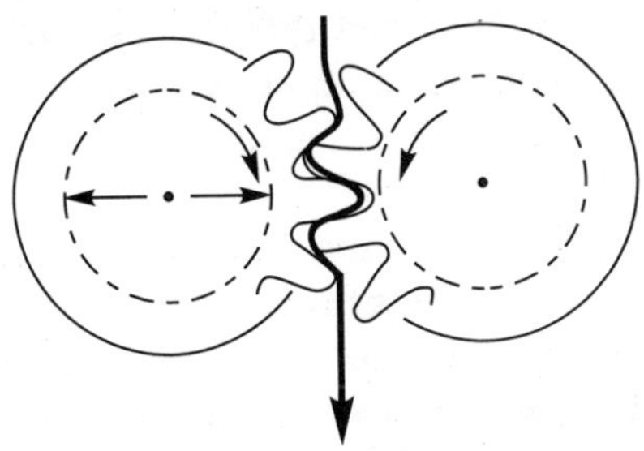

Fig. 12.3

The yarn is driven between the teeth of two tapered wheels. The yarn delivery rate is greatest when the toothed wheels are engaged to their fullest extent. The yarn metering is varied by the distance between the teeth, shape of tooth, engagement arrangement and depth of penetration. It is not a positive feed device because the yarn may slip between the teeth and the rate of yarn delivery changes continually.

12.5.1.2 Tape delivery device (Fig.12.4):

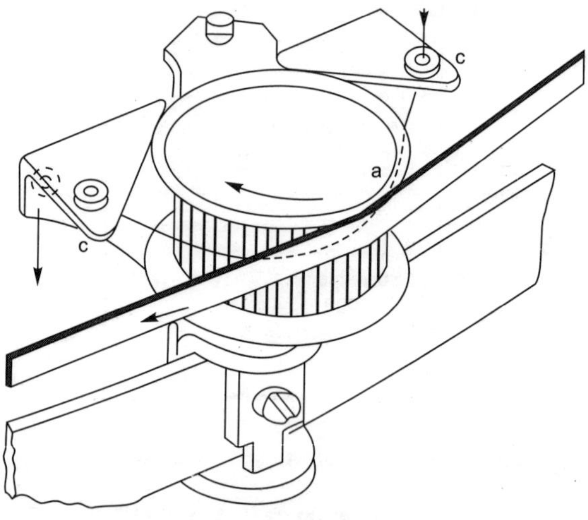

Fig. 12.4

At each feed, the machine is supplied with a smooth freely rotating wheel (*a*) and two guides (*c*) fixed at a certain distance. An endless tape passes around the full circumference of the machine, driving the freely rotating rollers. The yarn passes between the tape and the wheel (*a*). The tape is driven by a variable speed unit which is in turn, driven from the main machine drive. The variable speed unit enables the rate of delivery to be controlled. Fabrics knitted with this positive feed device are of much better quality. Tape frictional properties and variation in yarn tension are the disadvantages of the system. The device can only be used to knit structure that requires constant yarn feed at all the feeders in the machine.

12.5.2 Storage Feeding Devices (Fig.12.5)

The device works on the principle of constant yarn consumption per unit time per feeder with constant yarn reserve. Yarn input and output are equal and the process is simultaneous. This device is useful for the structures which need different amount of yarn to produce various courses that form the design. Also variation in input yarn tension can be avoided in this system. As the yarn is not directly fed to the knitting zone, unwinding tension variation due to package size changes can be avoided.

The yarn is drawn from the cone and wound onto a small coiler or cylindrical drum. The required amount of yarn is unwound from this drum when the needles form the loops. A certain amount of yarn reserve *i.e.*, constant number of coils are maintained in the drum. As the coils uncoil into the knitting zone, coiling automatically happens and the number of coils to be coiled will be equal to the number of coils uncoiled. Because of this arrangement, the yarn is delivered at a constant tension to the needles. The device handles very delicate yarns. Stop motion and indication light are also equipped with the device in the event of yarn breakages.

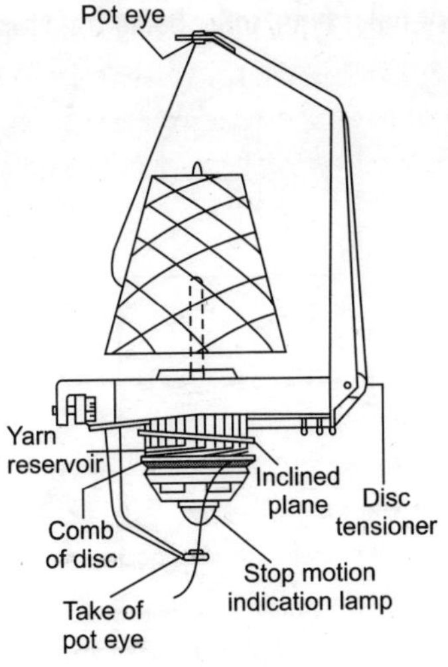

Fig. 12.5

12.5.3 Positive Feeding Devices

The device works under the principles of variable yarn delivery per unit time per feeder with variable yarn reserve. Yarn input and output are unequal and the process is not simultaneous. The yarn to be fed into the knitting zone is wound in the coiler. The number of coils will be a certain number at the starting. When this number of coils reach, the feeding to the coiler from the cone stops. Then the uncoiling starts and it goes until it reaches a certain number of coils which is the threshold value by the help of a sensor. As soon as the threshold reaches again the coiling starts as well as uncoiling. The process continues until it reaches the maximum number of coils. The prime requirement for the device is the coiling speed should be higher than the uncoiling speed.

Such positive feeding devices are of three types. Uniform feed and uniform tension are the advantages of these devices.

12.5.3.1 Positive feeder with rotating coiler type

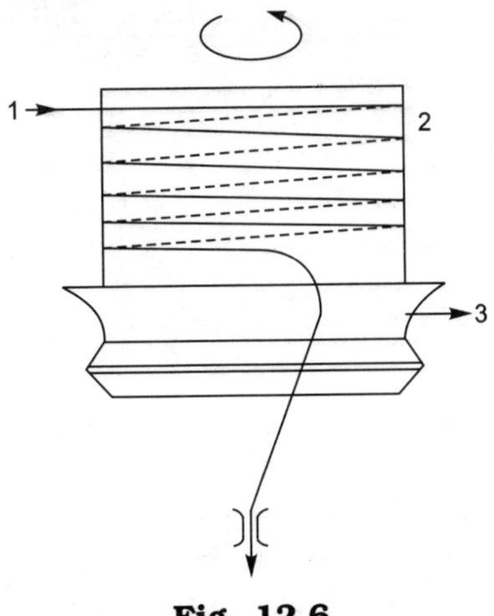

Fig. 12.6

Rotating coiler (2) with tangential take-up (1) at the top of the coiler and over end yarn let-off over yarn brake ring (3).

12.5.3.2 Positive feeder with stationary coiler type, coiler being from top to bottom

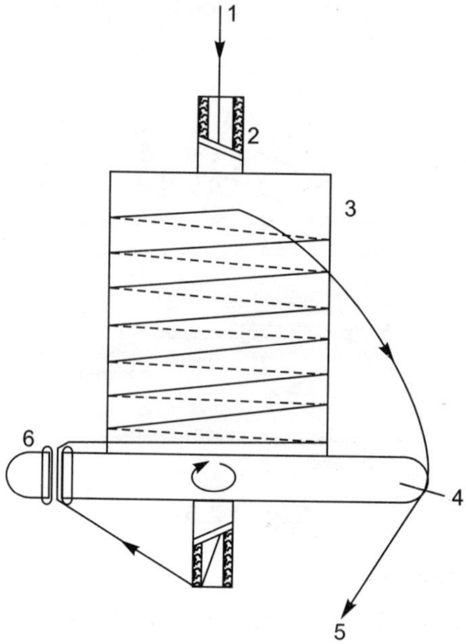

Fig. 12.7

Stationary coiler (3) with yarn take-up (1) from top through hollow axis (2) to bottom, tangentially yarn wind-on at bottom of coiler (3) through rotating disc (4) and yarn guide eye-let (6) and over end yarn let-off.

12.5.2.3 Positive feeders with stationary coiler type, coiling being from bottom to top

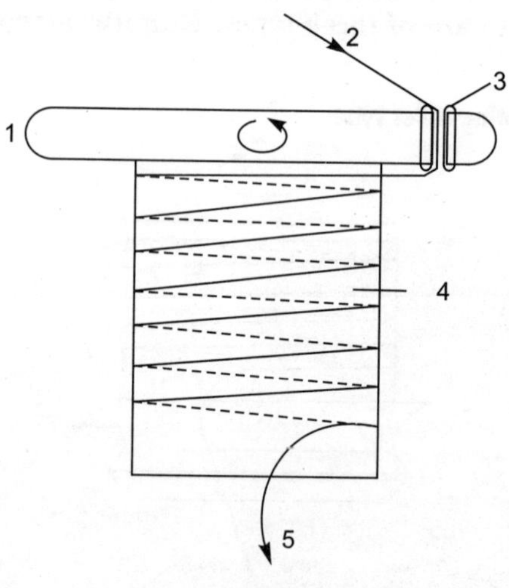

Fig. 12.8

Stationary coiler (4) with yarn take-up (2) over a rotating disc (1) and yarn guide eye-let (3) at top of coiler (4) and over end yarn let-off (5).

13

Circular Knitting Developments

13.1 INTRODUCTION

Today the consumption of the knitted goods have increased rapidly especially in outerwear. There is a strong consumer demand for garments, which combine comfort and fashion appeal and confirm to the more relaxed life style of today. The outerwear is a fashion sensitive area requiring a range of designs and colours, and the producer has to respond quickly and accurately to the changes in consumer taste while maintaining the product quality. To meet the above requirements the machinery manufacturers made striking development in knitting machine by implementing the aspects of increased and improved automation, better-built machines, improved electronics, aerospace metals in machine components and simplified operation. Versatility and the ability to change styles quickly were the main focus of most manufacturers of knitting machinery.

13.2 DEVELOPMENTS IN FRAME, DRIVE, BRAKE AND LUBRICATION

Frame: The knitting machines have three foot frames with base cross beams, for fabric take up, the upper frame with carrier plate having needle cylinder and dial housing, and the yarn bobbin carriers. The machines are manufactured as tall frames for high batch production and wide frames where the tubular fabric undergo as open width fabric take up.

Bearings: For operational reliability and quality of fabric produced, soft start and steady retardation down to zero - Fukahama's oil bath bearing systems are used which enables less damage and prolonged life.

Motors: Inverter motors are used to control the speed electronically by varying the frequency and for smooth start. Stop motors are also used for adjustment of cylinder and dial.

Brake: The machine should be stopped immediately for any disturbance caused due to yarn tension, yarn breakage, needle breakage or any other faults, without causing any fault in fabrics. The electromagnetic disc brakes are used for the purpose. They are fixed in the housing of the main drive armature and according to the force generated by the rheostat, which could be altered, brakes the machine when any impulse is received.

Oil bath gearing: As gearing is the most critical system in ensuring a stable and precision gaiting, all main gear systems are embedded in oil bath to lubricate and prolong the life span of the system. All gear systems are fluted simultaneously to minimize back lash between the dial and cylinder needles. This precision engineering concept enables knitters to conduct fabric changes with ease during gating operations and achieve immaculate plating (especially spandex application) during the knitting operation. Pulsonic Lubrication system is effectively used for the automatic control of lubrication routine and take care of any rise in the temperature. The pulsonic pressure oiler allows an exact injection of smallest oil quantities per impulse. Oil is delivered through a pump under pressure and a distributor provides up to twelve oiling points with oil. The oil flow is controlled electronically. Oiling intervals and oiling periods are synchronized by a connected machine revolution sensor. The supplied oil quantity can be set individually and separately for every single oiling point between 0.1 and 24 ml per hour. This improved device is claimed to be the only oiler, which keeps a pre-set oil quantity per outlet permanently constant.

13.3 DEVELOPMENTS IN FEED REGION

Some important developments have taken place in the arrangement of yarn packages, passage of yarn and method of yarn feeding to the knitting elements.

Side creel: In the latest machines the over-head creels are replaced by side creels due to the difficulty in accommodating more number of packages on the overhead, difficulty in the creeling of package and attending to broken yarns. With the side creel it is possible to provide more than one cone holder per feeder.

Side creels and telescopic aluminum tubes with self-cleaning system ensure clean knit room, thereby reducing the end breakage. Yarn paths have been designed for fewer deflection points, thereby reducing variability in yarn tension, yarn hairiness and lint generation. All surfaces that the yarn must pass over now have high tech coatings to reduce friction and to protect the yarns.

Lint trap creel: As a solution to the problem of contamination from the collected lint and dust in high speed knitting machines the lint trap creel has been introduced. An enclosure is provided for the creel with access for creeling the cones and cleaning the inside of the creel. The creel is provided with an integral humidifier, which maintains the RH of 80 – 85% inside the creel. The lint trap creel ensures (*i*) reduction in the amount of lint and static electricity and (*ii*) improvement in fabric quality and knitting efficiency.

Positive storage feeder: A fundamental factor affecting the knitting quality is the amount of yarn knitted into the course. So it is essential to control the exact amount of yarn, to produce a particular quality of fabric instead of allowing the needles to determine the amount yarn knitted into the course. The positive storage feeder system ensures that the yarns are delivered to the knitting elements uniformly at the required rate and tension. In this device, a common belt driven by the quality Assurance Disc (QAD) assembly drives the individual pulleys of the positive storage feeder. The yarn is wrapped around the pulley that is mounted on the same shaft of the driven pulley. The pulley is designed in such a manner that the wrapped yarn does not get entangled or slip. The number of wraps can be varied to suit the individual requirement. Each positive storage feeder is provided with stop motion LED indicator, which blinks in the case of yarn breakage for easy location of the point of breakage.

Mayer & Cie has introduced a new feeding device Coni EP yarn storage feeder with positive feed or negative feed capabilities. It makes use of totally new thread wheel with a capability to choose the wheel as a positive yarn feeder or as a negative feeder. With combined yarn feeding capabilities only one unit per feed is required instead of two. This is advantageous where both plain and jacquard are knit on the same machine. Special feeders are provided for elastomeric yarn insertion. IRQ positive feed is provided for LYCRA knitting and special thread guide to produce plated (Standard) and triple plated fabrics.

Striper: Colour stripe fabrics are all time fabrics of knitted fabrics. There are always models equipped with yarn strippers whether it is a single jersey or double jersey or plain or electronic jacquard ma-chine. To increase the scope of striping, machines were initially offered with mechanically operated striping. Due to the limitations of its size to be incorporated on modern high feeder density machines, machinery manufac-turers have recently introduced electro mechanical stripers. Selection of each striping finger is effected from a programmable microcomputer in which the design information is fed in. Solenoids above each striping finger are used to transmit striping instruction in accordance with the programme. It has an individual trapping and cutting of yarn at each finger. To produce striped fabrics each feeder is fitted with a4 or 6 colour stripper. Fabrics decorated with 4 to 6 colours with unlimited range of striping width can be produced.

13.4 DEVELOPMENTS IN KNITTING ELEMENTS

13.4.1 Needles

There are stresses developed at the inner arc of the needle hook by the knitting yarn and fabric take down tension. Two different needle types are offered to reduce such problems. The first type needle with standard configuration is designed to allow the hook to bend open before breaking. The second type needle, with special configuration works identically up to the elastic limit of the needle, but instantly breaks the hook just after passing the elastic limit. There is no plastic deformation.

Some manufacturers offer conical needles with pressed flat on four sides of the needle hook (Square cross section) compared to the round cross section of the conventional cylindrical hooks. This new hook configuration guarantees greater rigidity as well as higher elastic limit. They also reduce tensile stresses exerted on the bottom of the hook by up to 40% under identical degree of force. Also the reduced cross section in the tip area of the conical hook creates greater clearance for the thread between the needle and sinker than is the case with cylindrical hooks. This enables trouble free knitting of fancy yarns.

The new steel/plastic composite needles effectively combine the advantages of lowprofile meandershaped needles with those of fullshank needles. The cut-outs are filled with plastic, preventing dirt from collecting and eliminating the problem of contamination.

The latch part of the Latch needles are spring loaded so that the latch will not close the needle after knotching over the old loop. This will of course help to lower the risk of miss feeding during the next loop forming cycle. A break off notch has also been built for the needle, this being an indentation made to the butt of the needles which reduces the force required to break it when needed, as a measure to protect against cam and cylinder damage.Thin needles of 0.3mm are employed to produce silk effect and very fine gauge fabrics using fine yarns. This reduces the strain on the yarn. A hook closer is also employed with this to assist during knitting.

13.4.2 Sinkers

G-Sinkers came out with change in the metallurgy and smoothened surface to reduce the abrasion. The Mayer & Cie introduced a double sinker technology for plush and terry fabrics, by which large float stitches can be achieved. These sinkers, with one throat hold the fabric and with another knock over plaiting hold the yarn to form loops. Changes in the loop length can be brought by setting and by change of dimensions of the sinkers.

In order to reduce the strain on the yarn, Mayer & Cie has introduced a new technology called Relanit. With respect to needle, the sinker is given an up and down movement apart from the radial movement, it normally goes. So, the reduced pulling force on the yarn and also fewer deflection points result in greater productivity, fewer faults in the knitted fabrics, reliable processing, higher working speeds and optimum quality of the knitted fabric.

13.4.3 Cam systems

Cam system developments were directed towards good stability at increased speeds, consistent needle clearing, stitch forming and speedy changeover.The new cam segments are made of Perunal, a high tech aerospace metal which is characterized by outstanding dimensional stability at different operating temperatures, better dissipation of heat, non-oxidizing properties and less wear on the cams and needles. With the computerized numerical control (CNC) on special alloy steel the perfection of track for high speed needle movements is ensured. At the same time to increase the stability at increased speed and vibration free running of needles, 4 or 5 cam tracks are provided in the new cams, thereby resulting in the wide scope for self designing. Single knit machines for production of jersey, Pique, and lay in fleece are equipped with 5 tracks of cams. Double knit machines for production of rib, interlock rib and interlock based fabrics are equipped with 2 tracks j of cams on cylinder and dial.

Nowadays, the cams are designed with gentle angles using the latest computer CAD/CAM technology in the simulation of needle movements. Combining this with the overlapping cam joints concepts, which gently transfer the high speed needles from one cam to another. The high precision earn production facility, the needle wear and vibration are greatly reduced in achieving an accurate and stable knitting operation as shown in Fig. 13.1. With the introduction of twin cam tracks for sinker with two butts, the user can be assured of a more stable, high speed sinker motion while knitting apart from better sinker control, more exact loop formation; increase in the sinker life.

The introduction of Rotary Drop Cam allows easy changing of cylinder or dial cams from the outside of the cam boxes where no yarn rethreading is required during changeover. All cam changes are made externally without the need to remove needles or cam. To get accurate quick style changeover, Mayer & Cie has introduced the "Switching Robot". The robot automatically moves from cam section to change the dial selection to knit, tuck or miss. By this "Switching Robot" rapid, accurate and technician free style changeover is achieved.

Decorative stitches like cable or cabling stitches can be produced by loop transfer cams, by means of which loops can be transferred from cylinder needle to dial needle. Stitch length can be adjusted at each cam segment by turning clockwise or anticlockwise instead of adjusting on each cam segment. The latest machine offers central stitch cam adjustment which can be performed by a since computer input command or by the adjustment of a central bolt that adjusts the cylinder with respect to all the stitch cams in the same time. The stitch length also controlled by stepper motor. Stepper motor controlled stitch cams allow for gradual and partial fabric widening and heightening.

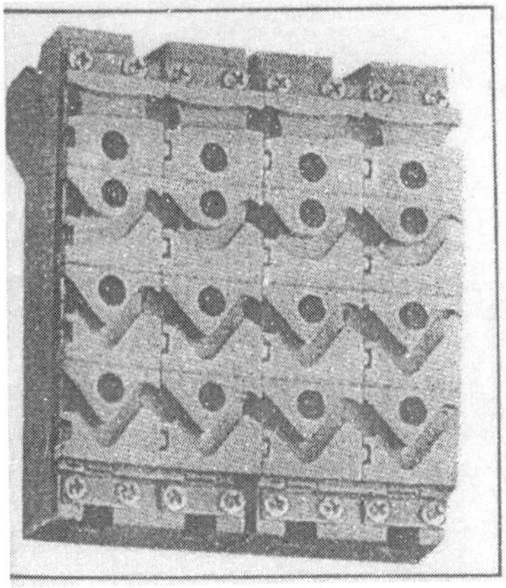

Fig. 13.1

13.4.4 Cylinder

Cylinder size and cylinder gauge can be changed easily and quickly. A full range of cylinder diameter is available from 15 to 34″. Gauge conversion can be carried out with no need to exchange cam parts and feeders. Diameter conversion is within the restriction of the frame size. Interchangeable mechanisms are available in the latest machine, as shown below.

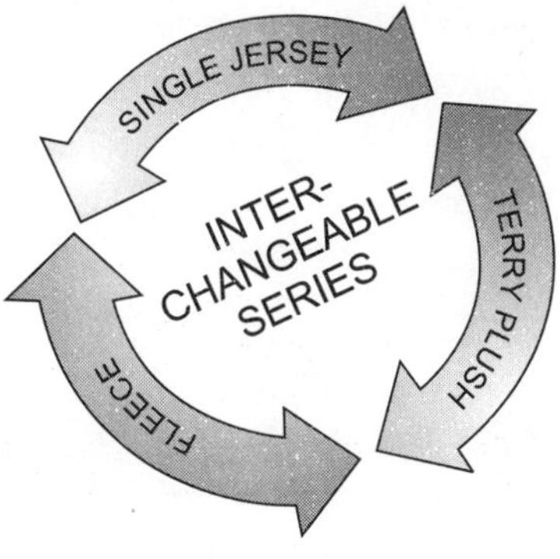

13.5 DEVELOPMENTS IN TAKE-DOWN REGION

Since the present day modern machines are high productive, it is very important to have mechanism for taking down the fabric more effectively. Takeup rollers are now being covered with coating, usually rubber, and have fully adjustable spacers capable of controlling fabrics over the full working width of the knitting machine. Spacer helps to reduce the distortion on the edge of the knitted tube. Consistent and reproducible takedown tensions are possible by computer control of a potentiometer.

Motorised take-up system, where the fabric roller is rotated by one torque motor with infinite control operated independently by a separate switch, is housed in the control box and can be adjusted to fabric tension. The system is adopted with inverter motor controller; which is programmable with one microprocessor provides a multifunction and high efficiency and ensures 100% starting torque as shown in Fig. 13.2.

Fig. 13.2

On the circular knitting machines the fabric is knitted in the form of a tube and is taken down as a sheet of fabric by the take-down rollers. In doing so the fabric passes through cadratex fabric spreader, which eliminates bowing of fabric and equalizes tension as in Fig. 13.3.

Another trend in machine design is to offer the option of extra large rolls. Most machinery manufacturers offer roll sizes up to 920 mm (36″) in diameter. This is important to both the knitter and the finisher. In knitting, machine downtime for doffing is reduced, knitting efficiency is increased and less labour is required. Dyers want long runs with a very few seams where rolls are joined together. For jet dying, the situation of one roll per fabric stand in the dyeing machine is ideal. For continuous bleaching ranges, the longer the roll, the better for reducing the number of seams required.

Fig. 13.3

With the introduction of Open Base Take-Up System the formation of central crease is eliminated. It consists of a system for the cutting and the opening of fabric tube on one side only before the fabric is rolled on to the roller. The width of this fabric roller is two times that of a machine with a standard base as it rolls the outspread fabric. The take-down is carried out on the fabric when it is ready outspread. The fabric edges are stretched by two rollers with worm screw profiles. The fabric is not pressed by the roller and the problem regarding the forming of creases is definitively solved as shown in Fig. 13.4.

Fig. 13.4

13.6 DEVELOPMENTS IN NEEDLE SELECTION

There is a major development in needle selection technique, viz 32 step mini jack needle selection, electronic needle selection and piezo system. In the piezo system a needle selector is composed of two ceramic plates. The two plates are stacked to each other. The principle of needle selection is that when one of the plates is bent by the effect of current and voltage, it makes the levers in the selector to move up or down to execute needle selection.

In the jack needle selection technique, levers in each needle selector are divided by two columns. Each column has 8 levers and each row has 2 levers. They are arranged in different levels to match the height of two butts on a pattern jack. Levers in the needle selector independently push forward the butt on a pattern jack and the pattern jack pushes intermediate jack up to achieve tuck or miss.

In the electronic needle selection technique, each feed is fitted with an electromagnet with pressers or springs. This selection requires only a minimum amount of energy. The slight displacement is then mechanically amplified by the intermediate clavettes, jacks, etc, in order to introduce the needles into the cam track of raising cams. A small magnet in polarized by a solenoid at the same frequency that the needles pass by the feed and according to the information given by the control unit. The electronic needle selection on the dial of a double knit machine allows for jacquard design on both face or back of the fabric and even new designs which result in textural and smooth motifs is possible. The other advantages of elec-tronic needle selection in both dial and cylinder are its quick, simple style and rapid changeover from jacquard to basic fabrics.

13.7 ELECTRONICS IN KNITTING

Electronics application has resulted in true control over the knitting parameters, quick change of fabric setup style changes and reproducibility. Machines of today are now equipped, not only with electronics to monitor basic machine functions, but also with electronics that measure machine parameters and stop the machine, if standards are not met. Many fully electronic machines are able to make style changes.

The following parameters are monitored with electronic machines:

- Operation of needle detectors
- Operation of latch openers
- Course and stitch length
- Yarn tensions
- The operation of quality wheels, tape feeders, takedown tensions
- Cylinder temperature, Oil pressures
- Defects such as barre, needle lines and others

The MCT matic monitoring system with Mayer & Cie controls yarn input tension, quality wheel settings, stitch formation and fabric take-down.

For quick style change, these systems allow fast and simple resetting of different fabric parameters. By keying or entering one command all the necessary settings for a given style, such as yarn input tension, stitch length, central adjustment of belt tension and quality wheel adjustment are carried out automatically. This results in more consistent fabric quality.

Electronic jacquard

Any picture can be reproduced using computer application. To reset the pattern, simply fix the module at the selection device and access the feeder number at the display. Every selector is then perfectly adjusted at the flick of a switch. The module switches automatically from one feeder number to the next. Apart from these applications, electronic needle selection, electronic stitch length control and electronic striping have improved design versatility and stable fabrics.

13.8 DEVELOPMENTS WITH REGARD TO FABRIC

Striped fabrics: Modern machines are produced with wide range of machine gauge. In latest machines, very fine gauge of 42 needles/inch has been introduced which enables use of finer count yarn for producing light weight fabrics. With the introduction of electro mechanical stripers, up to 6 striping fingers provided with each feeder has increased die design versatility. Electronic selection of each striping finger is effected from a programmable Microcomputer, which enables colour stripes of any range of width.

Jacquard fabrics: With the introduction of electronic jacquard knitting machine, any picture can be designed in or scanned into computer graphic system and stored in hard disk or floppy disc. It provides quick download, online production and convenient modification of the designs. The design capability depends on the imagination. One electronic controller simultaneously controls the needle selector and striper finger, thus the machine can produce not only jacquard fabrics, but also striped fabrics.

Terry fabrics: Some machines are equipped with descending sinkers of different specifications. These machines can knit terry fabrics of various thread height by changing different sinkers. The use of plating terry cam can produce thick velour fabric with upright loops. The use of ordinary terry cam can produce polar fleece fabrics with brushed loops on both sides of the fabrics. Elastic terry fabrics can be produced with the addition of the elastic yarns.

Double face terry fabrics: With a special notch in the sinker, the terry machine is able to knit two face terry loops in the fabrics. Length of the face loop is from 1.8 mm to 6.0 mm controlled by various sinker assemblies, and the length of the reverse loop is from 2.5 mm to 3.0 mm by adjusting the sinker cam. The products manufactured by this machine can replace the woven fabrics, including both towels, evening gowns, diapers for infants, pyjamas, beach robes and decorative fabrics.

Provision for the retensioning of terry loops produces a marked calibrating effect. The terry loops are uniform. Less material has to be removed during cutting. Loop length can be adjusted up to 0.3 mm with the graduated head for the knock-over sinkers.

Plated fabrics: Plating technique is the knitting of two threads in the same course selectively in the face and back. At least two feeders are combined to form one complete course with the threads fed to each of the system being formed into loops. The separate threads are plated in the order in which they are fed, *i.e.*, the first thread on the face and the second on the reverse. With each of the feeders capable of undertaking needle selection, which is combined with plating provides new design options in existing mesh structures.

Provision is also made for feeding elastomeric thread Lycra. When the elastic thread is plated in the back, terry like effects are produced on the reverse by the raising of the plating thread floats. But if those threads are plated, the non-plated loops on the face are raised away from the back in a three dimensional manner. These styles are also known as flexi mesh.

Plush or sliver knitted fabrics: High quality sliver is knit as to fabric to get plush fabrics. The 'fiber knit system' is reliable, simple to operate and capable of producing fabrics that are innova-tive with consistent quality. In this type of fabrics, almost all the fiber material is retained in the fabric with very little fiber lost in finishing, because the fibers are anchored twice into the ground of the fabric.

These fabrics are used for apparel, home furnishings, upholstery, automotive and industrial fabrics. Fiber staple from 25 to 35 mm is used. Fabric yield of 180 – 750 gm/m^2 is possible.

13.9 SPECIAL ADVANCEMENTS

Fig. 13.5 *Seamless Circular Knitting Machine*

The rate of fabric production on a circular knitting machine depends on the number of feeders and the rotational speed. Number of feed has been increased up to 108 feeds and 3.4 feeds/inch. The machine speed has been increased to 2.4 m/s.

Nowadays, circular knitting machines have been developed to produce seamless, complete garments that require no subsequent sewing whatsoever. Thanks to the development of one sophisticated technology referring to the circular system of fabric knitting, Santoni has reached a position in which machines can knit a ready-made garment without seams, *i.e.*, "seamless total garment".

The new class of machine SM8–VE as shown in Fig. 13.5, is a single cylinder, single jersey machine which produces garment pieces complete with inturned welt in a very wide range of stitch and pattern effects due to its individual electronic needle selection. Eight knitting feeders, four actuators with 16 levels and a range of standard diameters and gauges provide higher productivity, pattern versatility and the ability to produce seam free, body size briefs and other under garments which require a minimum make-up or at least considerably less than from normal open-width or tubular fabrics.

In its series, Santoni has now introduced a double jersey electronic circular knitting machine SM9 type for the production of underwear, outerwear, swimwear and medical wear, with continuous fabric with single welt and separation thread. The machine can produce any kind of rib and stitches, due to the individual selection of needles on both cylinder and dial, as well as in the transfer position.

The use of knitting equipments for technical textiles is another avenue towards building new design knitting machines. Apparel, technical, automotive, household, medical and surgical products such as vascular prostheses, heart valves etc.

A Circular rib knitting machine of Tritex International produce cuffs (for which flat knits are conventionally used) with 6 feeds, 72 × 72 needles and 4¼ inch diameter. The machine is precision engineered to high quality standards and is provided with the option of an elastic inlay facility.

Havry Lucas designed their interlock circular machine for the production of gusset material including cutting and rolling devices for the bands. Their fully electronic controlled circular knitting machine produces orthopaedic compression hosiery for both medical and sports applications. The electronically programmable machine controls the shape, dimensions and stitch length in steps of 0.01 mm. Interestingly, their plain knitting machine produce fabrics from both metal and plastic materials. These revolving cam box machines are used in filters, electronics shields, vibration fields (catalysator) and in aerospace and defence industries. Also, Bandaging materials are produced with 8 Ne yarn in single jersey machines. There are circular knitting machine with full electronic control of two, three or four colour jacquard patterns with individual needle selection for the production of cylindrical or shaped tubular supports for knees, ankles, elbows etc. The stitch length, shape and design of whole garment can be electronically programmed and controlled.

Circular knitting machines with fully automatic scanner which adjusts itself to the respective knitted surface are also becoming popular. The device efficiently controls longitudinal faults (such as broken needles, oil stripes, drag lines etc.), point faults (such as holes larger than 5 mm) and drag spots (such as oil spots). The scanner can be used on all types of circular knitting machines.

14

Calculations in Weft Knitting

14.1 PRODUCTION CALCULATIONS

Among the various factors those are to be considered in the manufacture of knitted fabrics, it is very important for the knitter to calculate the productivity of a machine in order to be able to schedule production and specify the delivery dates to the customer. Selected examples are given in this chapter to understand the methods of calculating production and efficiency of machines used for weft knitting. Productivity in weft knitting terms refers to the length of the fabric that comes out of the machine, the width of the fabric both single and double width and the weight of the fabrics produced in unit time. The following are the important parameters which decide the production calculations of circular weft knitting.

Machine Parameters

(*i*) Machine speed (rpm)
(*ii*) Machine diameter (inches)
(*iii*) Machine gauge (Needles/inch)
(*iv*) Number of feeders
(*v*) Machine efficiency
(*vi*) Number of needles

Yarn and fabric parameters

(*i*) Yarn count
(*ii*) Stitch length
(*iii*) Stitch density
(*iv*) Wales per inch
(*v*) Courses per inch

Gauge or cut is the number of tricks per circumferential inch of the machine. Though the machine gauge and yarn count are approximately interdependably, knitting a large range of yarn counts on a specific machine gauge is impossible. Similarly with a given yarn number, knitting different stitches is also limited.

The following simple formulae are used to calculate the knitting machine productivity:

(*i*) Fabric production in Yards/Hour is,

$$\frac{s \times f \times 60}{\text{CPI} \times 36}$$

(*ii*) Fabric production in pounds/hour is,

$$\frac{s \times f \times \text{N} \times l \times 60}{36 \times 840 \times \text{Ne}}$$

(*iii*) Fabric weight per linear yard (lbs) is,

$$\frac{\text{N} \times l \times \text{CPI} \times 36}{36 \times 840 \times \text{Ne}}$$

(*iv*) Fabric width (in inches) = Number of Needles/WPI

or Number of needles × Wale spacing

where, Wale spacing = 4 × yarn diameter

(*iv*) Fabric weight (*lbs*) per square yard

$$= \frac{\text{Weight per linear yard} \times 36}{\text{Fabric width in inches}} \quad \text{(for flat fabrics)}$$

$$= \frac{\text{Weight per linear yard} \times 36}{\text{Fabric width in inches} \times 2} \quad \text{(for tubular fabrics)}$$

where S = Machine speed in rpm

f = Number of feeders in the machine

N = Number of needles in the machine

l = Stitch length in inches

CPI = Courses per inch

WPI = Wales per inch

Example 1: *Calculate the production of knitted fabrics in terms of (i) Yards per hour, (ii) pounds per hour, (iii) Fabric weight per linear yard, (iv) Fabric width and (v) Fabric weight per square yard of a knitting machine running with the following data:*

Machine speed	–	20 rpm
Number of needles	–	490
Stitches per foot	–	72
Number of feeders	–	12
Yarn count	–	25 Ne
CPI	–	24
WPI	–	29

The calculations can be made by substituting the respective values in the formulae as given above. The machine efficiency also can be incorporated in this production calculation. Instead of directly by giving the stitch length, some-times the stitches per foot length is also given. This parameter is generally obtained by making two marks of 1 foot space in the yarn before knitting. After the yarn is knitted into fabric, number of loops between these two marks are noted down, which gives the stitches per foot length. From this the stitch length can be calculated as (as per the above example):

Stitch length $= \dfrac{12}{72} = 0.17$ inches

(*i*) Yards per hour $= \dfrac{s \times f \times 60}{\text{CPI} \times 36}$

$$= \frac{20 \times 12 \times 60}{24 \times 36} = 16.67$$

(*ii*) Pounds/hour $= \dfrac{s \times f \times \text{N} \times l \times 60}{36 \times 840 \times \text{Ne}}$

$$= \frac{20 \times 12 \times 490 \times 0.17 \times 60}{36 \times 840 \times 25}$$

$$= 1.59$$

(*iii*) Fabric weight (*lbs*) per linear yard

$$= \frac{\text{N} \times l \times \text{CPI} \times 36}{36 \times 840 \times \text{Ne}}$$

$$= \frac{490 \times 0.17 \times 24 \times 36}{36 \times 840 \times 25}$$

$$= 0.095$$

(*iv*) Fabric width $= 490/ 29$

$= 16.9$ inches (open width)

$= 8.45$ inches (folded width)

(*v*) Fabric weight per square yard

$$= \frac{\text{Weight per linear yard} \times 36}{\text{Fabric width in inches}}$$

$$= \frac{0.095}{16.9} = 0.0056 \textit{ lbs}$$

14.1.1 Fabric Length

The factor F is to be considered in calculating the linear length of knitted fabrics from the circular weft knitting machines. This factor is determined by the respective fabric structure. The factor F is taken into account the fact that a full course of loops is not always knitted at each feeder. For example, with plain knit fabric a complete course will be knitted at each feeder, therefore F = 1. However for a double jersey structure, such as interlock, each feeder only knits half a course of loops, therefore two feeders are required to knit a full course, so F = 0.5. For any structure, F can be calculated as,

$$\text{F} = \frac{1}{\text{Number of feeders required to knit one full cource of loops}}$$

Therefore, by incorporating the factor F,

$$\text{Fabric length in yards/hour} = \frac{s \times f \times 60 \times F}{CPI \times 36}$$

14.1.2 Fabric Weight

The constant C is important while calculating the production of knitted fabrics in terms of fabric weight. This constant also varies according to the fabric structure.

The constant C can be incorporated as

Fabric weight (*lbs*) per square inch

$$= \frac{WPI \times CPI \times l \times C}{36 \times 840 \times Ne}$$

Since the number of courses and wales per centimeter are measured on only one side of the fabric and double jersey structures have two sets of loops, the constant C has to be included in the equation to allow for the second set of loops. For single jersey structures, where there is only one set of loops, C = 1. For balanced double jersey structures, C = 2. For other structures, C has to be determined individually.

Example. 2: *Given a 14 gauge, 20 inch diameter, 40 feeder circular weft knitting machine operating at 25 rpm and 90 % efficiency, calculate the length, width and weight of the fabric that could be produced per hour using a two fold resultant yarn count of 12 Ne, Knitted to the following structures and specifications.*

(*i*) 1 × 1 Rib, loop length = 0.2 inch, CPI = 27.0 WPI = 20

(*ii*) Interlock, loop length = 0.12 inch, CPI = 37, WPI = 30

For 1 × 1 Rib, (F = 1, C = 2):

$$\text{Length (yards)} = \frac{40 \times 25 \times 60 \times l}{27 \times 36} = 61.73$$

$$\text{Width (inches)} = \frac{14 \times 20 \times 3.14}{20} = 43.96$$

$$\text{Weight } (lbs/\text{inch}^2) = \frac{27 \times 20 \times 0.2 \times 2}{36 \times 840 \times 12} = 0.000595$$

For interlock (F = 0.5, C = 2):

$$\text{Length (yards)} = \frac{40 \times 25 \times 60 \times 0.5}{37 \times 36} = 22.52$$

$$\text{Width (inches)} = \frac{14 \times 20 \times 3.14}{30} = 29.30$$

$$\text{Weight } (lbs/\text{inch}^2) = \frac{37 \times 30 \times 0.12 \times 2}{36 \times 840 \times 12} = 0.000734$$

14.2 OPTIMUM KNITTING CONDITIONS

Though the right choice of knitting machine for the knitting of appropriate knitted fabrics is essential, the selection of fibers and yarns are also important. Moreover the emphasis is to be given on optimum knitting conditions by which the knitting takes place to maintain the product uniformity. The following parameters are to be given due weightage during knitting.

(*i*) *Tightness Factor:* It is the ratio of the area covered by the yarn in one loop to the area occupied by that loop.

Tightness factor, $K = \dfrac{\sqrt{\text{tex}}}{l}$ where l is loop length in cm.

Tightness factor ranges from 11 (for slack fabrics) to 19 (for tight fabrics) and an average of 15 is preferable, which is optimum in general.

(*ii*) *Machine Gauge:* It is the number of needles per inch arranged on the needle bed. Yarn number and machine gauge are related as

(*a*) Yarn Tex $= \left(\dfrac{100}{G}\right)^2$

(*b*) Worsted count $= \dfrac{G^2}{10}$ where G = Needles/inch

(*iii*) *Input tension* $= \dfrac{0.1 \text{ gms}}{\text{tex}}$

(*iv*) *Loop length (l)* $= \dfrac{\sqrt{\text{tex}}}{K}$ cms, where K is tightness factor

(*v*) *Course length, L* = Loop length × number of needles per course.

Course length is the length of yarn consumed for full course during knitting.

(*vi*) *Run-in and run-in ratio:* Run-in is the yarn in take (inches per minute) at each feed for a given machine speed. It can be calculated by knowing the loop length, number of needles and machine speed.

Run–in = course length × rpm

Run-in ratio is the simple ratio between sets of feeders, feeding yarn at different run-ins. For a knitted structure produced with two feeders F1 and F2 which are supplying yarns at different run-ins.

Run-in ratio $= \dfrac{\text{Course length of F1}}{\text{Course length of F2}}$

In order to calculate the run-in for structures other than plain knits the following assumptions can be made.

(*a*) For rib gated structures, consider all the loops as plain loops

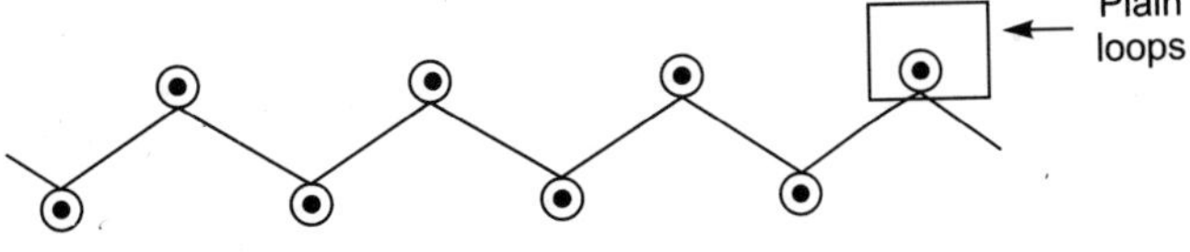

Fig. 14.1

(*b*) For interlock gated structures, loops are produced on both dial and cylinder and they are considered to be rib loops.

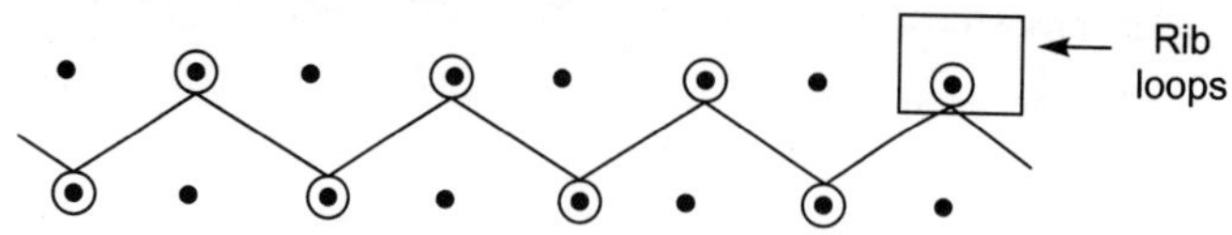

Fig. 14.2

(*c*) The knitted loops produced on dial needles only or cylinder needles only are considered to be plain loops.

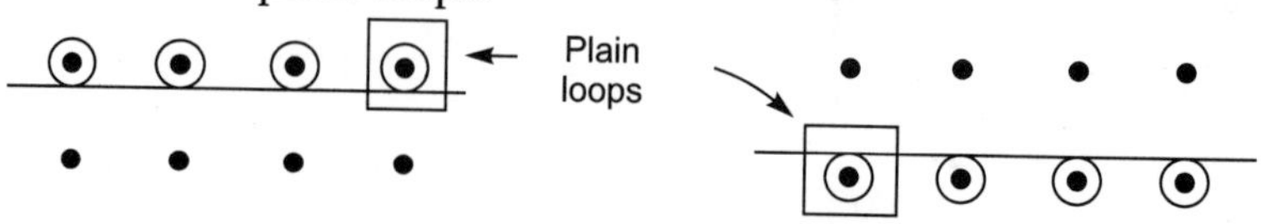

Fig. 14.3

(*d*) For both rib and interlock gated structures, the calculation for loop lengths of miss stitch and tuck stitches are,

$$\text{Miss stitch length} = \frac{1}{\text{Gauge}}$$

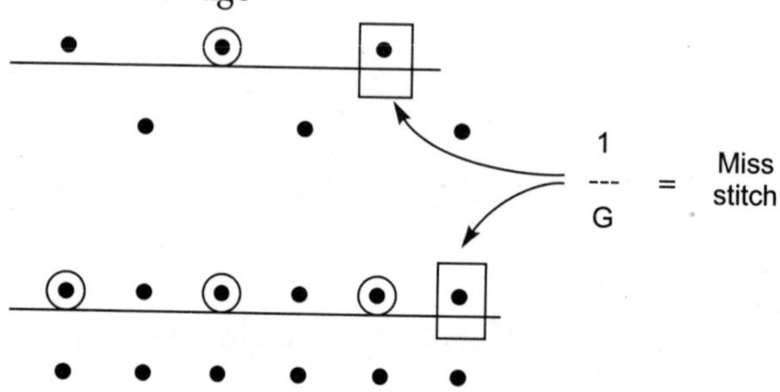

Fig. 14.4

Tuck Stitch length can be assumed as plain loop length in case of rib gaiting and rib loop length incase of inter lock gating.

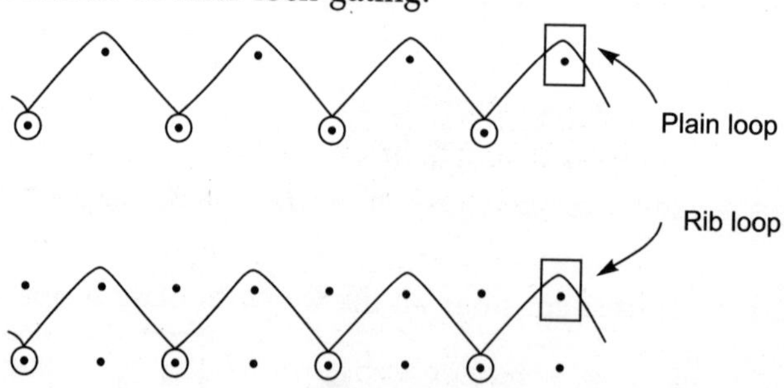

Fig 14.5

(*vii*) Fabric width (in cm) $= \dfrac{\text{Number of needles}}{\text{wales/cm}} = \dfrac{L}{kw}$, where L = Course length

(*viii*) Courses per centimeter,

$$\text{cpcm} = \frac{kc}{l}$$

(*ix*) Wales per centimeter,

$$\text{wpcm} = \frac{kw}{l}$$

(*x*) Stitch density,

$$S = \text{cpcm} \times \text{wpcm}$$

(*xi*) Fabric weight (gms/m^2),

$$\text{GSM} = \frac{ks \times \text{tex}}{100 \times l}$$

where $ks = kc \times kw$

where *kc*, *kw* and *ks* are constants for the geometry of plain knitted fabrics in various stable states, such as dry relaxed, wet relaxed and finished relaxed states.

or
$$\text{GSM} = \frac{S \times l \times T}{100}$$

where S = Stitch density (loops/cm^2)
l = loop length (mm)
T = yarn tex

Example 3: *For a circular plain knitting machine of 24 gauge, 30" diameter, 96 feeders and 35 rpm speed, find out the optimum knitting conditions.*

(*i*) $\text{Tex} = \left(\frac{100}{G}\right)^2 = \left(\frac{100}{24}\right)^2 = 17.3 \text{ tex} \approx 18 \text{ tex} \approx 32 \text{ Ne}$

(*ii*) Loop length: for a optimum knitting, tightness factor = 15

i.e., $15 = \frac{\sqrt{\text{tex}}}{l}$

Loop length $= \frac{\sqrt{18}}{15} = 0.28 \text{ cm}$

(*iii*) Input tension $= \frac{0.1 \text{ gms per tex}}{\text{tex}} = 1.8 \text{ gms}$

(*iv*) Course length = Loop length × number of needles
$= 0.28 \times \pi \text{ DG} = 633.4 \text{ cm}$

(*v*) Run-in = Course length × rpm
$= \frac{633.4 \times 35}{100} = 221.7 \text{ m/min}$

(*vi*) cpcm $= \frac{kc}{l}$, $\text{wpcm} = \frac{kw}{l}$

for finished relaxed states, assuming $kc = 5.5$ and $kw = 4.2$

$$\text{cpcm} = \frac{5.5}{0.28} = 19.5$$

$$\text{wpcm} = \frac{4.2}{0.28} = 14.9$$

(*vii*) Fabric width (at relaxed state) = Number of Needles/wpcm

$$= \frac{2262}{14.9} = 151 \text{ cm}$$

(*viii*) Fabric length (at relaxed state)

$$= \frac{\text{Number of feeders} \times \text{rpm} \times \text{time} \times \eta\%}{\text{cpcm}}$$

$$= \frac{96 \times 35 \times 60 \times 0.9}{19.5 \times 100} = 93 \text{ meters/hour}$$

(*ix*) GSM (at relaxed state)

$$= \frac{\text{cpcm} \times \text{wpcm} \times l \times \text{tex} \times \text{C}}{10} = \frac{19.5 \times 14.9 \times 0.28 \times 18 \times 1}{10}$$

$$= 147.48 \text{ gms}$$

Also, $\text{GSM} = ks \times \text{tex} / 10\, l$

where $ks = kc \times kw = 5.5 \times 4.2 = 23.10$

Therefore $\text{GSM} = \frac{23.10 \times 18}{10 \times 0.28}$

$$\text{GSM} = 148 \text{ gms/m}^2$$

Example 4: *It is required to knit a plain knit fabric to a finished weight of 150 gm/m² and a finished width of 160 cm on a 20 gauge machine with 30 inch diameter, 96 feeder and 30 rpm speed.*

Assuming the finished state constants $kc = 5.5$ and $kw = 4.2$,

Calculate,

(*i*) Optimum yarn tex
(*ii*) Optimum loop length
(*iii*) Optimum input tension
(*iv*) Run-in
(*v*) Length and width shrinkage

(*i*) Optimum yarn tex $= \left(\frac{100}{\text{G}}\right)^2 = \left(\frac{100}{20}\right)^2 = 25$ tex

(*ii*) Optimum stitch length

Assume optimum Tightness Factor, K = 15

i.e., $\text{K} = \frac{\sqrt{\text{tex}}}{l}$

Therefore $l = \frac{\sqrt{25}}{15} = 0.33$ cm

(*iii*) Optimum input tension $= \dfrac{0.1 \text{ gms per tex}}{\text{tex}}$

Therefore, Input tension = 2.5 gms

(*iv*) Course length $= l \times \text{needles} = l \times \pi DG$

$= 0.33 \times 3.14 \times 30 \times 20$

$= 621.7$ cm

(*v*) Run- in = course length × rpm

$= \dfrac{621.7 \times 30}{100} = 186.51$ m/min

(*vi*) $\text{cpcm} = \dfrac{kc}{l} = \dfrac{5.5}{0.33} = 16.67$

(*vii*) $\text{wpcm} = \dfrac{kw}{l} = \dfrac{4.2}{0.33} = 12.72$

(*viii*) Finished relaxed fabric width = Needles/wpcm

$= \dfrac{3.14 \times 20 \times 30}{12.72} = 148.11$ cm

(*ix*) Finished relaxed fabric length $= \dfrac{\text{Feeders} \times \text{rpm} \times \text{time} \times \eta\% \times F}{\text{cpcm}}$

$= \dfrac{96 \times 30 \times 60 \times 0.9 \times 1}{16.6 \times 100}$

$= 93.69$ mts/hr.

(*ix*) Length shrinkage and width shrinkage:

Given, finished width = 160 cm

Therefore,

Percentage width shrinkage $= \dfrac{160 - 148.11}{148.11} \times 100 = 8.02\%$

Given, finished weight = 150 gsm

i.e., $150 = \dfrac{\text{cpcm} \times \text{wpcm} \times l \times \text{tex.C}}{10}$

$= \dfrac{\text{cpcm} \times \pi DG/160 \times 0.33 \times 25 \times l}{10}$

Therefore, cpcm = 15.44

Finished length $= \dfrac{\text{feeders} \times \text{rpm} \times \text{time} \times n\%}{\text{cpcm}}$

$= \dfrac{96 \times 30 \times 60 \times 8 \times 0.9}{15.44 \times 100} = 100.74$ mts/hr

% length shrinkage $= \dfrac{100.74 - 93.69}{93.69} = 7.52\ \%$

$$\text{wpcm} = \text{Needles/finished width} = \frac{1884}{160} = 11.77$$

Width shrinkage $= \dfrac{12.72 - 11.77}{11.77} = 8.06\ \%$

Length shrinkage $= \dfrac{16.67 - 15.44}{15.44} = 7.51\ \%$

Example 5: *For a double jersey machine of 20 gauge, 48 feeder, 30 inch diameter, knitting acrylic spun yarn at a speed of 18 rpm, calculate the most suitable count (worsted, cotton count and tex), optimum input tension and optimum run in and run-in ratio for the following double knit structures.*

(*a*) Punto-di-roma

(*b*) Swiss double pique

(*i*) Optimum worsted count is given by

$$W_c = \frac{G^2}{10} = \frac{20 \times 20}{10} = 40 \text{ s worsted count}$$

For cotton count,

$$Cc = 2/3\ W_c = 40 \times 2/3 = 27 \text{ Ne}$$

For Tex, $\text{Tex} = 885.8/\text{Wc} = 22.0 \text{ tex}$

(*ii*) Optimum input tension:

$$T_i = \frac{0.1 \text{ gms per tex}}{\text{tex}} = 0.1 \times 22 = 2.2 \text{ grams}$$

(*iii*) Run-in and Run-in ratio: The optimum conditions for knittability occurs at tightness factor, K = 15.

(*a*) As the Punto-di-roma structure (Fig.14.6) is made up from both Rib and Plain knits only, the difference in Tightness K between these two units is approximately 20 %

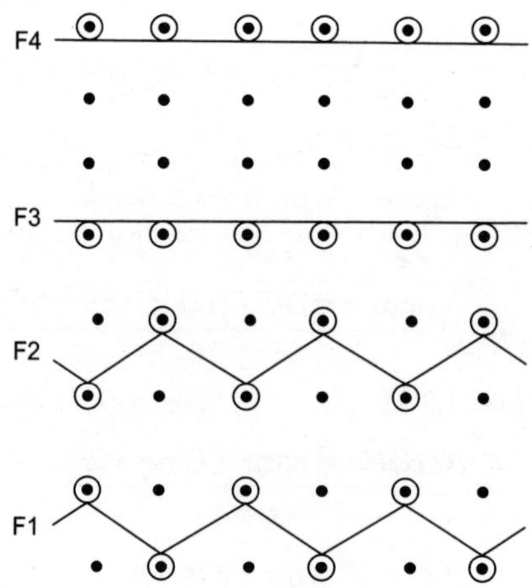

Fig. 14.6

i.e., $K_{plain} = 16.2$ and $K_{rib} = 13.8$

Now, the respective loop lengths are calculated as,

$$l_p = \frac{\sqrt{tex}}{K_p} = \frac{4.71}{16.2} = 0.291 \text{ cm}$$

$$l_r = \frac{\sqrt{tex}}{K_r} = \frac{4.71}{13.8} = 0.342 \text{ cm}$$

Referring the Punto-di-roma structure,

When knitting with dial and cylinder needles (feeder 1 and 2) the course length,

$$L_r = n.l_r$$

When knitting with dial or cylinder needles (feeder 3,4) the course length,

$$L_p = n.l_p$$

where n is the number of needles forming the loops

For Punto-di-roma, $n = N/2$.

where N is the total machine needles of dial and cylinder

i.e., $N = 2\,(30\pi 20) = 3772$

$$n = \frac{3772}{2} = 1886$$

Now course length, $L_r = 0.342 \times 1886$ cm

$L_p = 0.291 \times 1886$ cm

Therefore, Run-in $I_r = \dfrac{0.342 \times 1886 \times \text{speed}}{2.54 \times 12} = 381$ feet/min

$$I_p = \frac{0.291 \times 1886 \times 18}{2.54 \times 12} = 324 \text{ feet/min}$$

Therefore, Run-in ratio $= \dfrac{381}{324} = 1.2$

(*b*) As Swiss Double Pique is knitted with rib gaiting, all the loops are considered to be plain loops. *i.e.*, $K_p = 16.2$ and loop length, $l_p = \dfrac{\sqrt{tex}}{16.2} = 0.291$ cm

At the feeders (1,3) where dial and cylinder needles form loops, the total number of needles forming the loops are,

$$n + n/2 = 2829$$

Therefore, Course length produced with both dial and cylinder needles,

$$L_{d,\,c} = 2829 \times 0.291 \text{ cm}$$

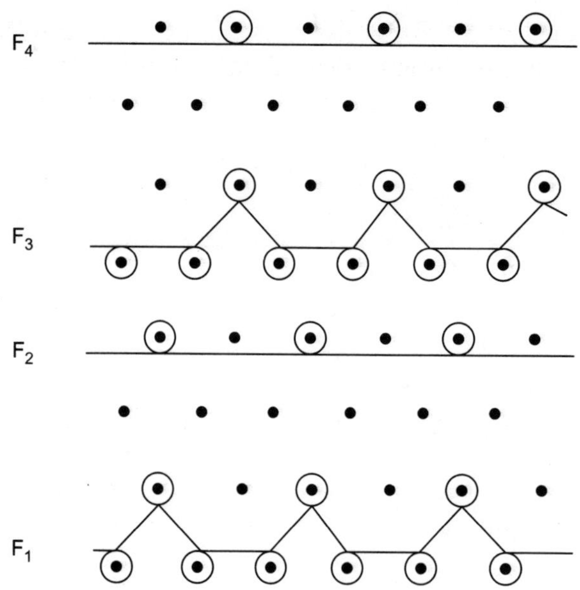

Fig. 14.7

At the feeders (2,4) where dial only form the loops, *n*/2 needles form knit stitches and the same *n*/2 needles form the miss stitches.

For miss stitches, $l = \frac{1}{G}$ inches $= \frac{1}{20} \times 2.54 = 0.127$ cm

$$L_d = \left(\frac{n}{2} \times 0.291\right) + \left(\frac{n}{2} \times 0.127\right) = 943\ (0.291 + 0.127) = 943 \times 0.418$$

Run-in, $$I_{d,c} = \frac{2829 \times 0.291 \times 18}{2.53 \times 12} = 486 \text{ ft/min}$$

$$I_d = \frac{943 \times 0.418 \times 18}{2.54 \times 12} = 233 \text{ ft/min}$$

Therefore, Run-in ratio $= \frac{486}{233} = 2.1$

14.3 ANALYSIS OF WEFT KNITTED FABRICS

To understand the history of the fabric as well as for its reproducibility, analyzing the fabric structure is important. Type of machine used and design principles can also be revealed by analyzing the knitted fabrics. For fabric analysis a piece glass (*i.e.,* a counting glass), a pair of scissors, ruler and calculator are needed. The following analysis sheet (Table 14.1) can be completed after having thoroughly analysed the fabric. The following procedure may assist in carrying out the analysis. These are only guidelines and it takes time and practice to acquire the skill in the analysis.

(*i*) *Fabric Name:* To find out to which type of the knitted structure the given sample belong viz., single jersey, rib purl, interlock etc. For single jersey fabrics, loops are seen at the face side and yarn lines are visible at the back as shown below:

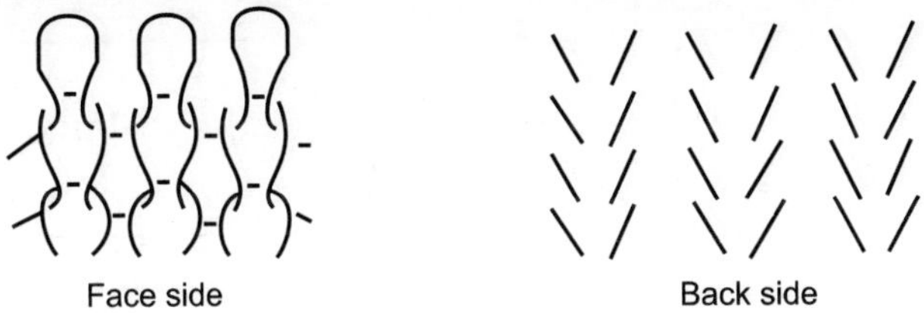

Fig. 14.8

Double jersey fabrics have similar appearance on both sides. By holding the fabrics horizontally, in such a manner to observe its cross section, the rib and interlock structures can be found by their cylinder and dial loop arrangements as given below:

Fig. 14.9

(*ii*) *Fabric Appearance:* To find out the technical face and technical back of the fabrics. The top of the fabric is the edge that was knitted last. The face of the fabric is always the side with the most knit stitches.

(*iii*) *Yarn Type:* The last knitted yarn is unraveled from the fabric and is observed for its types such as single, double, blend, milange, staple yarn, filament yarn, s/z twisted, 2/3 plyed etc.,

(*iv*) *Wales/cm and courses/cm:* With piece glass, the wales per inch can be measured and converted into per centimeter. It is advisable to analyze always the back of the fabric, and particularly for unbalanced fabric, the back is obvious.

$$c = \text{courses/cm and } w = \text{wales/cm}$$

(*v*) *Loop length:* Unravel 12 yarns from the sample and measure the total length, L_T cm (*i.e.*, L1 +L2 + L3 ...)

Weigh all the 12 yarns together, in grams (w_y)

Find out the average length, $L_{AV} = \dfrac{L_T}{12}$

Count the number of wales occupied by L_{AV}

Now, the average loop length = L_{AV}/No. of wales

(*vi*) *Grams per square meter:* Cut a piece of 10 cm × 10 cm from the fabric sample

Find out the weight of the sample, w

Now, $\quad\quad \text{GSM} = w \times 100$

(*vii*) *Yarn Tex:* Using the total yarn length and weight as found above,
Yarn number can be calculated as,

$$\text{Yarn Tex} = \frac{w_y \times 10^5}{L_T}$$

(*viii*) *Tightness Factor:* Using the values of tex and loop length as found above, tightness factor can be calculated as,

$$TF = \frac{\sqrt{\text{tex}}}{l} \quad \text{where } l = \text{loop length in cms}$$

(*ix*) *Fabric construction:* By carefully analyzing the fabric using a pick needle, the full design repeat of the structure, number of feeders used, needle gating etc., can be found. The repeat can be drawn either as point paper design (symbolic notation) or yarn path diagram (diagrammatic notation).

To prepare the fabric for the analysis, first a wale line is to be marked at the centre of the fabric sample, which is the starting point. Using a pick needle, highest stitches are to be pulled out, which is the yarn that was knitted last. Ensure that one yarn is pulled at a time; continue the deknitting until a yarn is reached that goes completely across the fabric. Now the sample is completely prepared for analysis. Now start deknitting the yarn that was last one knitted. Starting from the marking line, observe the type of stitches (knit, tuck, miss) the yarn is made into at each wales positions. Record them in a point paper from right to left (Fig. 14.10) as the yarn is deknitted from right to left. The next yarn is deknitted and recorded in the same manner. This process is repeated until two repeats are seen on the paper.

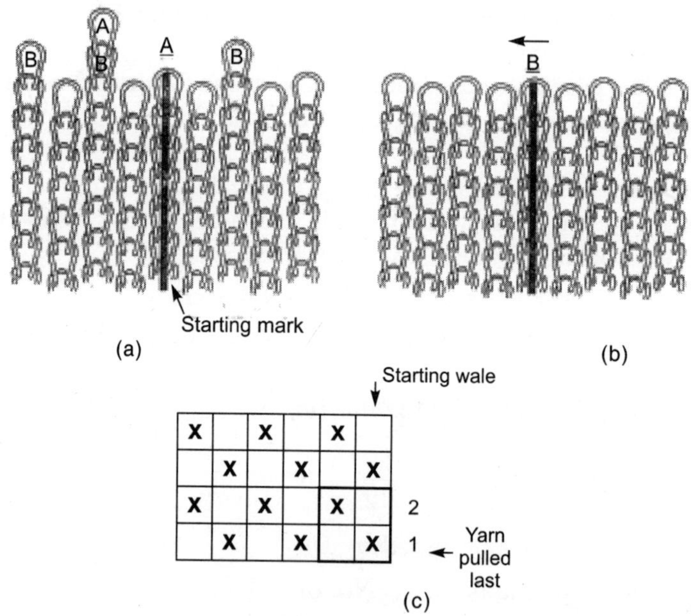

Fig. 14.10

The fabric construction can also be shown as a yarn path diagram (diagrammatic notation) by analyzing the fabric sample as shown below for a double knit swiss pique fabrics. The diagram represents full repeat of the design structure, number of feeders used and needle gaiting.

(*x*) *Run-in Ratio:* Run-in-ratio $= \dfrac{\text{Course length of feeder F1}}{\text{Course length of feeder F2}}$

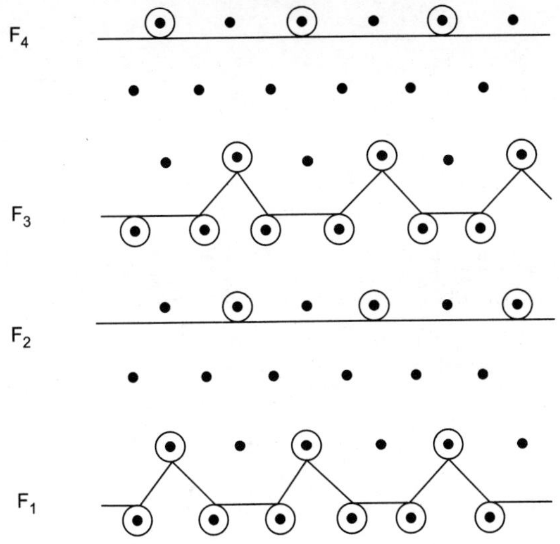

Fig. 14.11

The Table 14.1, given below enables to record the knitted fabric particulars after the analysis.

Table 14.1

Fabric Name :	Fabric appearance :	Yarn Type :
No. of courses per 10 cm :	No. of wales per 10 cm :	Length of yarn per wale :
Fabric construction		Machine Gating :
		Average length :
		Wt of unroved yarn :
Tightness factor :	Yarn tex :	Average loop length :
Area of sample :	Fabric GSM : (From sample)	Affix fabric sample :
Wt. of sample :		
Machine gauge :	Fabric GSM : (By calculating)	

15

Finishing of Knitted Fabrics

15.1 INTRODUCTION

Knitted fabrics are commonly used for under garments, sports wear and to a limited extent as casual wear. One peculiar quality of knitted fabrics is that it can be easily deformed and will loss its shape when stretched. For these fabrics which are meant to be used as under garments and sports wear, the main requirement is that the material should be absorbent so that sweat can be absorbed easily. Knitted fabrics have a natural ability to recover from creases. In the case of casual wears, some amount of creasing is required to enhance the appearance and serviceability of the fabric. One of the most common finishing is to apply a softening agent to the fabric.

There are different finishing processes for cotton and polyester knitted fabrics. In order to increase their performance and value, the knitted fabrics are subjected to certain finishing treatments such as softening, bio-polishing, anti-microbial finish, raising, emerying, compacting, calendering, soil and stain release finish, UV-protect finish etc., for cotton knitted fabrics and in addition to the above the anti-static finish and thermosetting for polyester knitted fabrics. Finishing with anyone of the above process is mandatory now-days as per their end use requirements. The chemicals required for finishing woven and knitted fabrics are almost similar. But the method and intensity of application differs. Both exhaust techniques as well as continuous application techniques are followed to impart the above finishes.

15.2 FINISHING OF COTTON KNITTED FABRICS

15.2.1 Softening

Softeners are one of the important finishing agents which give softness to the fabric. It gives not only softness but also handle, drape, cutting and sewing qualities to the materials.

Method: In discontinuous tubular fabric finishing this process is carried out in overflow machir s during bleaching for white goods or in the last vat in dyeing or impregnation during balloon squeezing. If the fabric is going to be dyed, softening can be performed in the rinsing bath after dyeing.

Chemicals: In the softening process mostly non-ionic and cationic softening agents are used. Cationic softening agent gives very effective results. For white goods non-ionic softeners are used. Acetic acid is used in the bath to maintain pH at 5-5.5.

Application: The material is treated in overflow machine with a softener and acetic acid at 40–50° C for 20–30 minutes or it may be continuously applied using a balloon padding machine taking the softener and acetic acid in the balloon padding bath.

Effect: It gives soft feeling to the surface of the fabric and impart more pleasant handle.

15.2.2 Crease-resist Finish

The object of this finish is to keep the fabric flat and smooth and free from undesirable creases. It is purely a chemical and permanent finish.

Method: There are two methods for crease-resist finish for knitted fabrics : (*i*) Impregnation with crease-resist agents, the drying and condensation in one stage (*ii*) Application of crease-resist finish by spraying crease resist agents at the entrance of the stenter.

Chemicals: Formaldehyde resins (DMDHEU), catalyst, acetic acid, citric acid, softener, wetting agent are used.

Application: The fabric is impregnated with the above chemicals at pH4 using a padding mangle with an expression 65% – 75% (liquor uptake), then it is dried at 110 – 130 °C(to leave 10 % moisture). After that the fabric is cured at 150 °C for 3–4 minutes during which the condensation polymerization takes place.

Effect: The crease resist finishing, in addition to stabilizing the fabric is providing unshrinkability by reducing the swelling of the fibres.The swelling value for normal cotton fibers is 42%–45%, after crease resist treatment this value decreases to 18%–24%. That is why the crease resist treatment also imparts unshrinkability to the fabric.

15.2.3 Peach Finish

To create peach skin like surface on the fabric which give soft, smooth feel and provide a more formal look while maintaining comfort and a casual feel.

Method of Application: Knitted fabric is passed over a rotating high speed rollers covered with emery cloth or sand papers. Emerying, sanding, sueding, raising, peaching are some of the descriptions of the same finish. By adjusting the angle of contact with the rollers, the grain size of emery used, pressure and temperature, the intensity of the sanding can be varied. By making first rollers using coarser emery than latter ones, we will get the best effect.

Effects: Peaching softens fabric handle and gives a greater degree of warmth to the fabric. It improves moisture retention properties with good wash fastners for synthetic, natural fibers or composites. Peach finish results in higher bursting strength, higher compressibility, higher friction and higher wicking. But flammability is somewhat poor.

15.2.4 Bio-polish Finish/Enzyme Finish/Anti-pilling Finish

Processing of cellulose knitted fabrics and garments with cellulase enzymes is generally referred as bio-polishing or bio-finishing.

Method: Enzymes are applied on knitted fabrics using overflow machine.

Chemicals: Cellulase enzyme, acetic acid to maintain pH 5-5.5.

Application: The above chemicals are applied with fabric at pH 5–5.5 at 50–55°C for 20–60 minutes depending on the effect required.

Effect: The protruding fiber ends are removed from the surface of the cotton there by giving a soft and polished surface.

15.2.5 Calendering

To upgrade the fabric handle and to impart a smooth silky touch to the fabric.

Method: There are two types of tubular knitted fabric calendars: (*i*) Steaming glazing (decatising) calendar. (*ii*) Decatising machine with finishing felt covered iron. Another surface smoothing process is ironing with a pressing machine.

Application: Subsequent to the steaming process the fabric is passed between the calendaring cylinders under a requested pressure and heat according to the kind of fabric between 130 °C at 4 atmospheres and 150 °C at 6 atmospheres. The heat is kept stable by thermostatic adjustment. The fabric is the cooled on a conveyor band with steam a temperature up to 150 °C and with oil a temperature up to 210 °C can be obtained. Operating speed is 60 meters per minute. The two passages of the fabric can be done in one single passage through the machine.

Effect: The calendaring process also ensures the return to their original lengths of loops which have been lengthened due to the tensioning effects of processes.

15.2.6 Compacting

The object of compacting is to prevent shrinkage of knitted fabric,and it is a perfect shrinkage control system for cotton and blended tubular knitted fabrics. Tubular Compactor control the shrinkage of tubular knitted fabrics, 3% to max. 0%.

Method of Application: The fabric is wetted and steamed on the compacting machine. The fabric is first stretched width wise and, during that time, it is over fed length wise. Then it is shrunk length wise in the compressive shrinkage unit. At the delivery end the fabric is cooled upon cooling drums and wound on to a roll. A maximum over feed of 40% may be given. Operation speed is up to 35 meters per minute.

Effect: It is a mechanical finishing process performed to secure the size stability for cellulosic materials such as cotton, viscose staple fiber, linen, rayon. It is mainly for imparting unshrinkability to knitted fabrics.

15.2.7 Raising

The fibers are raised to a certain level and the fibers are made to protrude from the surface of the fabric uniformly.

Method of Application: Fabric passes through first group of raising cylinders, counter pile and pile. Then through the second group of cylinders, two counter pile and finally through third group of cylinders and two pile.

Effect: A pile surface is conferred to the knitted fabric by a mechanical raising process. It gives the fabric a hairy, bulky and insulating structure. Two changes take place on the fabrics during the raising operation: (*i*) A lessening in the strength of surface fibers occurs, since they are damaged during raising. (*ii*) A certain amount of fiber is lost. It is desired that the amount of loss doesn't exceed 10% in heavy fabrics and even less in lighter goods.

15.2.8 Emerying

Emerying is a kind of pile forming process like raising.

Method of Application: The emerying operation is performed on open width fabrics rather than tubular knitted fabrics. There is only one single emerying cylinder which is covered with emery dust or with emery paper as an abrasive matter. The fabric is brought into contact with this fast rotating cylinder and abraded. As a fashion effect, it is done in denims and gabardines.

Effect: With emerying, the surface hairs of the fabric are separated from each other and a short piled surface similar to velvet is obtained. The effect on the surface of the fabric can be fine or coarse depending upon the coarseness or fineness of the emery.

15.2.9 Shearing

The object of shearing is to remove the fiber neps on the surface of the fabric or to cut them in a given height uniformly.

Method: Machines with single shearing unit are used for single face operations, and machines with two shearing units are preferred in order to finish the shearing in a single passage on both faces.

Application: There are a shearing bed, a lower knife and an upper knife. When the fabric passes by the shearing bed, hairs which have become perpendicular to the fabric surface are grabbed by the stationary lower knife and cut by the revolving upper knife.

Effect: The shearing operation removes the fiber naps on the surface of the fabric or cut them in a given height uniformly. In cotton finishing, shearing is generally performed after raising. In this way the pile is cut in a uniform height.

15. 2.10 Antimicrobial Finish

The object of this finish is to protect the cellulosic knitted fabrics from the microbial attack such as bacteria and fungi.

Method: Films or coatings that have direct surface contact with the fabrics enables activity against bacterial growth. Pad–dry–cure process is practiced for fixation of anti-bacterial additive onto fabric with the help of cross-linking agent.

Chemicals: Sodium salt of heterocyclic compound, pentachlorophenyl laurate and non-ionic emulsifying agent. Aqueous solution of sodium ortho phenylphenate, an organic compound containing nitrogen with no halogen, combination of heterocyclic organic compound, derivatives of bezimidazole is used.

Application: The above chemicals are applied in a padding mangle, then the fabric is dried and cured at 150 °C for 5 minutes.

Effect: It provides protection from the microbial attack such as bacteria and fungi which may lead to health hazards.

15.2.11 UV Protection Finish

The object of this finish is to protect the fabric from UV attack.

Method: By exhaust application like reactive dyes.

Application: The fabric is applied with UV protective finish chemicals along with salt and soda ash.

Effect: It protects the fabric from UV attack.

15.3 FINISHING OF POLYESTER KNITTED FABRICS

Finishes used in polyester and polyester blended knitted fabrics are:

Anti static finish, softening finish, thermosetting, compacting, emerying, raising (if desired) and shearing.

15.3.1 Antistatic Finish

The object of this finish is to prevent the static charge generation. Hydrophobic fibers are tend to develop static charge.

Method of Application: All softeners have anti-static properties and in certain softeners the anti-static property is superior. The softeners to be used are resistant to washing. The silicon elastomeric softeners which form a film upon the goods make the fabrics resistant to anti-static materials and are permanent.

Effect: Static electricity is reduced on polyester goods.

15.3.2 Thermosetting

The object of this finish is to give heat setting to the synthetics and blended fabrics.

Method: Thermosetting of polyester is done using stenter in a continuous manner.

Application: It is performed with hot air at 160 °C to 180 °C for 30 seconds. Use of saturated steam lowers the setting temperature and improves the handle.

Effect: In preliminary setting, shrinking of goods and formation of crease marks are prevented, in final thermosetting final form and final sizes are given to the material.

15.3.3 Compacting

Method of Application: The fabric is wetted and steamed on the compacting machine. The fabric is first stretched width wise and, during that time, it is over fed length wise. Then it is shrank length wise in the compressive shrinkage unit. At the delivery end the fabric is cooled upon cooling drums and wound on to a roll. A maximum over feed of 40% may be given. Operation speed is up to 35 meters per minute.

Effect: It is a mechanical finishing process performed to secure the size stability for cellulosic materials such as cotton, viscose staple fiber, linen, rayon. It is mainly for imparting unshrinkability to knitted fabrics.

15.3.4 Raising

Method of Application: In the raising process both the main cylinder and the small cylinders turn upon their axis. The turning direction of the cylinders is the reverse of that of the fabric and the main cylinder. The main drum turns in the clockwise direction together with the fabric. The clothed raising cylinders turn in the anticlock wise direction and the wires penetrate into the fabric surface and raise the fiber ends to the surface of the fabric.

Effect: Raising effect is obtained by bringing to the surface of the fiber ends one by one by the raising card wires.

15.3.5 Emerying

It is a similar method to raising and gives to the fabric suede like handle. It is the process where the fabric is passed through one or more revolving cylinders covered with emery paper. With emerying the hairs on the face of the fabric are separated from each other and suede like, very short piled surface is obtained.

15.3.6 Softening

Method: In discontinuous tubular fabric finishing, this process is carried out in overflow machines.

Chemicals: Silicone softeners in emulsion form is used.

Application: Impregnating the above chemicals with an expression of 60%-70% and then dry at 120 °C to 130 °C.

Effect: It gives soft feeling to the surface of the fabric and impart more pleasant handle.

15.3.7 Shearing

Shearing of polyester knitted fabrics is performed for two main purposes: (*i*) To give an even pile height to raised fabric. (*ii*) In loop pile fabrics to obtain a cut pile surface.

Method of Application: In shearing machines, there are a shearing bed, a lower knife and an upper knife. When the fabric passes by the shearing bed, hairs which have become perpendicular to the fabric surface are grabbed by the stationary lower knife and cut by the revolving upper knife.

Effect: It is performed in order to remove the fiber naps on the surface of the fabric or to cut them in a given height uniformly. A birds eye view of various finishing treatments for cotton and polyester knitted fabrics have been discussed. With reference to polyester knitted fabrics certain finishing treatments such as Anti-static, thermosetting are exclusive.

Warp Knitting

16

Warp Knitting

16.1 INTRODUCTION

The another interesting segment of the knitting industry is the warp knitting. Unlike weft knitting, most of the warp knitting machines are open width/flat type. As the name implies, the loop formation is warp wise *i.e.*, vertically upwards. This unique feature of loop continuity in upward direction makes the warp knitted fabrics more special with respect to their characteristics, production and applications. As they resemble weaving machines, originally these were referred a warp knitting looms. Though the machine initiation has started very long back, in the middle of the twentieth century only the major developments in the manufacture of warp knitting machines has taken place. The warp knitting machines have gained their importance due to the advent of man-made synthetic fibers such as nylon, polypropylene, polyester and acrylics. The major credit in the development of these machines goes to Mr. Karl Mayar of Germany who had identified a wide range of its interesting applications in the areas such as apparel, home furnishings, industrial and technical textiles. His Karl Mayar company is the leading warp knitting machinery manufacturer in the world.

16.2 COMPARISON OF WEFT KNITTING AND WARP KNITTING

The weft knitting and warp knitting are the two distinct fabric forming systems having loop structure as the common feature. Though both are unique in their respective planes, a comparison between them with respect to their knittability, patterning, fabric characteristics, production, economics and applications may throw a better understanding of them.

(*a*) In weft knitting, the loop formation takes place course wise, *i.e.*, loops are formed one after the other in each course in horizontal manner. Hence the knitting is sequential. For a machine having 20 feeders, twenty courses are produced per revolution. In warp knitting, the loop formation is wales wise, vertically. *i.e.*, thousands of loops are formed simultaneously per course per knitting cycle.

(*b*) As a simple case, it is sufficient to feed one yarn and to produce a weft knitted fabric. Where as in warp knitting multiple ends based on the number of needles in the machine is required to produce the fabric.

(*c*) In weft knitting, the yarn supply is in the form of cones or cheeses. The yarn supply capacity depends on the number of feeders in the machine. In warp knitting, the yarns are supplied in beam form, which is in the form of sections.

(*d*) The yarns used in the weft knitting are mostly of spun type. Yarn lubrication in the form of waxing is required for those yarns to control the yarn hairiness. The warp knit yarns are mostly of filament type, which requires lubrication in the form antistatic oiling.

(*e*) Preparation of sectional warp beam is an important operation for warp knitting. Whereas for weft knitting, the yarn preparation is simple.

(*f*) The products of weft knitting are mostly confined to limited applications, mostly apparels. The fabric dimensions are fixed with respect to their diameter and machine gauge. The outlets of warp knit fabrics are unlimited such as inner and outer wears, industrial, technical and domestic. The use of filaments enhanced the scope for their wide opportunity.

(*g*) Change of patterning is difficult in weft knitting, as cams at every feeder positions are to be changed. In warp knitting, pattern changes are simple as they do not require special attachments.

(*h*) The weft knitted fabric stretch both in width wise and length wise with respect to their type of structure. The warp knit fabrics usually stretch in width wise direction.

(*i*) The weft knitting machines are simple by their elements and mechanism. The fabrics are cheaper to produce. The warp knitting machines are more complicated by their knitting elements as well as mechanism. The manufacturing cost is also more for them.

(*j*) The dimensional stability of warp knitted fabrics is superior to the weft knits. Each loop of warp knit consists both a loop part (called as overlap) and a diagonal straight yarn part (called as under lap) which makes the fabric more stable.

(*k*) The superior hand values of weft knitted fabrics make them more suitable for children and ladies apparel compared to warp knits.

(*l*) Consistent and more uniform fabric quality is possible with warp knitting compared to its counterpart.

(*m*) Economical advantages are with weft knitting with respect to their machine cost, floor space and yarn type. Normally, costlier and stronger yarns are required with warp knitting.

(*n*) Simultaneous knitting and high speed operation make the warp knitting machines more productive compared to the weft knitting.

(*o*) Conversion of fabric into garment is advantageous with warp knitting due to the simple cutting and sewing operation. They are better tailorable compared to weft knitted fabrics.

16.3 CLASSIFICATION OF WARP KNITTING

The classification of warp knitting machines is based on a number of factors such as the type of needles, number of guide bars, machine speed and machine complexity. All the three needle types, viz., latch, spring beard and compound needles are used with warp knitting machines. The needles are not independent for their action, but are mounted on a common needle bar. Number of needle bars as well as their width also takes part in classifying the warp knitting machines. Increase in number of guide bars increase the machine complexity; there by reduce the speed, which also influences the machine classification. In general the warp knitting machines are broadly classified into two major groups viz., Tricot and Raschel. Conventionally, the distinction between these two was with respect to their needle types used. Raschel machines were equipped with latch needles and Tricot machines with spring beard needle. However, with the advent of compound needles, the same old classification is no longer possible as compound needles replace both the spring beard and latch needles in modern warp knitting machines.

Though both the Tricot and Raschel machines use sinkers, the sinkers used for Tricot knitting machines control the fabric throughout the knitting cycle. Whereas in Raschel knitting machines, the sinkers are only used to ensure that the fabric stays down when the needles rise. Due to this reason, high takeup tension is required for Raschel machines to pull the fabric tightly downward from the knitting zone at an angle of 160°. On Tricot machines, gentle take down at right angles from the knitting zone is sufficient.

Both the Tricot and Raschel machines use multiguide bars. Though the maximum number of warp beams and guide bars are four with conventional Tricot machines, the majority of the Tricot machines use only two guide bars. Guide bars to a maximum of 48 are possible with Raschel machines.

Both the Tricot and Raschel machines are equipped with single set of needles. A warp knitting machine equipped with two sets of spring beard needle is referred as the Simplex Machines. Similarly a two-needle bar Raschel machine is equipped with two sets of vertically mounted latch needles. The fabric is removed from the needles vertically downwards between the two needle beds.

In Table 16.1, the two major classes of warp knitting machines are being compared for their functional elements.

Table: 16.1 Tricot vs Raschel Machines

Functional Aspects	*Tricot*	*Raschel*
Needle Type	Spring beard or Compound	Latch
Machine Gauge	Finer (expressed as needles per inch) 28-32 NPI	Coarser (Expressed as needles per 2 inches)16–8 NPI
Yarn type used	Filament (mostly)	Spun (mostly)
Number of guide bars	Less (2 to 4)	More (6–48)
Fabric complexity	Simple	Complex
Machine speed	High speed (1200 cycles per minute.	Slow speed (600 cycles per minute)
Machine width	Wider (168 inches)	Narrow (< 100 inches)
Fabric take-down	At 160° from the knitting zone	At 90° from the knitting zone

16.4 YARN REQUIREMENTS

Unlike weft knitting, in warp knitting, a wide range of yarns are used. Todays warp knitting is not only with the rayon, nylon and polyester areas, but has expanded to fibres such as carbon, glass, metals, aramide, polypropylene, polyurethane and polyolefins. The exploration of warp knitting machines into new markets and end uses in building materials, geo textiles, erosion control and reinforcement fabrics, automotive, medical as well as home and apparel applications necessitated the usage of the above mentioned new fibres.

Table 16.2 shows a wide classification of warp knitting machines.

Table: 16.2 Classification of Warp Knitting Machinery

- Warp Knitting Machines
  - Tricot machines (spring beard and compound needle)
    - Single needle bar
      - 2-5 bars lignerie, shirts dresses
      - Weft insertion 2-3 bars bed linen, curtains, lining
      - Cut Presser blouses, curtains, dresses
      - Pile Terry 3- 4 bars bed linen, bath robes, gowns
    - Double needle bar
      - Simplex (Gloves, lingerie)
  - Raschel machines (Latch needle)
    - Single Needle bar
      - Standard Rachel 2-9 bars high speed
      - Weft insertion 1-10 bars curtains, Industrial furnishings
      - Jacquard lace up to 78 bars (fall plate) electronically controlled jacquard lace
      - Special version up to 18 bars (fall plate) weft insertion curtains
      - Lace 12 - 48 bars (fall plate) curtain lace, foundation garments.
    - Double Needle bar
      - Face to face Plush (jacquard) 6-7 bars
      - Cords trim bandages 4-8 bars
      - Special version split film fruit and vegetables bags
      - High Pile fabrics 6-bars
      - Sacks Tubular nets

The overall construction of the warp-knitting machine is shown in Fig. 16.1. The warp threads fed from the warp beams are knitted into fabric and wound into batching device. The knitting elements are driven by the main shaft.

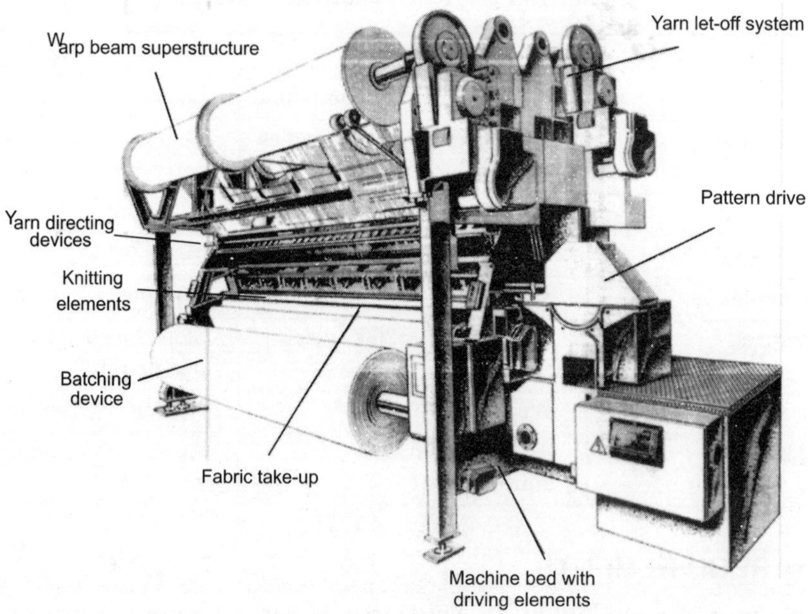

Fig. 16.1

17

Functional Elements of Warp Knitting

17.1 LOOP STRUCTURE AND TERMS (Fig. 17.1)

Fabrics of the warp kintting are formed by intermeshing of loops vertically down the length of the fabric in contrast to weft knitting. All the needles to the complete width of the knitting area are fed with atleast one yarn in line with the direction of fabric production. As mentioned earlier, both the spun as well as continuous filaments can be used in the production of warp knitted fabrics. A simple warp knitted loop structure in shown in Fig. 17.1. As mentioned in Chapter 2.1., the vertical line of loops (*i.e.*, wale) and the horizontal line of loops (*i.e.*, course), the loop portion (*i.e.*, overlap) and the diagonal floats of yarn (*i.e.*, underlap) are seen in the Fig. 17.1. The Warp knitted fabric structure has dissimilar appearance on the technical face and technical back side.

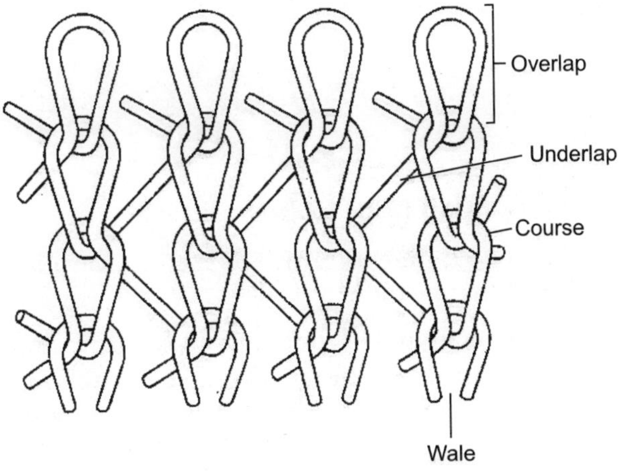

Fig. 17.1

The diagram is the technical face side of the fabric in which the loops (*i.e.*, overlaps) have the clear visibility. At the back side of the fabric, the diagonal line of yarns (*i.e.*, underlaps) run right and left throughout in a zigzag manner. These underlaps play an important role in influencing the pattern effects. The length or extent of these underlap floats and their direction of running cause a variety of design possibilities in warp knitting.

17.1.1 Overlap and Underlap (Fig. 17.2)

The loop structure of warp knitting is made into two parts. The first part is the actual loop itself which is formed by wrapping the yarn around the needle and drawing through the old loop. This part of the structure is called as overlap. In warp knitting the loops are called as laps, as the guides lap the warp yarn around the needle to form a loop. The guides from the back of the needles pass through the needle spaces to their front, wrap yarn over them and return back to their back position. By this movement of guides with respect to the needles, a loop is formed at the front (i.e., hook part) of the needle. Therefore, overlap always takes place at the front of the needle, with one needle space.

The second part of the loop structure is the length of yarn connecting the loops, which is called an underlap. After the overlap is completed, the guides at the back of the needles move to the next overlap positions for the next cycle of operation. This part of the yarn run at the back of the needles is termed as underlap. In warp knitting cycle, the overlap always precedes the underlap. The guide displacement with respect to the needle is restricted to one in the case of overlaps and one and above in the case of underlaps. The underlaps cross and cover more wales on its way, with the results that the fabric becomes heavier, thicker and denser.

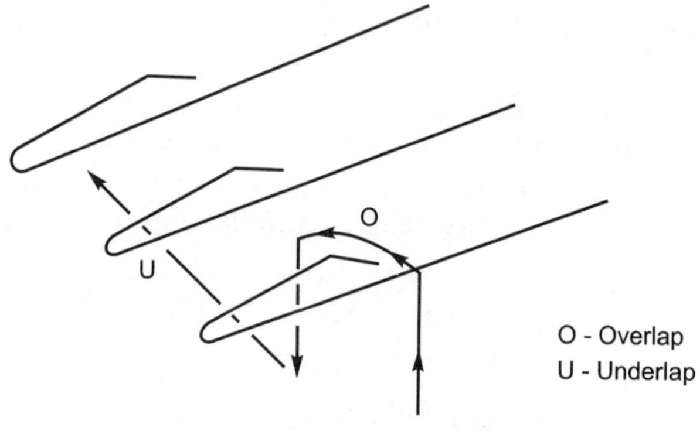

Fig. 17.2

17.1.2 Closed lap and Open lap (Fig. 17.3)

In the production of overlap, two different forms are used in warp knitting, based on the way the yarns are wrapped around the needles.

A closed lap in produced when an underlap follows in opposite direction to the overlap and thus laps the thread around both sides of the needles. This is the most common lap, used in most of the warp knitted structures. Closed laps are heavier, more compact, opaque and less extensible.

An open lap is produced either when the underlap is in the same direction as the overlap, or it is omitted so that the next overlap commences from the space where the previous overlap finished. The open lap is used when special needs arise or when technical limitations are imposed.

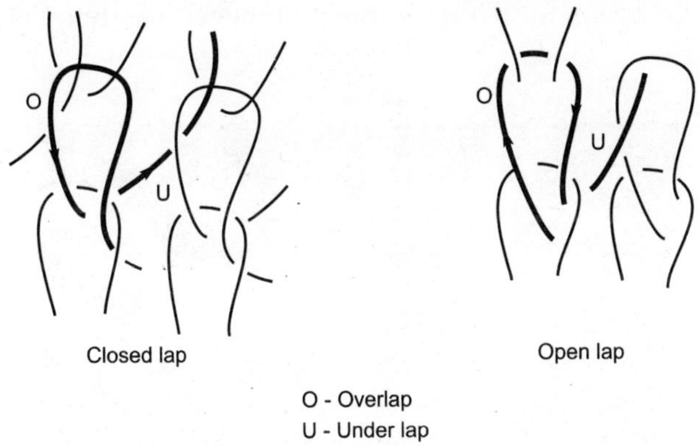

Fig. 17.3

17.1.3 Stitch Density and Stitch Length

The stitch density of warp knitted fabric is defined as the total number of loops in a square area. The area is usually a square inch or square centimetre and the density is obtained by multiplying the number of courses by the number of wales in that area.

The stitch length of warp knitting is different to that of weft knitting. As the warp knit stitch consists both the loop part and the diagonal cross link, the stitch length can be calculated as the sum of both overlap and underlap portions.

17.2 WARP KNITTING MACHINE ELEMENTS

Majority of the warp knitting machines are straight needle bed type, not circular. In a way they resemble the weaving loom by their appearance. The yarns are fed from warp beam and the knitted fabric is collected in open width form, not the tubular.

Both the Tricot and Raschel type machines have almost common functional elements even though they are differing by certain machine features as discussed in the previous chapter. The following are the important functional elements of warp kintting

1. Needle bar
2. Sinker bar
3. Guide bar
4. Presser bar
5. Warp beam
6. Pattern wheel
7. Chain links
8. Latch wire
9. Trick plate

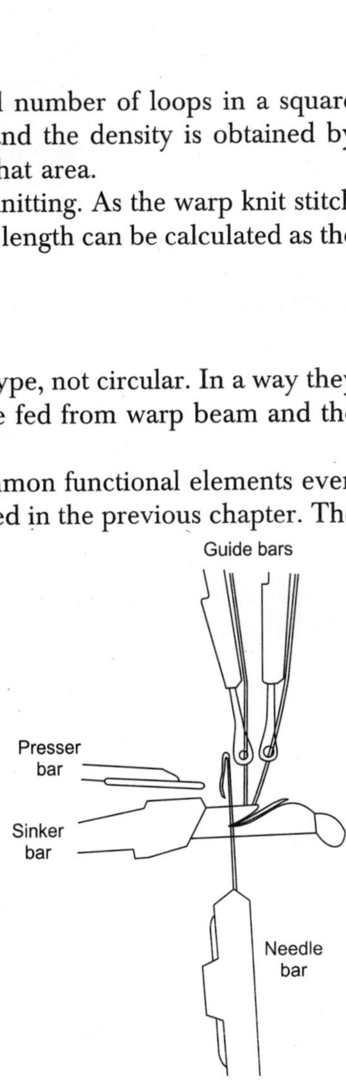

Fig. 17.4

Fig. 17.4 shows a simple view of the arrangement of various knitting elements in a warp knitting machine. The warp yarn supply beams are at the back of the machine placed at an elevated level. The guide bars are at the front of the machine, close to the operator. Behind the guide bar are the needle bar, sinker bar and presser bar as shown in the Fig. 17.3.

Plate 17.1 shows the arrangement of the knitting elements in a Six-Guide Bar-Latch Needle Warp-knitting Machine.

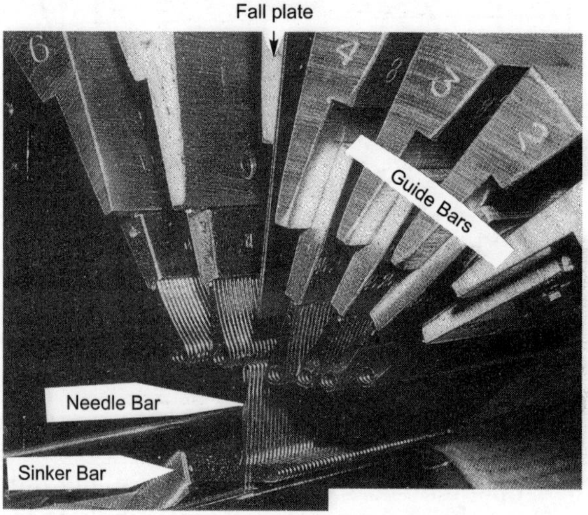

Plate 17.1

The hooks of the needle are facing towards the back of the machine. In the diagram, only the single elements of the respective needle bar, sinker bar and guide bars are visible. (Fig. 17.4)

When one element is illustrated, it is assumed that there exists a whole set of similar elements which cannot be seen because they are all exactly aligned.

17.2.1 Needle Bar (Fig. 17.5)

The Warp knitting machines use any of the three basic types of needles viz. latch, spring beard and compound which depends on the type of the machine such as Tricot or Raschel. Unlike weft knitting, in warp knitting the needles are not independent. The needles are mounted individually or in leads, and are clamped to a common bar, called as needle bar which extends to the full width of the machine. The space between the needles decides the gauge of the machine. The needle separation is normally equal to that of the guides and must also be accurately controlled. The movement of the needles is simultaneous in which all the needles collectively move up and down together as the needle bar is lifted and lowered. This needle bar movement is by means of a cam or crank drive.

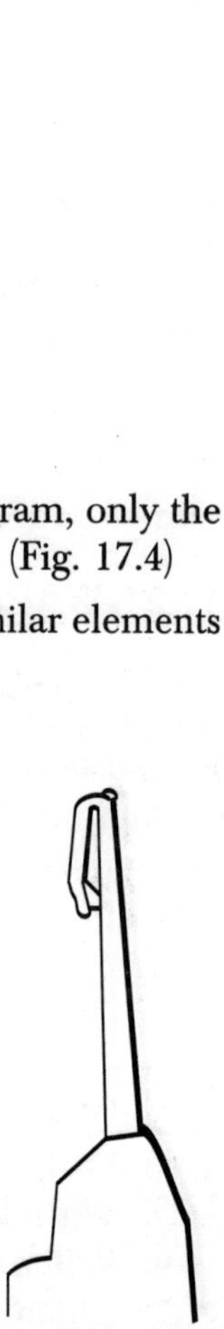

Fig. 17.5

17.2.2 Sinker Bar (Fig. 17.6)

The sinkers are placed in the spaces between the needles. Similar to the needles, the fine, thin steel sinkers are mounted in leads which are clamped to a common bar called sinker bar. The bar runs to the complete width of the machine. Unlike needles, the sinkers are mounted horizontally, equal to the gauge of the needles. The main objective of the sinkers is to impart control over the fabric during the loop formation. The forward and backward movements of the sinker bar are given by a cam or crank drive.

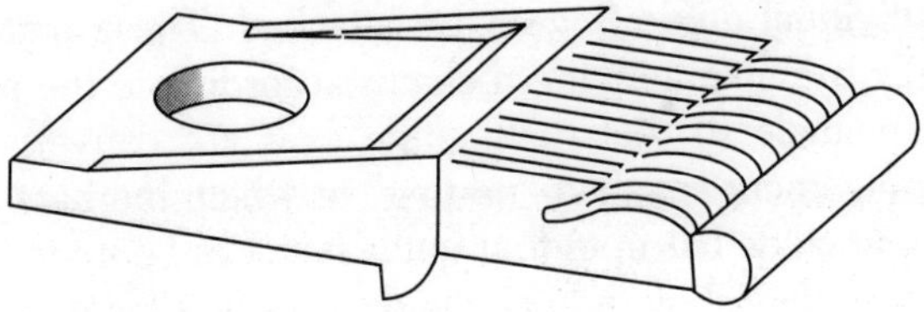

Fig. 17.6

17.2.3 Guide Bar (Fig. 17.7)

The objective of the guide bar is to present yarns to the needles to enable them to carryout their loop forming knitting action. Unlike weft knitting, in warp knitting, the needles are not catching the yarns, instead yarns are supplied to them by the guides of the guide bar.

The guide bar consists of thin flat steel guide strips mounted collectively together. Several guides are usually mounted in leads of one inch width to ensure accurate guide separation. These leads are then connected to a horizontal bar that make the complete guide bar assembly. The guide bars runs to the full width of the machine. The guides hang down from the guide bar and there are holes drilled at the end of the guides to facilitate threading through them. Each guide in the guide bar is usually provided with a single end of yarn. Warp knitting machines are usually equipped with two or more guide bars. Machines having single guide bar are uncommon.

The gauge of the guide bar is equal to that of needle bar as well as sinker bar. With the machines having more number of guide bars, the guide bars are numbered from the front to the back, *i.e.,* guide bar at the front of the machine and near to the operator is called as front guide bar and the lastone is called as the back guide bar. This is the case with Tricot machines. However, with Raschel machines, the numbering is reversed. Each guide bar has its own warp beam.

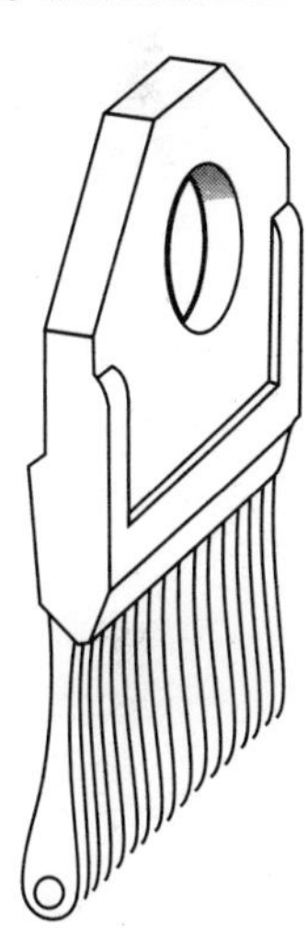

Fig. 17.7 *Individual guides*

The threading of the yarns through the guide bars is of different styles viz. full set threading, part set threading and half set threading. If all the guides in a guide bar are threaded without leaving empty, it is a full-set threading. If half of the guides are threaded *i.e.,* threaded alternately or if the threading is at random based on the pattern requirement, they are known as half set threading and partset threading receptively. In a 2-guide bar machine, if both the guide bars are full-set threaded, then each needle will receive two warp threads and each loop of the fabric, will, therefore have two yarns.

Tricot machines have less number of guide bars (2 to 4), mostly fully threaded where as Raschel machines are equipped with large number of guide bars (as much as 70), mostly partly threaded which allow the greater patterning capability. These multibar Raschel machines are equipped with two types of guide bars, the first type of guide bars are similar to tricot knitting, having two or three guide bars in most cases that construct the ground fabric.

The second type of guide bars also referred as pattern bars are used for pattering purpose. These bars usually require only one thread for each patterning repeat, so that only a few yarns are threaded across the full width of such a bar. These pattern bars are designed as narrow, light weight metal strips

(Fig. 17.8) on to which individual guide fingers are attached. These pattern bars are only partly threaded, driven individually by separate pattern chains according to the pattern. Being only partly threaded they are set at an angle so that their guide eyes are converged, aligned in the same displacement line. This arrangement is called "nesting" in which the bars are set in groups (nests). Each nest is considered as one guide bar upto four guide bars can be nested together and they swing as a single guide bar. However they shog independently in such a manner that the guides of bars in the same nest cannot cross or approach within two needle spaces of each other (Fig. 17.9 and 17.10).

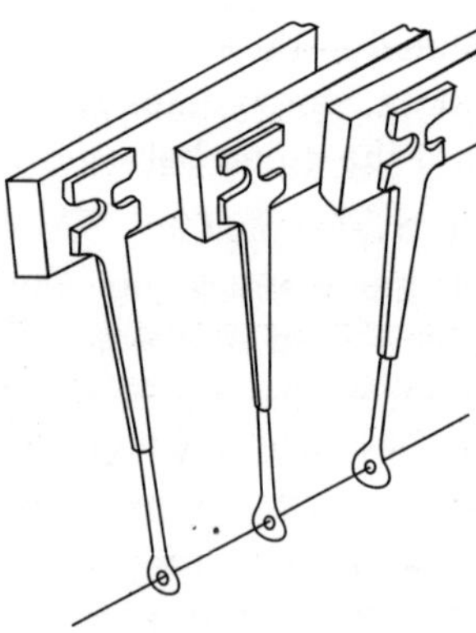

Fig. 17.8

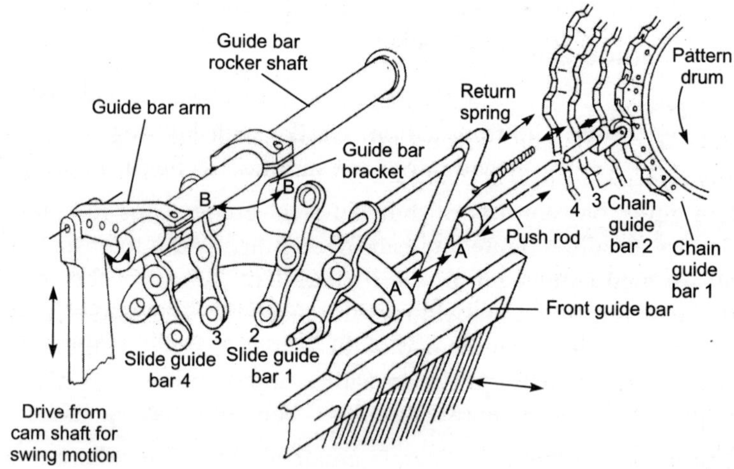

Fig. 17.9

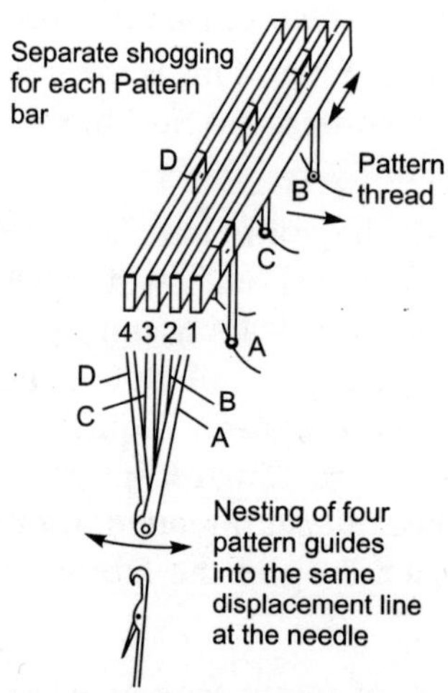

Fig. 17.10

On a 42 bar lace Raschel machine for example, there will be 13 displacement lines. The first two lines are conventional guidebars for knitting the ground. The next two lines are the two nests each of two pattern bars and the remaining nine lines are the nine nests, each of four pattern bars.(*i.e.*, $1 + 1 + 4(2 \times 2) + 36(9 \times 4) = 42$).

The guide bars undergo two kinds of movement during their knitting cycle for the complete loop forming action. Lateral movements on the hook side and the reverse side of the needle. These are the two lapping movements, one wrap the yarn around the needle at its front side called the overlap and the other make diagonal cross links at the back of the needles called the underlap.

The to and fro action. *i.e.,* the forward and backward movement of the guide bar makes the guides to pass through the needle spaces and carry their yarn to the opposite side of the machine. This swinging action of the guide bar is obtained from a common rocker shaft (Fig. 17.9), and the drive to the later is derived from the main cam or crank shaft in a similar manner to the needles and sinkers.

These side way movements of the guide bars are controlled by a pattern chain, mounted on a pattern drum and located at one side of the machine. These sideway movements occur parallel to the needle bar and are called as shogging movements. As the guide bars are not rigidly connected to the swing assembly, they are allowed to move freely in a lateral way.

17.2.4 Pattern Drum (Fig. 17.11)

As mentioned earlier, out of the two types of the guide bar movements viz. swinging, i.e., the guides passing through the needles to and fro and shogging, *i.e.*, the sideways movements of guides parallel to the needles the later type of movements, are facilitated by pattern drums. The pattern drum consist of an endless chain made up of pattern chain links wound over its periphery. The different heights of the chain links give different displacements to the guide bars. Pattern wheel is a circular disc which doesn't have chain links and its periphery is cut or stepped into different heights. The heights of the steps decide the lateral displacement of the guide bars. Both the pattern wheels and pattern drums can be used with warp knitting machines.

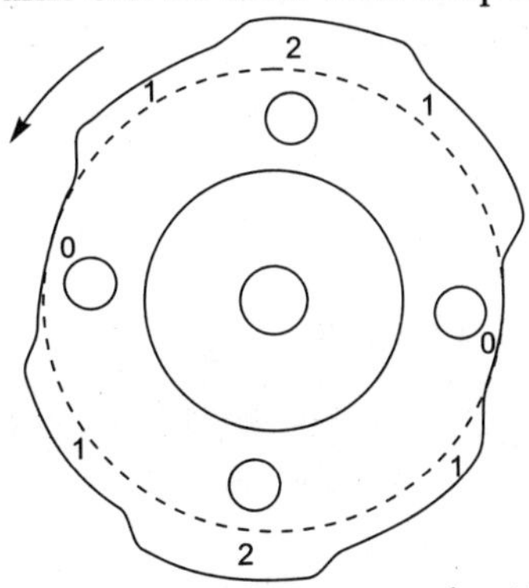

Fig. 17.11(a) *Pattern drum mounted with pattern chain links*

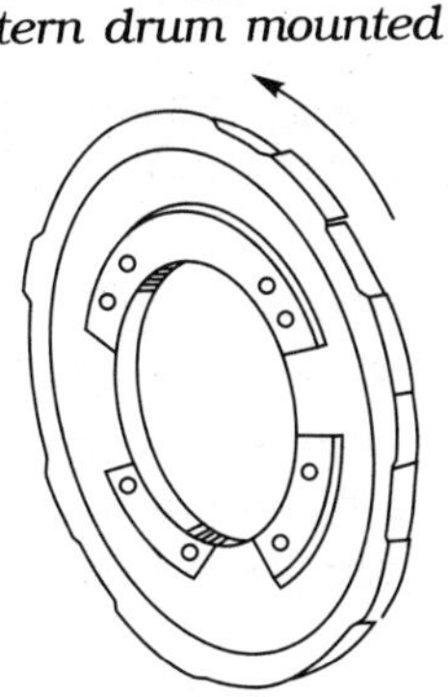

Fig. 17.11(b) *Pattern wheel*

17.2.5 Chain Links (Fig. 17.12)

The pattern chain links are the Y shaped metal blocks having fork at the front and tail at the back. These links are connected one into other as an endless chain and wound over the pattern drum. When the drum rotates the different height of the links facilitate the pattern drum to shift the guide bars to different lateral displacements. Links of shortest height displace the guides to one needle space and of larger heights provide the guides for more needle space displacements.

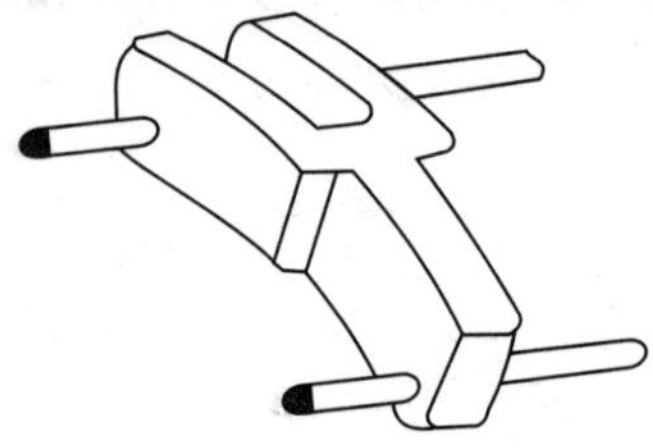

Fig. 17.12

17.2.6 Warp Beams (Fig. 17.13)

Like weaving, in warp knitting also, the yarns are supplied to the needles in the form of warp sheets from warp beams. The number of beams used on a machine is normally equal to the number of guide bars of the machine.

Each individual warp sheet is usually supplied from its own beam, which may consist of several section beams.

For the convenience of handling, a number of section beams may be attached to a beam shaft to achieve the full width of the warpsheet, for example, a warpsheet 84 inches wide might be supplied from a full width beam having two beams each of 42 inches wide or four beams each of 21 inches wide.

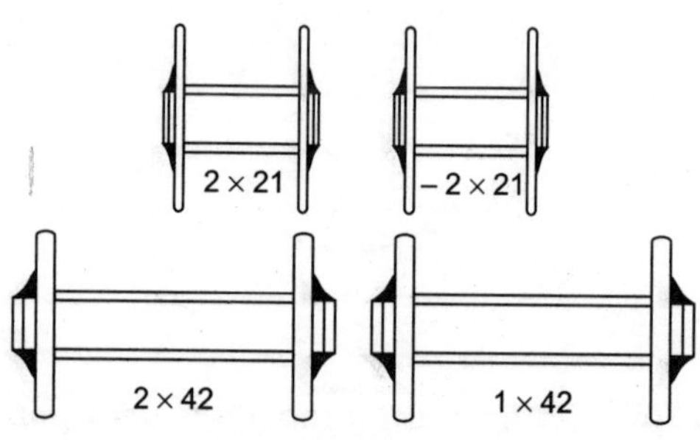

Fig. 17.13

17.2.7 Presser Bar (Fig. 17.14)

With the Tricot machines which are using spring beard type of needles, the presser bar is used. The presser bar runs to the full width of the machine and moves forward and backward like the sinker bar. The movement is obtained from a cam or crank drive, similar to the needle bars. The presser bar, during the loop forming action, presses the beard of the needle into the groove or recess, there by close the needle hook to trap the yarn.

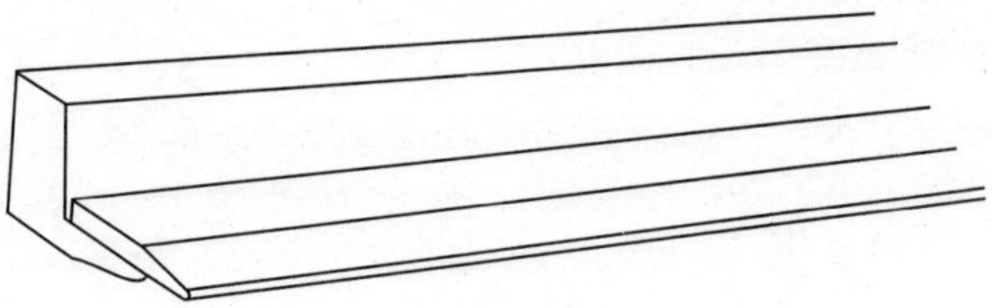

Fig. 17.14

17.2.8 Trick Plate (Fig. 17.15)

This can be considered as a needle bed of Raschel warp knitting machines, similar to that of weft knitting needle bed, since the needles are placed in tricks cut into it.

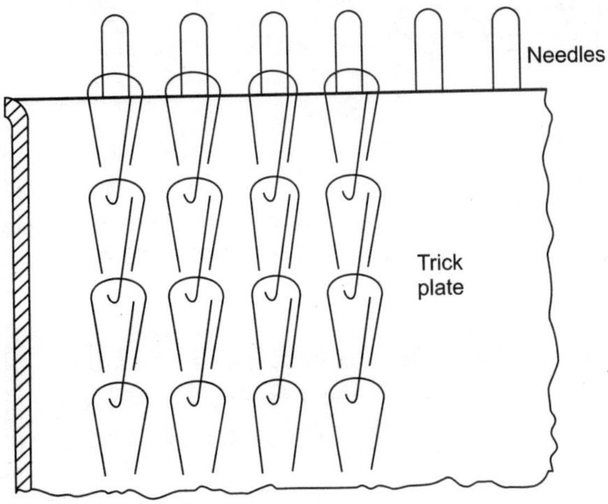

Fig. 17.15

It is placed to the full width of the machine. The lateral oscillation of the latch needles at faster running conditions are restricted by the trickplate when the needles move up and down. The top edge of the trickplate is designed in such a manner to ensure perfect knock over operation during the loop formation.

17.2.9 Latch Wire (Fig. 17.16)

The Raschel machines working with latch needles are fitted with latch wires. It is a steel wire placed to the full width of the machine, parallel to the needles. After the needles clearing operation due to the momentum, the latches may return back quickly and close the needle hooks. With closed hook, the needle cannot receive a new yarn and thus the loop formation is hindered. The latch wire placed close to the needles stops the flick back of latches and keeps the hooks in open conditions.

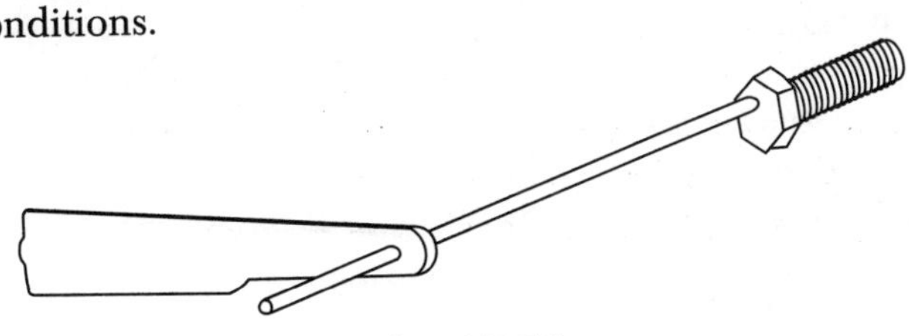

Fig. 17.16

18

Patterning in Warp Knitting

18.1 INTRODUCTION

In weft knitting, the patterning is based on the independent selection of knitting needles for various stitches such as knit, tuck and miss or float. But in warp knitting, the needles are not to be selected for loop formation, but the guides move around the needles, wrap yarns over them and form the loops. Therefore, the patterning is based on the lateral movements (*i.e.,* shogging) of the guide bars at the front and back of the needles *i.e.,* the overlap and underlap. Pattern wheels or the pattern chain links wound over the pattern drums control the lateral movements of the guide bars.

18.2 SWINGING

As mentioned earlier, the guide bars undergo two types of movements, viz. swinging and shogging. The swinging is the to and fro-movement of the guide bar so that the guides pass through the needles front and back. This movement is derived from a rocket shaft, which is driven by the main shaft of the machine, via. Levers, pivots and linkages (Fig. 17.8). The two swinging movements produce the two side limbs of the loop structure. The forward and backward swinging movements are equal in their magnitude. It is possible to achieve a longer or shorter swinging movement by connecting the leverage arrangements of the pushrod in different positions.

18.3 SHOGGING

Shogging is the lateral movement of the guide bar at the front as well as back of the needles. Shogging takes place in parallel to the needle bar. Shogging at the front of needle is overlap and at the needles back is underlap. The overlap is restricted to only one needle space displacement to wrap a yarn over the needles hook where as the underlap can be extended to more needle spaces to impart diagonal cross links at the back side of the fabric. So, shogging takes place twice during each knitting cycle. During underlapping, the guides are displaced from the needles previously wrapped, to the needles which will be wrapped in the next knitting cycle. This extent of underlap decides the structure of the fabric.

18.4 PATTERNING MECHANISM

The shogging movements of the guide bar mainly influence the patterning in warp knitting. So the patterning mechanism in warp knitting is aimed at controlling the guide bars for their overlap and underlap operation. The patterning mechanism essentially consist of the pattern wheel or a pattern drum wounded with pattern chain links.

18.4.1 Pattern Wheel

The mechanism (Fig. 18.1) consists of a pattern wheel whose periphery is stepped into different heights. The step heights decide the displacement of guides with respect to needle spaces. A height of no.1 represents one needle space coverage by the guides; no.2 represents two needle spaces and so on.

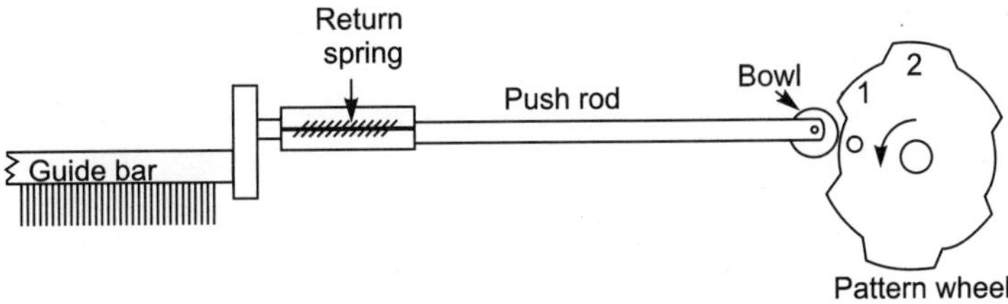

Fig. 18.1

A bowl which is rigid over the pattern wheel is connected to the guide bar by a push rod and a return spring. As the pattern wheel rotates, based on the contour provided by the various step heights of the pattern wheel, the bowl is shifted to different extent (from the centre of the pattern wheel), which in turn gives the lateral movements to the guide bars. The return spring is always having a pressing tendency over the guide bar towards pattern wheel. The pattern wheel is precisely pre-cut for a certain design. For any change of design, the wheel is to be replaced according to the pattern repeat. Though the pattern wheel provide accuracy and smooth running at higher speeds, they are economical only for long production runs of some common simple repeat structures. For fancy structures, frequent change of pattern and long pattern repeats, the shogging movements are carried out with pattern drums by the assembling a chain of reusable pattern links. As two lateral movements (viz., overlap and underlap) are required for the one complete course repeat, the minimum number of pattern wheel steps to make a fabric is four. The circumference of a pattern wheel may contain the repeats 8, 4, 2, or 1 times. If the pattern repeats exceeds 8 or if they could not be contained within the circumference of the pattern wheel, (*i.e.*, non divisible) then the pattern drum mounted with endless pattern links is used.

18.4.2 Pattern Drum and Chain Links

The pattern drum assembly consists of an endless chain made up of pattern links of different heights which is wound over the periphery of the drum. The pattern chain links have been accurately ground to provide a smooth and accurately timed side ways movement of the guide bars. The chain links available to produce the shogg movement are of four types. (Fig. 18.2) viz. A – straight, B – ground at leading, C – ground at trailing, D – ground at both leading and trailing. They are also numbered as 0,1,2,3,4 and so on with repsect to their heights. Height 1 will bring the guide bar closer to the pattern drum and vice versa. Each successive links will move the guide bar one needle spaces further than the previous number.

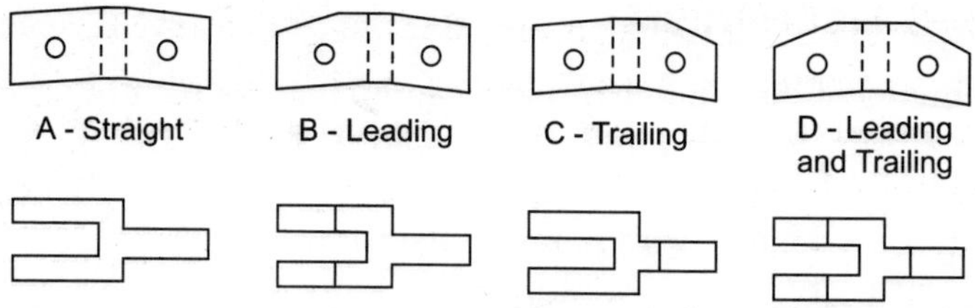

Fig. 18.2

The links have a front fork and a back tail and an endless chain is formed by connecting the fork end of a link to the tail end of the preceding one. The links have two holes, into which steel pins are inserted to connect the links into a chain. The links are slightly arched to fit the surface of the pattern drum. The links are made to fit a certain machine gauge and to avoid confusion, the machine gauge and the height of the link are stamped on to the link. The ground or slope of the links at the leading and trailing edges ensure smooth displacement of the guide bars without any jerk. During the rotation of the pattern drum, the fork end of the link leads.

While assembling the pattern chain links, the following points are to be observed (Fig. 18.3)

- The link preceding to the ground down link, should be straight at the back.
- The link succeeding the ground down link, should be straight at the front.
- Overlap is usually produced with one single variation in height from one chain link to next.
- Underlap is produced with one or two variations in height over three chain links.
- Special care must be taken to ensure right position of the chain corresponds to the movements of the knitting elements.

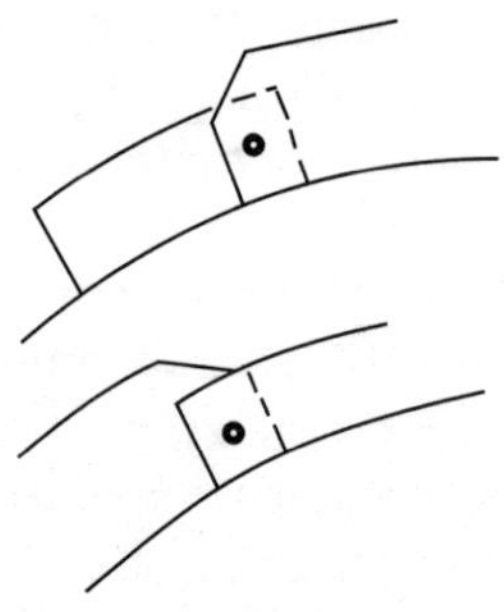

Fig. 18.3

If the chain is placed one link forward or backward, the guide bar shogs its underlap on the hook side of the needle, wrapping more than one needle and causing damage to the knitting elements.

Tricot machines which are functioning with less number of guide bars (mostly 2 guide bars), perform mostly with pattern wheels for their shogging movements. With Raschel machines, many different types of pattern mechanisms are used, depending on machine type and purpose. Raschel machines working with less number of guide bars use a pattern mechanism similar to that of Tricot machines. However, the Raschel machines, when equipped with more pattern guide bars along with their ground bars, use another type of mechanism which operates the guide bars through special shogging levers. Both the Tricot and Raschel multibar machines are very commonly equipped with a mechanism which combines both the principles. This combined mechanism operates the ground guide bars and pattern guide bars by pattern wheels and pattern drums respectively.(Fig. 18.4)

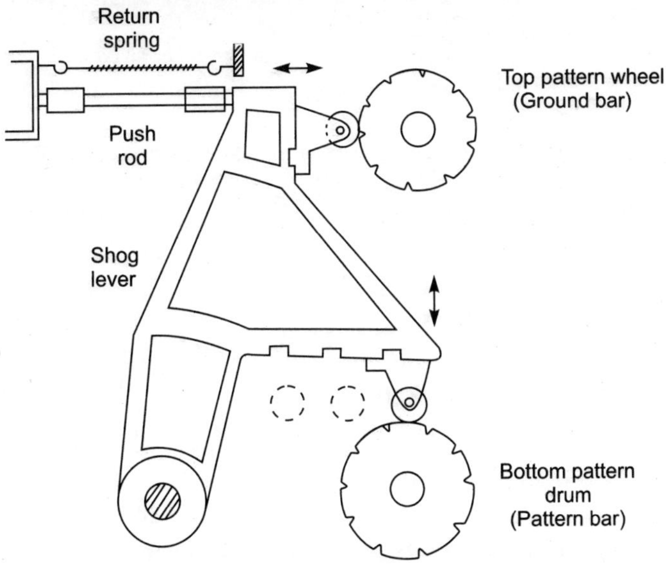

Fig. 18.4

This combined mechanism offers the following possibilities:

- The top pattern wheel control the ground guide bars, thus produce the ground structure of the fabric. The bar shogs twice during each knitting cycle, thus produce overlap and underlap. Two chain links per knitted course are required to produce the two shogging movements.
- The bottom pattern drum control the pattern guide bars. Those bars, as they are meant for special stitch movements, (say for example laying-in) do not wrap yarn around the needles, and only underlap shogging movement is required. So only one chain link per course is necessary. The chain therefore is only half the length it would otherwise have been.
- The shogging lever placed in between the pattern guide bar and the bottom pattern drum magnify the movement given by the pattern link. The leverage of the shogging lever increase the lateral movement of the guide bar in such a manner that a small movement produced by the pattern link will be doubled by the shogging lever and hence more needle spaces can be covered.
- The follower roller of the shogging lever that rest on the bottom pattern drum can be positioned in different arrangements. This enables different leverage ratios to be obtained. By doing so, machines of different gauges can be operated, using pattern links of only one gauge (Fig. 18.5).

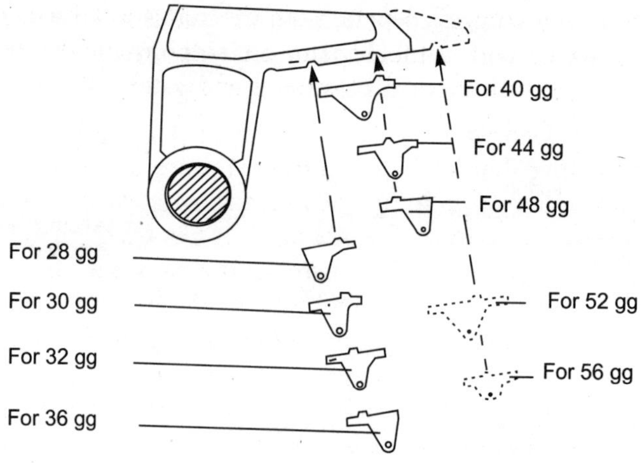

Fig. 18.5

18.5 MULTIBAR PATTERNING MECHANISM

The patterning mechanism of multibar machines, which operates both the ground guide bars and pattern guide bars is illustrated in Fig. 18.4.

The top pattern drum which is in line with a guide bar operates at two links per course and is thus used for ground bars. This produced the ground fabric. A pattern wheel can also be used instead of pattern drum for the ground bars. The bottom pattern drum operates a shog lever which is fulcrummed at its lower end. To the upper end of the shog lever, pattern guide bars are connected by their push rod. Each pattern guide bar will be having its own shog lever and push rod. The bottom pattern drum controls the movement of the pattern bars only. As these pattern bars are meant for pattern effects (for example, laying-in), only an underlap shogging movement is required. To produce such designs, the pattern bar operate at one link per course and thus inserts underlaps.

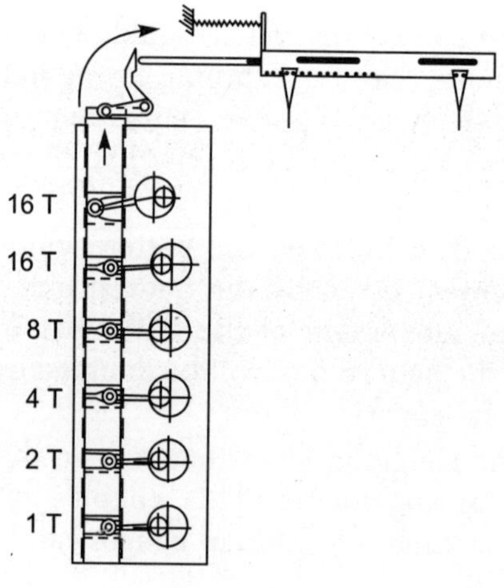

Fig. 18.6

An advanced electronically controlled patterning mechanism, known as "SU" or summary drive, opened a new area in the multibar patterning, in which the traditional chain link mechanism is replaced. The mechanism is illustrated in Fig. 18.6.

Every pattern bar is equipped with such a "SU" unit. Each unit is made up of six eccentric cams, mounted on six rotating shafts. The eccentric cams are not fixed to the shaft, and so do not rotate while the shaft rotates. There are two electromagnets situated on each side of the cam and they are energized one at a time. When an electrical signal is transmitted by a microprocessor, one of the two electromagnets locks the cam into the shaft, so that the cam rotates with the shaft. The eccentric movement moves the push rod forward and the roller in front of the push rod forces the segmented vertical bar to move upwards. This vertical movement is transformed at the top of the segment bar into a horizontal shogging movement. A signal by the microprocessor to the other electro magnet again locks the eccentric cam on to the rotating shaft, pulls the roller out of the segmented vertical bar, shortens it and shogs the guide bar by horizontally in the opposite direction.

Each unit, within a set of six cams, when activated, results in a different shogging motion. As denoted in diagram the cams, from the bottom to the top, when inserted or withdrawn, moves the guide bar to different needle spaces, such as 1, 2, 4, 8 and 16 respectively. By inserting or withdrawing any of these cams simultaneously, a shogging movement of any number of needle spaces can be achieved upto a total of 47.

For example, to displace the guide bar by 11 needle spaces to the right, the first, second and fourth eccentric cams are inserted. *i.e.,* $1 + 2 + 8 = 11$.

From this position, in order to displace the guide bar by 5 needles spaces to the left, the third eccentric cam is inserted, while the first and fourth are withdrawn. i.e., $11 + 4 - 1 - 8 = 6$.

18.6 NOTATION

In weft knitting, the representation of the fabric structures are based on the notation of the basic stitches such as knit, purl, tuck and miss. In warp knitting unlike weft knitting, as there is no variation in the lifting of needles (which are lifted to only fixed heights), and the guides are lapping yarn over the needles, the notation are based on the movement of guide bars with respect to the needles. *i.e.,* the overlap and the underlap movement. As already mentioned, the overlap is only one needle space displacement of the guide bar and the underlap is more needle space displacement.

The warp knitted structure is represented by plotting the lapping movements of the guide bar on a point paper (Fig.18.7), where each dot represents a needle in the needle bar, and each row of the dot represents a course. Successive courses are read upwards on the paper from the bottom to the top. As the guides position themselves in the space between the needles, the position between the vertical column of points *i.e.,* the needle spaces being numbered from the right to left starting from 0. This is because the patterning mechanism is on the right side of the machine. It is assumed that the needles are facing towards the back of the machine and the guides start moving from the front side of the machine to the back.

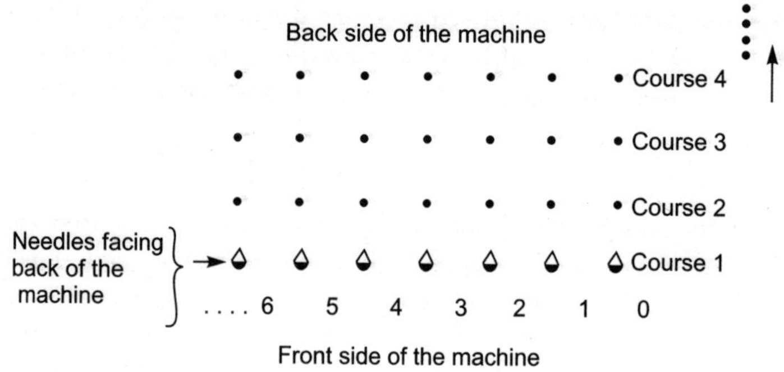

Fig. 18.7

During loop formation, the guides make four kinds of movements.

- Starting from the front of the machine, the guide swing through the needle space to the back of the machine.(*i.e.,* hooks side).
  It makes a lateral shogging movement in front of the needle, which is one needle space displacement (*i.e.,* overlap).
- The guide again swing back to the front of the machine through the needle space.
- It makes another lateral shogging movement at the back of the machine when the loop is cast off. This movement is to place the guide to a new needle position for the next knitting course to take place.

Pattern chain notation

In representing the guide bar lapping movements, the overlap movements are indicated by dash (-) with the needle spaces that the guide bar has moved from and to. The underlap movements are indicated by a slash (/) which is drawn between two overlap representations.

In the lapping diagram shown below in (Fig. 18.8), the dotted lines represent the swinging movements and the fully drawn lines represent the shogging movements. The first overlap was caused when the guide bar moved from space 1 to space 0, which would be noted as 1 to 0. The second overlap was caused when the guide bar moved from space 0 to space 2, noted as 1-2. By joining the two overlaps, the under is completed automatically, as the underlap was caused when the guide bar moved from space 0 to space 1, which is noted as 0 /1 and then from space 2 to space 1, noted as 2/1. The full chain pattern notation would therefore be 1-0 /1-2/1-0/1-2 and so on repeating to form the structure. Usually one repeat of the pattern is normally represented, i.e., 1-0/1-2. In order to activate the guide bars as per the above repeat, the pattern chain is formed from individual links. For the two courses repeat 1-0 /1-2 notation, the links used for the pattern chain for one full repeat would be as shown in the Fig. 18.9.

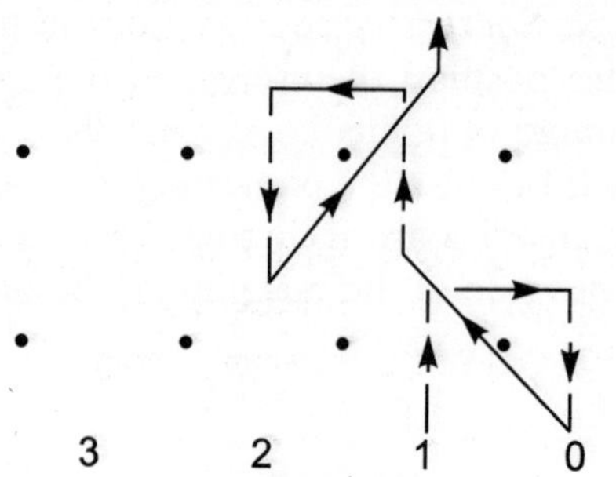

Fig. 18.8

A single step in link height requires two chain links for overlap (i.e., 1 to 0 and 1 to 2) and underlap (0 to 1) movements per stitch as shown in the Fig. 18.9. It can be noted that the guide bars change their shogging direction for overlap and underlap based on the trailing and leading grounds of the chain links.

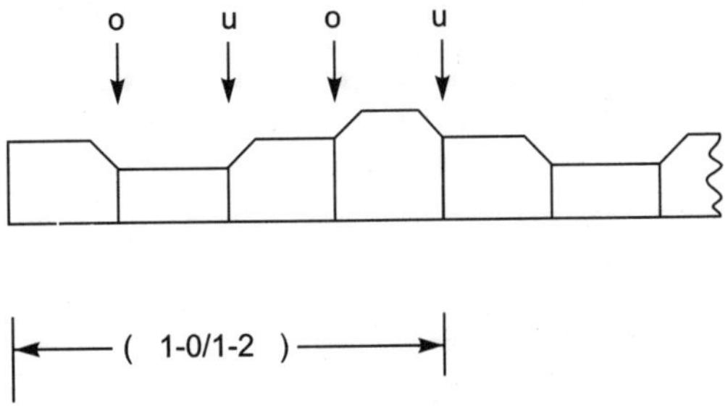

Fig. 18.9

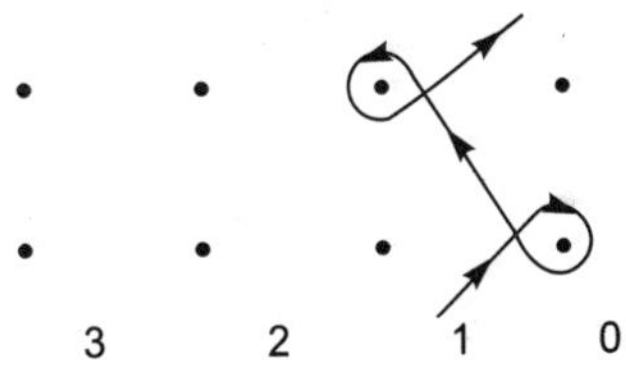

Fig. 18.10

This pattern forming links would be repeated until all the links have been included to fill the pattern drum, which normally requires 48 links in total. With two links per course, 24 courses will therefore be required for one revolution of the drum.

The final version of the lapping diagram will be shown in the form of curved lines as shown in Fig. 18.10 having considered the similarity of yarns knitted over the fabric structure.

The overlaps and underlaps are represented by the letters O and U respectively.

While plotting the guide bar lapping diagram, the following basic rules are to be observed:

- For fabrics composed of more than one guide bar, lapping movements of each guide bar must be represented separately.
- The starting of the lapping diagram of all the guide bars during knitting, must be plotted from the same course.
- Numbering of the needle space should correspond to the position of the pattern mechanism. For machine with patterning mechanism on their right side, which is more common, the numbering is from right to left.

In order to provide smooth displacement to the guide bars and to obtain minimum vibration, modern machines use three links for each knitting cycle. This is because the guides spend very little time on the hook side of the needles during overlap, when compared to the underlap. The three links per course movement distributes the time allocated for each shogging movement in a better way, *i.e.,* one third for overlap and two thirds for underlap. It is mentioned earlier, that the overlap consist one needle space and the underlap consist several needles space. Therefore two links available for the underlap allow a long shogging movement to be distributed between them. Thus a third intermediate link allows time for the underlap, provides smooth running condition. For the same two-course repeat 1-0 /1-2 notation discussed earlier in (Fig. 18.10), a new setup of pattern chain links by incorporating intermediate links is shown in Fig. 18.11. Under such cases, for 48 link pattern drum, with 3 links per course, 16 courses can be produced per revolution of the pattern drum.

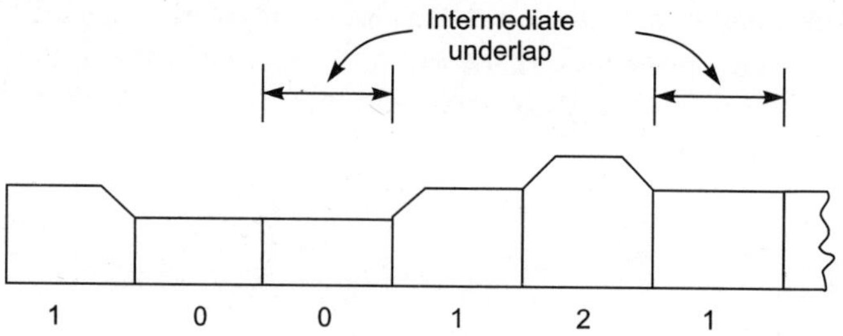

Fig. 18.11

Another example of pattern chain links for a two-course repeat 1-0/4-5 notations, having 2 links per course movement is illustrated in Fig.18.12 (a) & (b).

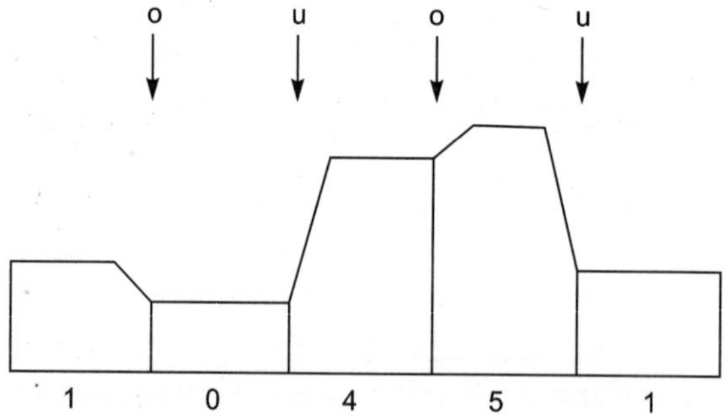

Fig. 18.12 (a) *Two links per course chain arrangement*

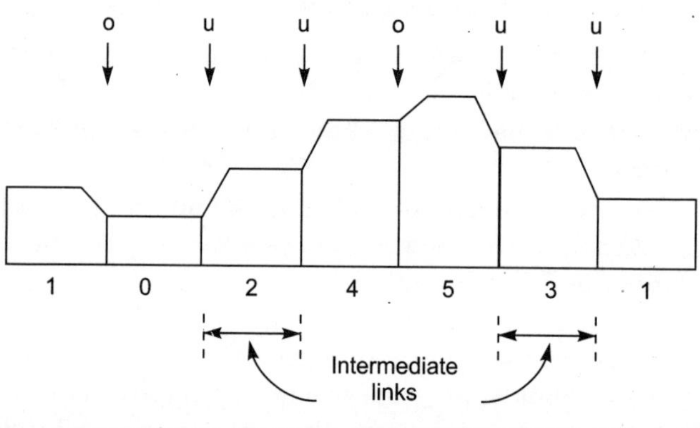

Fig. 18.12 (b) *Three links per course chain arrangement*

The chain link scheme for the pattern 1-0/4-5 with three link arrangement as it is mounted over the pattern drum is shown in Fig 18.12 (c).

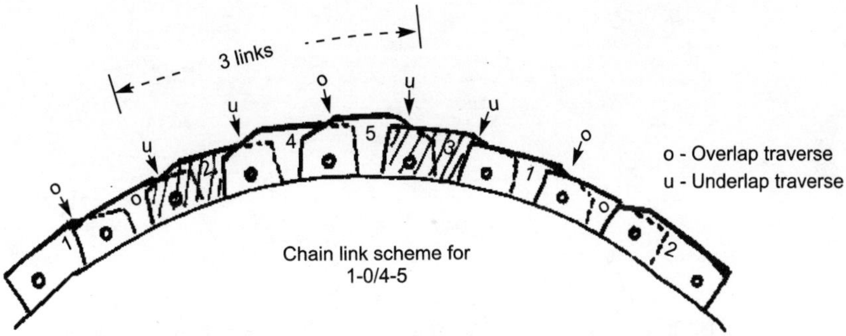

Fig. 18.12(c)

The conventional ground edge links may not give sufficient smooth working conditions with high speed machines. In such cases, special links, called profile links are used. In these links, the grinding to the edge of the link is curved and not straight (Fig. 18.13).

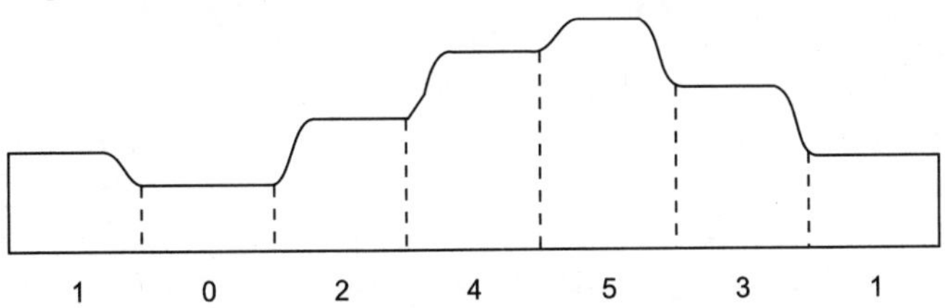

Fig. 18.13

19

Tricot and Raschel Machines

19.1 A COMPARISON

The two classes of machines used for the production of warp knitted fabrics are Tricot and Raschel. There are a number of fundamental design differences between Tricot and Raschel machines. Tricot machines use both the bearded and compound needles, with the later becoming extremely popular since its shorter and smoother needle movement has enabled increased production speeds to be achieved. Usually two guidebars are used on Tricot machines, though upto five guidebars are possible. They are used for the production of apparel and household fabrics, particularly when bearded needles are used to process continuous filament yarns.

Raschel machines use latch and compound needles are usually coarser in gauge than the Tricot machines. Raschel machines can have upto 78 guidebars and they are capable of processing both staple fibre and continuous filament yarns for furnishing, Industrial and jacquard fabrics. As mentioned above, the Tricot machines widely use less number of guidebars (3 or 4) and the use of multibar machines is no longer popular. The multibar tricot machines were constructed to meet a specific requirement of warp knitters.

Both the Tricot and Raschel machines differ by way of their take up principle also. In Tricot knitting, since the amount of yarn in each loop is always set by the let-off mechanism, the take-up affects mainly the shape of the loop and tensions in the yarns. Also, in Tricot take-up the main objective is just to collect the fabric produced in the knitting zone. Whereas in Raschel knitting, the fabric take-up mechanism determines the fabric quality and appearance. The loop forming process on a Raschel machine imparts high take-up tension. This difference leads to different construction principles of take-up mechanism in Tricot and Raschel machines. The Tricot takeup mechanism is positioned further away from the knitting zone, while the Raschel one is positioned as close as possible to the needles, pulling the fabric very tightly downwards.

19.2 TRICOT MACHINE DESCRIPTION

Fig.19.1 shows a simplified section view of a Tricot warp knitting machine. The Figure reveals some unique characteristics of the machine in terms of the relative position of the warp beams, the number of guide bars used and the angle at which the fabric is taken up from the needles.

The warp supply beams are placed at the back of the machine and the fabric collection is at the front. The fabric is drawn away almost at right angles to the needle bar. The warp beams are accommodated in an inclined arc towards the back of the machine with the top beam supplying the front guide bar and the bottom beam supplying the back guide bar. The warp sheets are taken through their tension rails (sley combers) situated at the front of the machine.

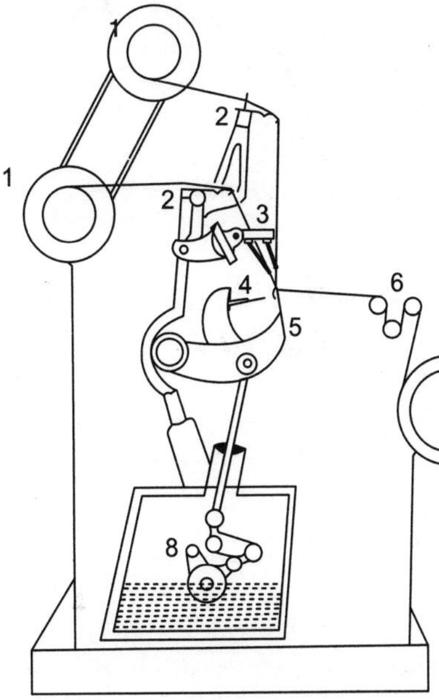

Fig. 19.1

The needle bed consists of vertically mounted needles in a bar. The gauge of the machine is expressed in needles per inch (28–40NPI). Mechanical attention to the knitting elements is carried out at the front of the machine as the beams prevent access to the back. Due to the placement of warp beams and guidebars, the threading to the guidebars usually commence with the back bar. This threading sequence makes the guidebars to be numbered from the back towards the front of the machine (Fig.19.2). The majority of Tricot machines employ only two guidebars.

The knitting zone of the machine comprises all the important elements such as needle bar, sinker bar, presser bar, guide bar and their synchronicity in their driving is of great important. Each of the different elements are driven separately and independently from the machine bed through cam (or) eccentric drive.

The needle bar in which the needles are set is driven up and down by means of cams in conventional and eccentrics in modern high speed machines. The eccentrics are connected to the needle bar by a crank assembly and the combined movement produce the desired effect.

The horizontal movement of the sinker bar can also be obtained from the crank/eccentric assembly. The vertical movement of the eccentric cam is transferred into horizontal movement to the sinker bar by a crank mechanism through a push bar.

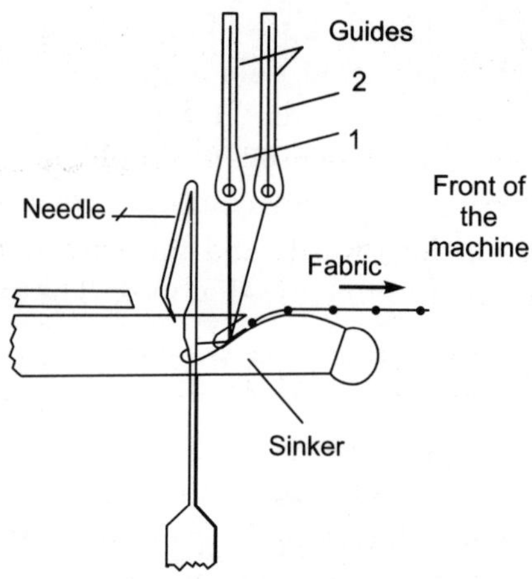

Fig. 19.2

The drive to the guide bars is with reference to its compound movement, which consists of a swinging movements-through the needles and shogging lateral movements-parallel to the needles.

The swinging movement is generated by a mechanism similar to that as above. The vertical movement created by the driving assembly is transformed into the swinging movement of the rocker shaft that carry the guidebars. The lateral shogging movements of the guidebars is generated by the patterning mechanism which is situated on the side of the machine. The pattern mechanism is driven by the main shaft and the pattern drum rotates in a constant ratio to the speed of the machine. The pattern drum chain link numbering is 0, 1, 2, 3, 4, etc., generally with three links per course.

The yarn let-off consists of a length regulated-mechanically stepless variable regulating device, constantly driven from the main shaft by an intermediate shaft. Constant let-off speed is guaranteed eventhough the diameter of the warp beam is constantly decreasing.

Fabric take up rollers are driven by the rotating movement of the mainshaft of the machine through change gear and warm gear assembly. The surface of the rolers is covered with a special material which has a high friction coefficient so that any slippage between fabric and rollers is eliminated.

19.3 KNITTING CYCLE OF TRICOT MACHINE (BEARDED NEEDLE)

The knitting action for one knitting cycle is carried out by the combined operation of the basic knitting elements such as bearded needle, presser, sinker and guide. One revolution of the drive shaft gives one cycle of the operation which is explained below in Fig. 19.3.

Position 1**:** It is the start of the knitting cycle. The needles are lifted to 2/3rd of their full height from knock over. The guides are at the front of the machine and sinkers are in the forward position, holding the loops. The presser is withdrawn.

Position 2**:** The guides swing through the needles to the beard side and make a lateral shog in front of the needles for one needle space. The needle sinkers and presser remain idle (about 45° of rotation).

Position 3: The guides swing towards the front of the machine through the needles and the loop part (overlap) is completed.

Position 4: The needle raised to their maximum lift and cause the thread to slip from the hook portion to the needle stem.

Position 5: The needles move downwards and the yarn is trapped into the hook of the needles. The presser bar move forward to close the hook. The sinkers move backward.

Position 6: As the sinkers withdraw, the upward curve of their bellies lands the old overlaps into the closed beards. As soon as the fabric loop is landed on to the beard, the presser is withdrawn.

Position 7: The pressure is withdrawn and the needles continue to descend, pulling the hook through the old loop and hence knocking over occurs. The guides make lateral shogging at the front so that the underlap takes place. The guides move to their new position at which they can swing for the next cycle.

Position 8: The sinkers move forward to hold down the fabric loops and push them away from the ascending needles which are rising to the rest position.

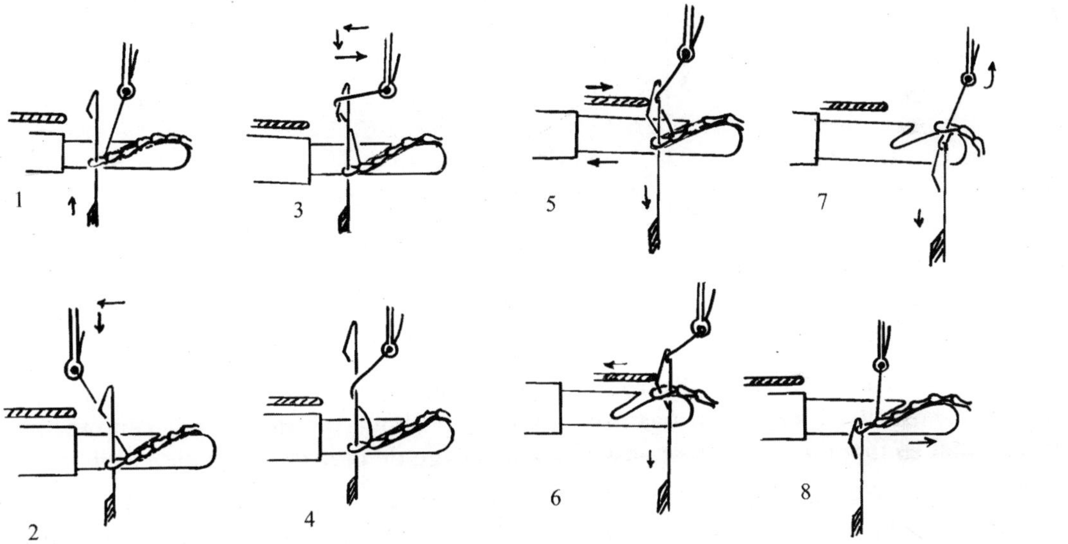

Fig. 19.3

19.4 RASCHEL MACHINE DESCRIPTION

A simplified view of a Raschel warp knitting machine is shown in Fig.19.4. The Figure shows the arrangements of various warp knitting elements of the knitting zone, supply beams, take down device and drive. The gauge of the machine is generally expressed in needles per two inches. (i.e., 36 gauge means 18 needles per inch). The chain links of the machine are usually numbered in even numbers 0, 2, 4, 6, etc., generally with two links per course.

Unlike the Tricot machines, the warp supply beams of the Raschel machines are placed above the machine. This enables easy accessible at the front for fabric inspection and at the back for mechanical attention to the knitting elements. The guide bars are numbered from the front towards the back of the machine. More number of supply beams can be accommodated in Raschel knitting so that atleast four 32 inch diameters beams or large numbers of small diameter pattern beams are engaged. Larger number of warp beams necessitates larger number of guidebars with Raschel machines. The warp sheets are taken through tension rails which are the tension compensating devices.

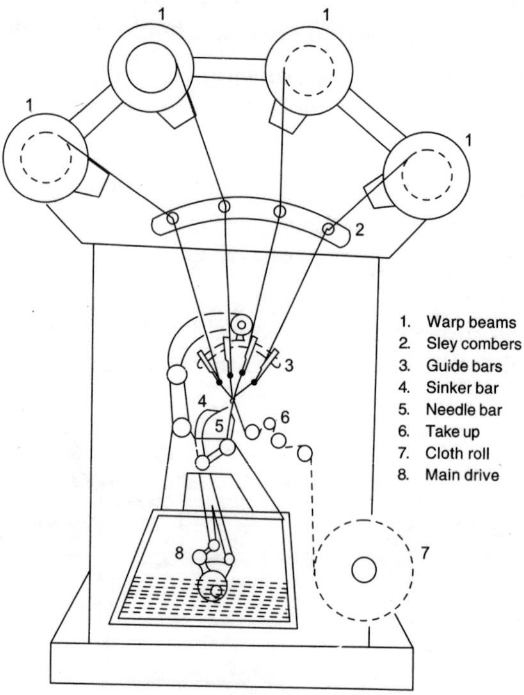

Fig. 19.4

The fabric is drawn downwards from the needles at an angle of 120-160 degrees by a series of take-down rollers. The warp beams are arranged above the needle bar centred over the rocker-shaft so that the warp sheet pass down to the guide bars on either side of it.

The needle trick plate supports the fabric and assist the knock over operation. Raschel sinkers perform only the holding down of the loops while the needles rise. As the take up mechanism take care of the fabric tension and clearing of the needles, the function of the Raschel sinkers are only of secondary importance. So they move away towards the back of the machine for the rest of the knitting cycle.

Different driving arrangements such as eccentrics and pot cam are used in modern Raschel machines. Usually, eccentrics are used to drive high speed machines with a lower number of guidebars, while pot cams are used to produce a more complicated movement for machines equipped with greater number of guidebars running at a lower speed. Drive in combination of both eccentrics and pot cam are also available.

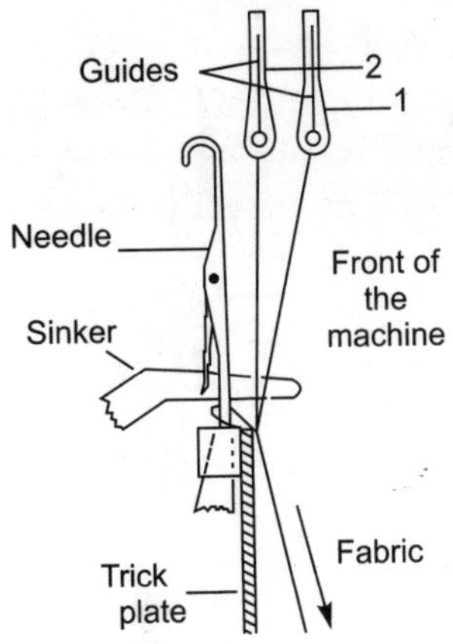

Fig. 19.5

Two following rollers, viz., the lift and counter rollers follow the contour of the pot-cam, give vertical movements to the needle bar through a push bar. In some Raschel machines, the needle bar is driven horizontally also, in a counter movement to the swing of the guides. This horizontal movement is used to shorten the guide bar swing so that the machine speed can be increased. For this horizontal movement of the needle bar, another pot-cam assembly is used. Similar driving arrangements are used to operate the sinker bar horizontally from the pot cam assembly.

Similar to that of Tricot machines, the guide bars of the Raschel machine have compound movement, viz., swinging and shogging. The swinging movement is obtained from the pot-cam through the push rod, whose up and down movement is transformed into the to-and-fro swinging movement via., leverage connections. The shogging movement of the guidebars is produced by the patterning mechanism. As the Raschel machines use two types of guide bars viz., ground bars for the ground structure and pattern guide bars for design effect, the machines are equipped with patterning mechanisms for the above two combinations.

In Raschel multi-guide bar machines, apart from the positive let-off for main supply beams, the spot beams which supply the partly-threaded pattern guide bars are completely negatively turned. These light weight beams turn easily and are placed with small weights at one end to ensure balanced operation.

As mentioned earlier, in Raschel machines, to impart tension during take-up, the fabric is pulled downwards very tightly at 120-160 degree to the needle bed. Also the take-up mechanism is placed close to the needles unlike Tricot knitting.

19.5 KNITTING CYCLE OF RASCHEL MACHINE (LATCH NEEDLE)

Figures. 19.6 shows the main elements involved in the loop formation of a latch needle Rachel warp knitting machine. The various diagrams illustrate the sequence of events in one knitting cycle. The Rachel needles have longer latches than the weft knitting needles in order to ensure perfect wrapping of yarn well within the hook portion.

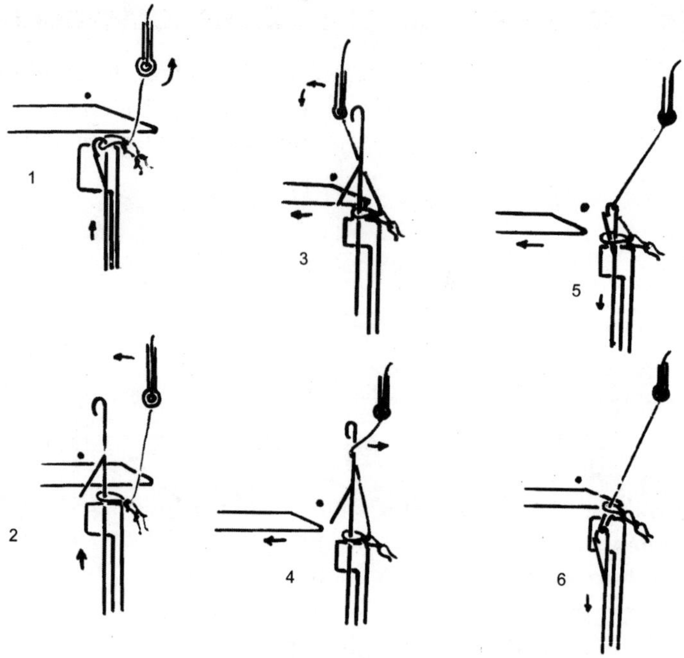

Fig.19.6

Position 1: It is the holding down position in which after completing a course of loops, the needle's head is just in line with the top edge of the trick plate. The guide bar at the front of the machine complete its underlap shog in readiness for the next course. The sinker bar moves forward to hold the fabric down.

Position 2: At this clearing position, the needle bar rises to its full height. The latches get opened and the old loop is at the stem of the needle. The flip back of the latches is prevented by the latch wire. The guide bar start swinging towards the back of the machine.

Position 3: The sinker bar starts to withdraw and the guidebar swing to the back of the machine and then shog on the latchside of the needle for the overlap.

Position 4: The guide bar makes return swing to the front side of the machine through the needles. The warp yarns completely wrapped around their respective needles.

Position 5: The needle bar descends and the old loop closes the latch, trapping the new overlaps inside. The guide bar movement terminates.

Position 6: The needle bar descends downwards further, its head passes below the surface of the trickplate, drawing the new loop through the old loop which is cast -off. The trickplate supports the old loop and the sinkers advance towards the trickplate. The underlap shog of the guide bar starts.

19.6 KNITTING CYCLE OF RASCHEL MACHINE (COMPOUND NEEDLE)

The loop forming action of compound needle is carried out by its combined action along with other knitting elements such as guide bar, sinker bar and trickplate. Figure.19.7 shows the various stages in the knitting cycle for a compound needle warp knitting machine.

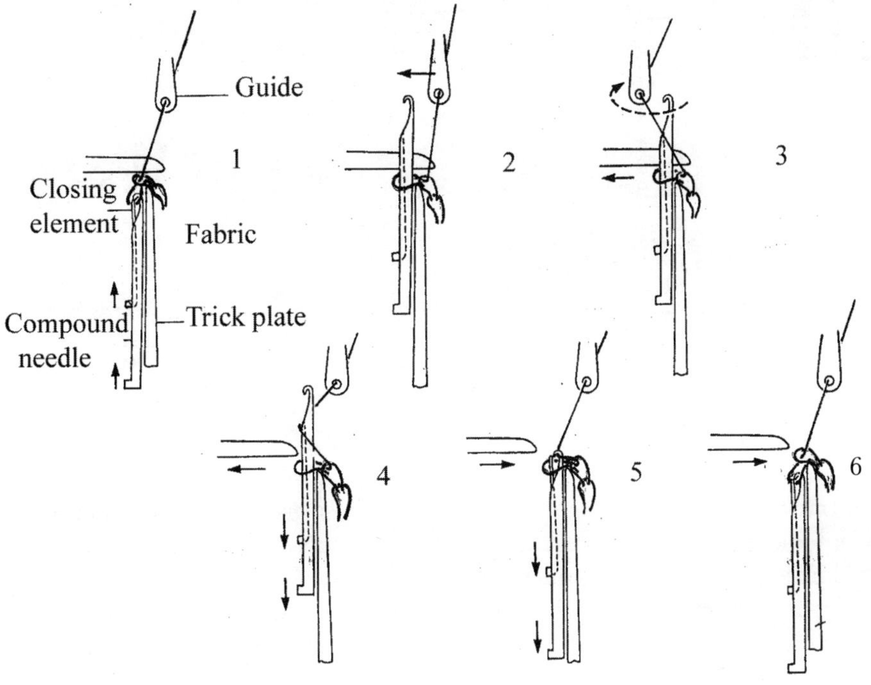

Fig.19.7

Position 1: The needles are at their lowest level in their knock over position. After completing the previous set of loops, they are about to be lifted. The sinkers are in their advanced position.

Position 2: The needles have risen to the full height. The movements of the hook and tongue are timed in such a manner that the hook position gets opened. The guide then swing towards the back of the machine.

Position 3: The sinkers start to withdraw. The guides shog in front of the needles at their hook side for one needle space and swing back to the front of the machine. An overlap is produced.

Position 4: The sinkers continue their withdrawal movement and the needles descend.

Position 5: The hook part of the needle is closed and the newly wrapped yarn is trapped in. The guides shog sideways to reposition themselves for the start of the next course and the underlap is completed.

Position 6: The needles descend to their knock over and a new course of loops are produced.

20

Principal Stitches of Warp Knitting

20.1 INTRODUCTION

Among the various factors influencing the warp knit structures, such as machine gauge, loop type, guide bar lapping, guide bar threading, types of fibres and yarn, number of guide bars, number of needle bars etc., the basic lapping movements of the guide bar are very important. Most warp knitted fabrics are produced by a few guide bars, each one knitting a simple and basic lapping movements. These lapping movements can be regarded as stitches of warp knitting. This chapter describes such basic stitches of warp knitting. The chain notations for the stitches are given in the respective diagrams.

20.2 TRICOT STITCH (Fig. 20.1) 1 AND 1 LAPPING

If the threads are lapped by the guide bars round the two neighboring needles, in alternating course sequence, a tricot stitch is produced. It is also referred as "1 and 1" lapping, which implies an underlap of one needle space and one overlapped needle. An overlap followed by an underlap in the opposite direction produce a closed lap type tricot and an overlap followed by an underlap in the same direction produce an open type tricot stitch. This lapping movement is used in the production of single bar fabrics, which have restricted commercial value. The fabric produced with tricot stitches are unstable and also leads to loop inclination.

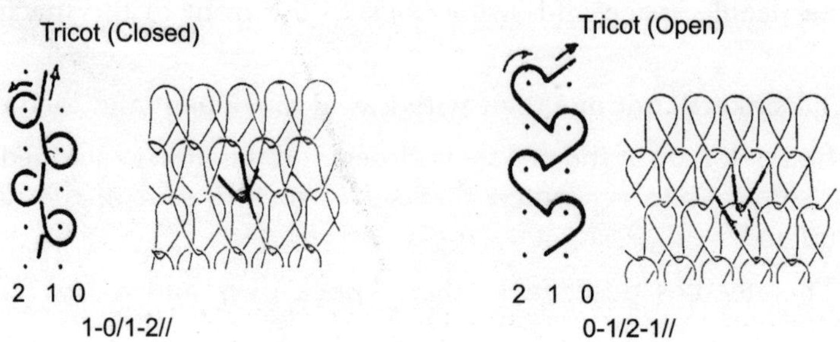

Fig. 20.1

20.3 PILLAR OR CHAIN STITCH (Fig. 20.2)

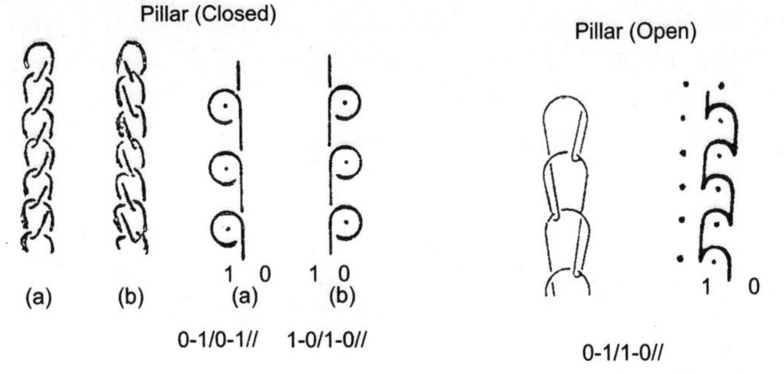

Fig. 20.2

A pillar stitch is formed when a needle is being lapped continuously by the same guide. Since the guide bar does not lap their adjacent needles, there are no side way connection and no fabric is formed. It is not possible to construct fabrics only with pillar stitches, unless atleast one more underlap forming guide bar is employed. Different net fabrics, constructed of chain lapping movements, are produced in great quantities by the Raschel industry. The pillar lapping movement can be closed or open or can be a combination of closed and open laps. Having no side way under laps, yarn consumption of the guide bar is relatively very small. This, in addition to stable construction lengthwise, make the pillar stitch very popular for the production of certain fabric types.

20.4 IN-LAY STITCH (Fig. 20.3)

Fig. 20.3

If there is no overlap and only underlap exist, the stitch is known as in-lay or laying-in. Here, some of the guide bars do not knit the yarn into the fabric; instead, the yarn ends, threaded in these guide bars, are only inserted into the fabric *i.e.,* they are laid-in. To inlay the yarn into the fabric, special lapping movements are required. The laid-in yarn can be extended over one or several wales. Using colored and specialty yarns, the designers can use this technique to enlarge his range of designs. The inlay coloured yarns threaded with back guide bar are trapped by the stitches formed by the front guide bar.

20.5 BLIND STITCH (Fig. 20.4)

It is a miss-lap stitch in which there is no overlap as well as underlap. Here the guide bar swing continuously between the same two needles and then runs between two wales and floats freely on the technical face of the fabric. This miss-lap principle is widely used with multi-bar Raschel machines. The guide bars produce this blind stitch work in conjunction with other guide bars which produce the ground of the fabric. The blind lap is held thoroughly by other binding elements of the structure.

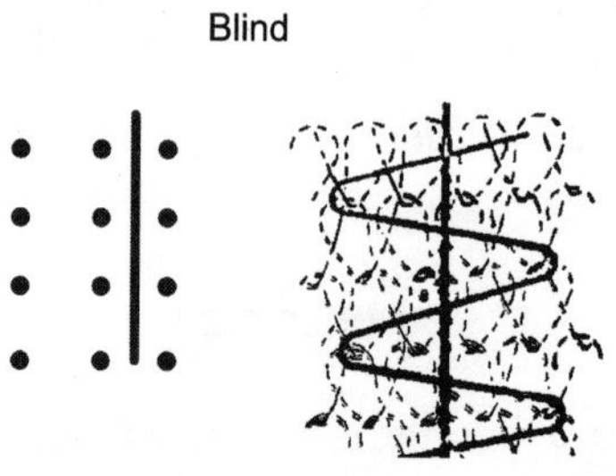

Fig. 20.4

20.6 2 AND 1 LAPPING (Fig. 20.5)

By increasing the underlap produced by the guide bar to one more needle, a 2 and 1 stitch is produced. Longer shogging movements causes the yarn to lie more horizontally in the fabric, thus increase the widthwise stability. The guide bar consumes more yarn so that the fabric is heavier and has a better covering factor. The lapping movements of both open and close are possible.

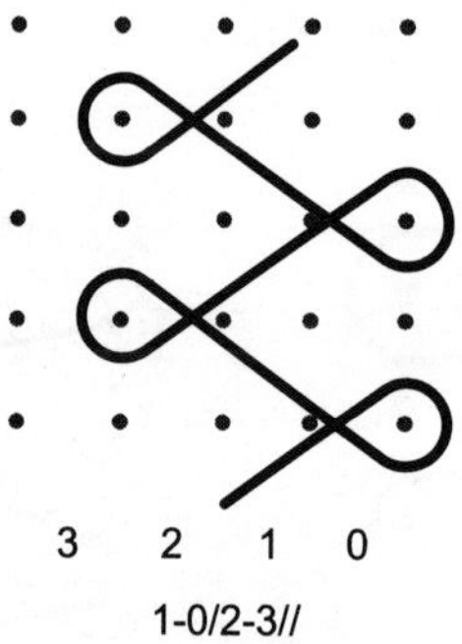

Fig. 20.5

20.7 LONGER LAPPINGS (Fig. 20.6)

By increasing the underlap shogging movement further by one or more needles spaces, 3 and 1, 4 and 1 lappings can be produced. They can be produced with closed and open lapping movements. As more yarn is used for this lapping movement, widthwise stability,fabric weight and fabric cover are increased.

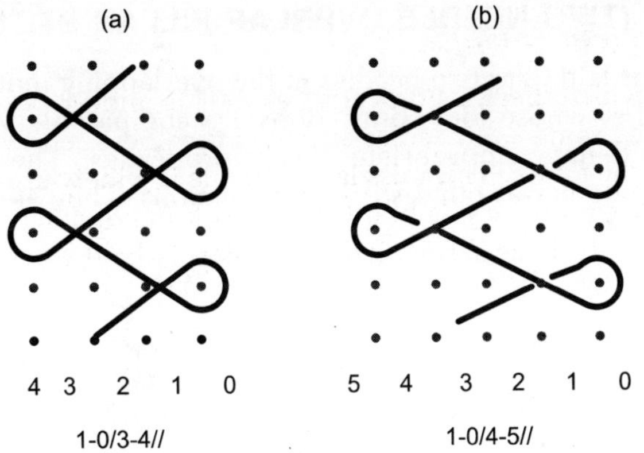

Fig. 20.6

20.8 ATLAS STITCH (Fig. 20.7)

The guide bar moves in one direction continuously for a few knitting cycles and wrap yarn over the needles. After a predetermined number of courses, the guide bar moves in opposite direction and wrap yarn for the same number of knitting cycles. This lapping can be of both open and closed type. Atlas lapping is often used with guide bars threaded with coloured yarns. With two guide bars moving in opposite direction, this atlas lapping produce attractive diagonal or diamond shapes. In case of closed atlas, after a number of knitting cycle, at the point of return (*i.e.*, return-row), the lap is open and again further, the closed laps continue. Vice-versa is the case for open atlas.

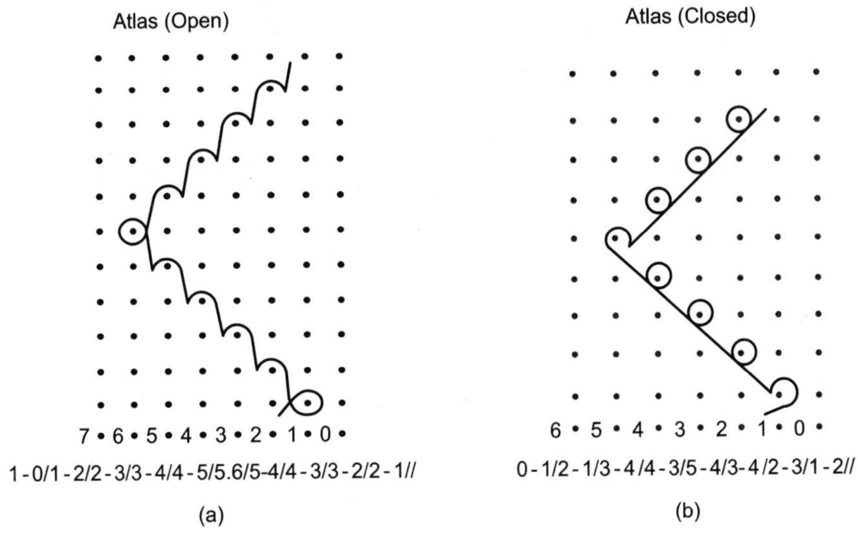

Fig. 20.7

20.9 TWILL STITCH (TWO NEEDLE OVERLAP PILLAR STITCH) (FIG. 20.8)

In this stitch, the thread is laid over two needles in the overlapping motion. This produces two stitches next to each other whose foot connections are comparable to those of weft knitted fabrics. The yarn should have appropriate elastic properties. These two needle overlap constructions are not as popular as other stitch constructions. They are used for heavy fabrics produced with one guide bar.

Twill (closed)
0-2/0-2//

Twill (open)
2-0/0-2//

Fig. 20.8

20.10 TWO NEEDLE OVERLAP SEMI-TRICOT STITCH (FIG. 20.9)

The stitch shows alternatively one wale with straight stitches and one wale with slanting stitches. In fabric of such stitches, the closed stitches are generally distorted whereas the open stitches are straight and rather uniform in appearance.

Closed (2-0/1-3//)

Open (3-1/0-2//)

Fig. 20.9

21

Structures of Warp Knitting

21.1 SINGLE BAR FABRICS

The warp knitted fabrics which are produced by a single guide bar of the machine in which all the yarns of the guide bar follow the same repeated action are known as single bar fabrics. The fully threaded single guide bar warp knitted fabrics have little commercial importance, due to their lower cover, lack of stability and limited design effects. Also the structures undergo loop distortion due to loop inclination as in each course there is a definite loop tilting either to the right or to the left depending upon the movements of the guide bars (Fig. 21.1) some of the structures will split easily if only slightly damaged. Solid and stable single guide bar fabrics, to some extent can be produced if the wales are connected together at every course. This is achieved by causing each warp thread to knit on a different needle at each successive course.

Fig. 21.1

The various stitches (Chapter 20) formed by the lapping movements of the guidebar can be used in combination with one, two or three guide bars to produce some popular two bar or three bar fabrics.

21.2 TWO-BAR FABRICS

The warp knit fabric production with two guide bars gives a large pattern potentiality than with single bar fabrics. These fabrics are most popular among all warp knitted structures and are produced mainly with two-course repeat cycle with a change of direction lap at each course. Although these structures are produced with Tricot machines, the Raschel machines also can be used as well. The two-bar fabrics have their commercial importance, better dimensional stability and strength compared to single-bar fabrics. Use of continuous filaments is common in most of the cases. As two guide bars are used, generally two warp beams are required to knit. Based on pattern requirement, the two-bar fabrics can be produced with full-set threading and part-set threading of the guidebars.

21.2.1 Two-bar Full Set Threading

In a simple knitting, each needle within the knitting width receives at least one warp thread at each course. In the case of full set threading, all the guides of both guide bars are fully threaded and hence each needle will have two warp threads and each loop of the fabric will be therefore having two yarns. In the two course repeat cycle, there will be a change of direction lap at each course. Each guide bar contributes a thread to every over lap, and the two under-laps can be clearly distinguished as they lap to a different angle, extent and direction. Generally under normal conditions, the threads of front guide bar (FGB) tend to dominate the face as well as back of the fabric *i.e.,* the threads of the back guide bar (BGB) are sandwiched between the face loops and back underlaps of the front guide bar. The two yarns supplied to the needle can be laid across the needle beards in the same or opposite directions, depends on the relative movements of the guidebars. If laying is in the same direction, loop distortion and inclined loops similar to single bar fabrics occurred. If laying is in the opposite direction, loop stability and vertical loops are formed. In two bar fabrics, the face of the fabric shows purely vertical wales having similar appearance of plain knit fabrics, and at the back, underlaps run diagonally right and left, crossing between the wales. The following basic features are applicable with reference to the two bar structures.

- The loops will lie straight in the fabric, if the underlaps of the two bars move in opposite direction.
- The loops will lie at an angle in the fabric, if the underlaps of the two bars move together (in parallel). The direction of loop inclination depends on the direction of movement of under lap.
- If the front guide bar makes longer underlaps and the back guide bar makes shorter one, the fabric will have widthwise elasticity.
- The longer underlaps of the front guide bar given lustrous effect at the technical back of the fabric.
- If the back guide bar makes larger underlaps and the front guide bar shorter, and both the bars move in opposite direction, a rigid stable structure is produced. Here the yarn movement is restricted as the back guide bar underlaps are trapped by the shorter underlaps of the front guide bar.
- The underlaps of the front guide bar will appear at the top of the technical back of the fabric.

The two-bar full set threaded Tricot warp-knitting machine is shown in Plate No.21.1

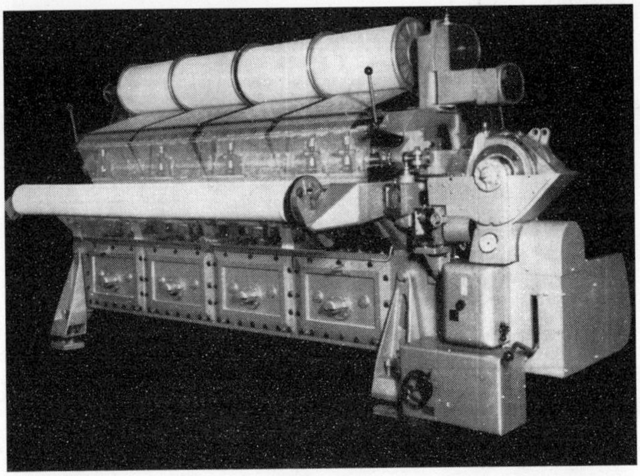

Plate No. 21.1. *Tricot-Two Bar Warp Knitting Machine*

21.2.2 Two-Bar Full-set Tricot Fabrics

As mentioned earlier, two bar full set warp knitted fabrics are produced mainly with Tricot machines and in this part, certain commercial fabrics are discussed with their lapping movements. Two separate warp beams feed the two fully threaded guide bars. The movements of the guide bars for the underlappings control the fabric elasticity, rigidity, stiffness and softness. The direction and magnitude of front and back guide bars produce various two bar knitted structures. If the magnitude of front guide bar and back guide bar is same, but the direction of movement is opposite, a full-Tricot structure is produced. If the movement of the guide bars are in opposite direction, but the front guide bar make longer underlaps than the back guide bar, a locknit fabric is produced. Whereas, if the back guide bar make longer underlaps than the front guide bar, reverse locknit, queen's cord and shark-skin structures are produced. If the front guide bar makes longer underlaps than the back guide bar, but both the guide bars move in the same direction, a loop-raised structure is produced.

21.2.2.1 Full-Tricot (Fig. 21.2) : Both the guide bars move in opposite direction but with equal magnitude of movements. The loops lie straight and the technical face of the fabric has similar appearance to single jersey. The back side shown diagonal lines of underlaps running right and left. The chain notations for the front guide bar is 1-0/1-2 and the back guide bar is 1-2/1-0. Due to shorter underlaps, the fabric is light in weight. Though the fabric is commercially very attractive, the fabric split very easily as each wale is connected only to the adjacent wales.

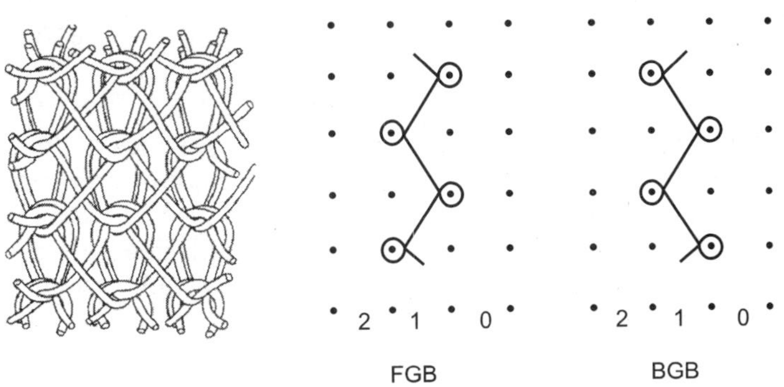

Fig. 21.2

21.2.2.2 Locknit (Fig. 21.3) : The guide bar move in opposite direction with longer underlaps of the front guide bar. The pleasant feel and considerable elasticity make the fabric more suitable for ladies lingerie. The non-split nature makes the locknit the most popular of all warpknits. As the fabric contract widthwise, the final width of the fabric may only be 2/3 of the needle bar width. The amount of shrinkage depends mainly on knitting conditions, yarn type, yarn tension etc. The chain notations for the front guide bar 1-0/2-3 and the back guide bar is 1-2/1-0.

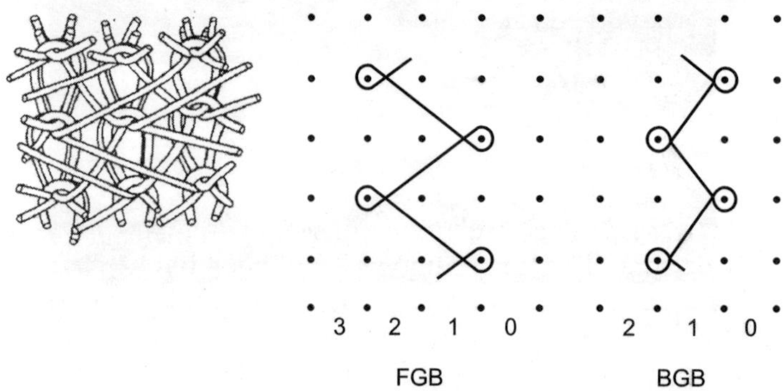

Fig. 21.3

21.2.2.3 Satin (Fig. 21.4) : The increased lapping movements of the front guide bar results in a structure called satin. Like woven satins, the warp knitted satins have longer floats at the backside of the structure and these longer floats give more lusture and feel to the fabric. Due to longer underlaps of the front guide bar, the fabric shrinks considerably immediately after leaving the needles. At the same time, it is elastic and very comfortable to wear. The chain notation for the front guide bar is 3-4/1-0 (or) 4-5/1-0 and back guide bar is 1-0/1-2.

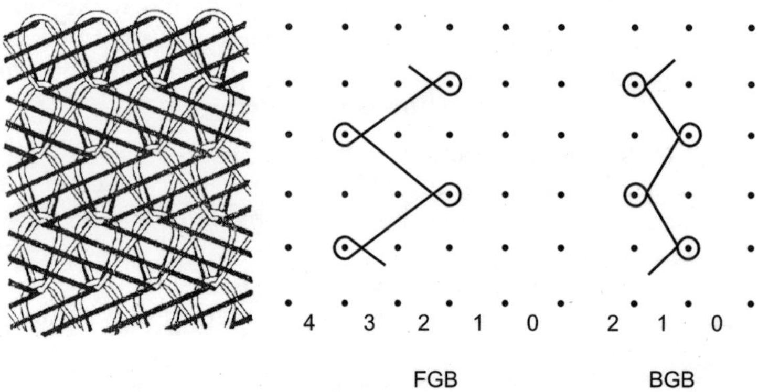

Fig. 21.4

21.2.2.4 Reverse Locknit (Fig. 21.5) : As the name implies, the lappings of the front and back guide bars are exactly the reverse of the locknit structure. The structure is semi-stable and still shrinks considerably on leaving the knitting zone. The longer underlaps of the back guide bar are trapped under the underlaps of the front guide bar and the fabric will therefore become rigid with less elasticity. The chain notations for the front guide bar is 1-2/1-0 and back guide bar is 1-0/2-3. The fabric is suitable for shirting.

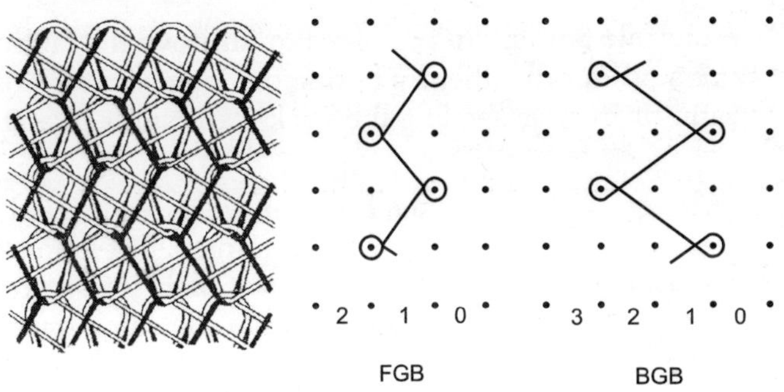

Fig. 21.5

21.2.2.5 Shark-Skin (Fig. 21.6) : It is a reverse version of sati with longer underlaps of back bar and shorter underlaps of front bar. The longer underlaps of back bar are trapped by the front bar, which restricts the shrink ability of the fabric. As the name implies, the surface of the technical back is rough, the fabric is very rigid with maximum shear resistance. It is a shirting fabric having less lusture than other warp knit fabrics. The chain notations are front guide bar 1-0/1-2 and back guide bar 4-5/1-0.

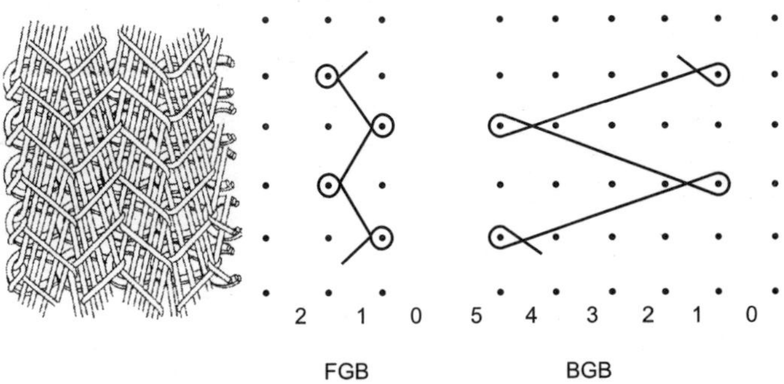

Fig. 21.6

21.2.2.6 Queen's cord (Fig. 21.7) : The fabric is more stable than the shark-skin due to minimum shogging distance of front guide bar and longer underlaps of back guide bar. The vertical pillar stitches of the front bar, trap the longer underlaps of the back bar, thus restrict the movements of the threads. The fabric is less extensible with less shrinkage on finishing. Therefore the find width is very similar to that of knitted width. It is a very popular shirting fabric. By threading coloured yarns with front guide bar, a pin-stripe effect is produced. The chain notations for the front bar is 0-1/1-0 and back bar is 1-0/3-4.

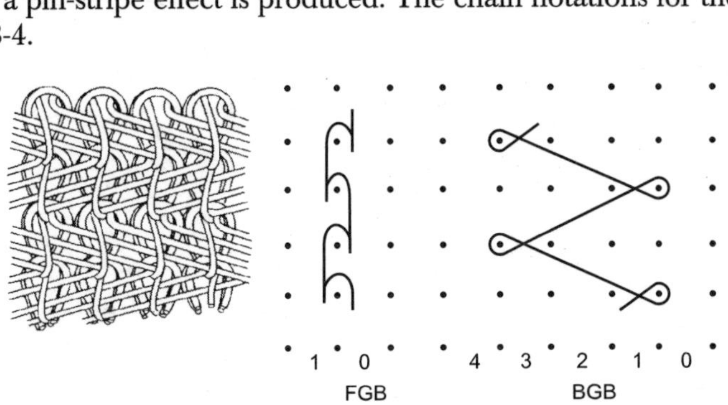

Fig. 21.7

21.2.2.7 Pile Fabrics : Two bar pile fabrics are produced to improve the fabric appearance or thermal properties. They can be produced either by raising the longer underlaps of the front bar in a construction similar to satin or by over feeding the back warp yarns in a construction similar to reverse locknit.

For brushed fabrics, the longer underlaps of the front bar are raised by card clothed rollers during the finishing process. And after a few such stages, the pile is formed. The loop raising operation makes the wale lines to become vertical. By shearing the pile, velour effect is also produced. It is important that the lapping movements of both bars are carried out in the same direction, so that the long underlaps of the front bar can be easily pulled with no resistance from the back guide bar underlaps. The chain notations are front guide bar 1-0/4-5, and back guide bar 1-0/1-2 (Fig. 21.8a).

The loop pile fabric is produced by over feeding the back warp yarn. The excess yarns of the back guide bar protrude between the underlaps of the front guide bar and form pile on the technical back. The lapping movements of both the guide bars are carried out in the same direction. The chain notations are (Fig. 21.8. b) front guide bar 1-0/1-2, back guide bar 1-0/2-3.

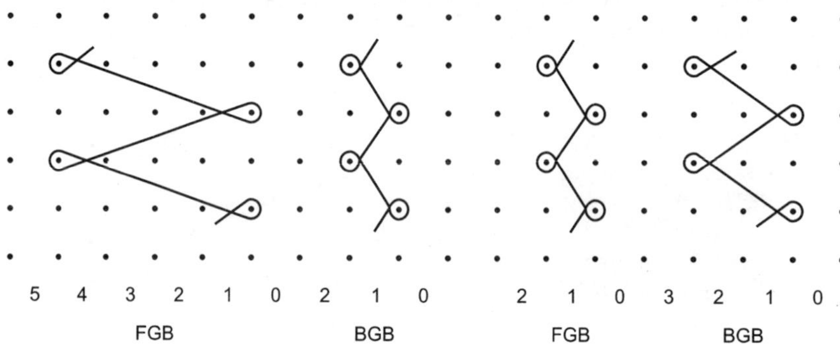

Fig. 21.8 (a) & (b)

21.3 THREE-BAR FABRICS (Fig. 21.9)

The three bar machines are much limited with spring bearded needles compared to the two bar types. The three bar machines produce fancy effects for decorative and curtain fabrics. The demand for such machine is very limited and there are very few structures available for apparel purposes.

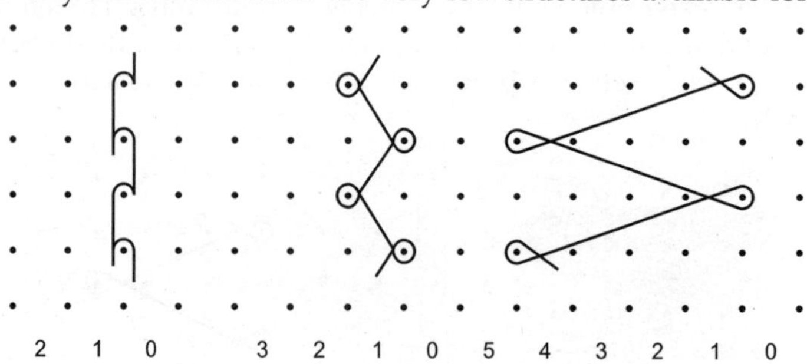

Fig. 21.9

The Three-Bar Tricot warp knitting machine is shown in Plate No. 21.2.

Plate No. 21.2 *Tricot three bar warp knitting machine*

In most of the three bar structures, two guide bars knit a regular pattern and the third guide bar knit a regular pattern and the third guide bar, which is usually the front guide bar forms a chain stitch throughout, gives a straight pillar effect. The back and middle guide bars produce a normal shark skin fabric (viz. BGB 4-5/1-0 and MGB 1-0/1-2) while the front guide bar, which is partly threaded, produces a pillar stitch (viz. FGB 0-1/1-0). If this chain stitch is done with coloured yarn at an interval of five or six threads, an attractive striped shirting can be produced. Also the chain stitches by the front guide bar increases the stability of the shark skin structure, which is comparable to that of woven structure.

21.4 FOUR-BAR FABRICS

The four bar fabrics with spring bearded needles are rare in use. The machine speed reduces with increase in the number of guide bars. For example, the speed of two-bar machine is 1000 cpm, three-bar machine is 800 cpm and four bar machine is 600 cpm. In four bar machines, a firm foundation is done with two guide bars and the other two guide bars are used to produce design or ribbed effect on the basic fabric. Narrow width Angel lace patterned fabric

Plate No. 21.3 *Tricot four-bar warp knitting machine*

is one such example. In three or four guide bar structures, usually a laid-in thread is used, which laid into the basic ground structure, but does not knitted into loops. The Four Bar Tricot warp knitting machine is shown in Plate No. 21.3.

21.5 RASCHEL FABRICS

The Raschel machines originally adapted the name 'Raschel' for the production of the lace fabrics. Though lace was the original product, today the Rachel machines produce fabrics for a variety of end uses, such as nets, curtains, outerwear, upholstery and elasticated fabrics. The machine varieties are much greater than the Tricot and are built with 2 to 78 guide bars with machine gauge of 3 to 32 NPI on both compound and latch needle types.

21.5.1 Net Fabrics

Net fabrics are one of the major products of Raschel industry. Net fabrics are used for a variety of end uses such as shade nets, fish nets, wrap nets, mosquito nets, packing nets and sports nets. Formation of pillar stitches and interlacing/bridging them at different styles are the principles of net fabrics production. Net structures are broadly classified into the following types.

- Nets in which vertical pillars are formed by one set of yarn and second set of yarn bridges the gap between each two wales horizontally.
- Nets which are formed by interlacing pillars. The bridging is made by the so that inclination and distortion of the wales, so that different shapes of openings are obtained.

21.5.1.1 Vertical Pillar structures : In this type of net structures, the guide bars are fully threaded. In a simple construction (Fig. 21.10), the front bar chains constantly and the back bar is laying in. To improve widthwise stability, the inlay is made at least more than one needle space.

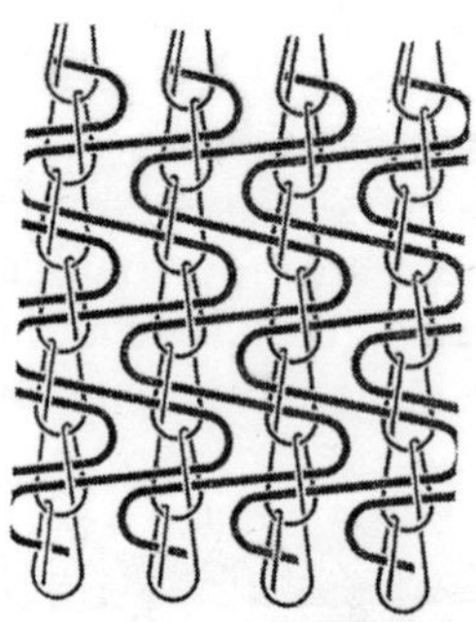

Fig. 21.10

The Marquisette net structure is one of the most popular structures formed in this way (Fig. 21.11). The structure is produced with two or three fully threaded guide bars. In a three bar structure, the front guide bar is constantly chaining with open lap, while the back guide bars inlay yarns. The inlaid yarns follow the pillar for a few courses and only then they bridges the vertical pillars.The chain notations are:

Front bar: 1-1-0-0/0-0-1-1/1-1-0-0/0-0-1-1/1-1-0-0/0-0-1-1
Second bar: 2-2-2-2/0-0-0-0/1-1-1-1/0-0-0-0/2-2-2-2/1-1-1-1
Back bar: 0-0-0-0/3-3-3-3/2-2-2-2/3-3-3-3/0-0-0-0/1-1-1-1

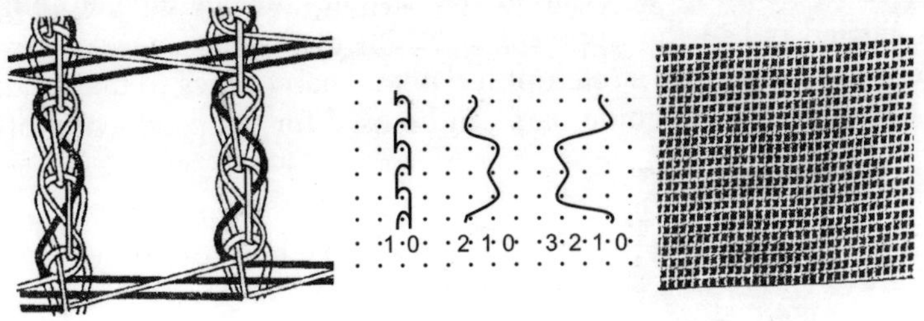

Fig. 21.11

The marquisette structures are very popular in the production of net curtains.

Voile is the other type of straight vertical pillar net structure. This is also produced with three guide bars. The front bar produce chain, and the fully threaded inlay bars move right and left to make horizontal side connection (Fig. 21.12). Pallet wrapping net is the net structure produced of this type which is used to wrap and protect bulky shipping consignments such as cotton bales and stacks of corrugated paper boxes. Pallet nets are made with polyethylene or polypropylene tapes, also with chain loops as slack as possible. Similar construction with different yarn types are used for geo textiles, agricultural textiles and other technical fabrics.

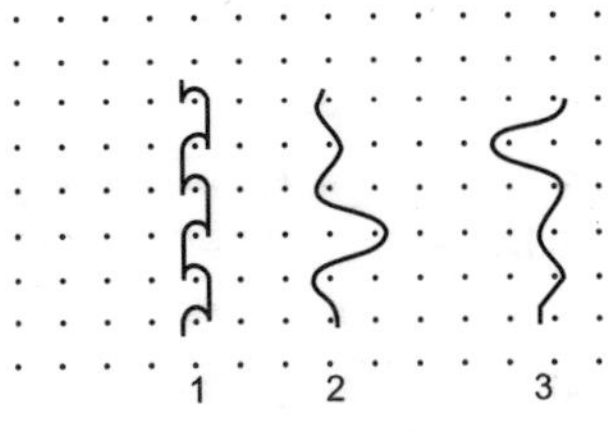

Fig. 21.12

21.5.1.2 Interlacing Pillar structures : In these net fabrics, as a simple construction (Fig.21.13) two guide bars produce pillar stitches and the net structure is formed by interlacing the pillars after a predetermined number of courses. No special yarns are necessary to connect the pillars. The guide bars are threaded in a sequence 1 in and 1 out.

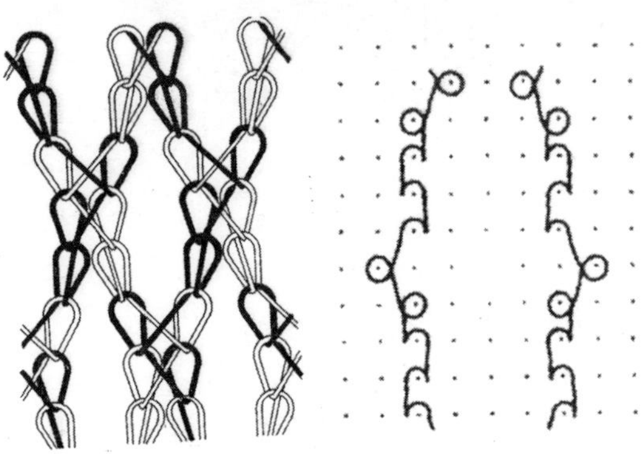

Fig. 21.13

Both the guide bars constantly make pillars with their own yarn on the same needle and after certain courses they shog in opposition to one needle space. During this course, each of the guide bars drawn its new loops through the old loops of the adjacent yarns. After this connection, the guide bars are shogged to their original position and resumes the chaining lapping movement. The next connection is carried out in the opposite direction to the first and so the diamond shaped openings are formed. The widthwise stretch of the net is many times to that of the knitting width. Raschel machines with two to eight guide bars can be used for the production of this type of nets.

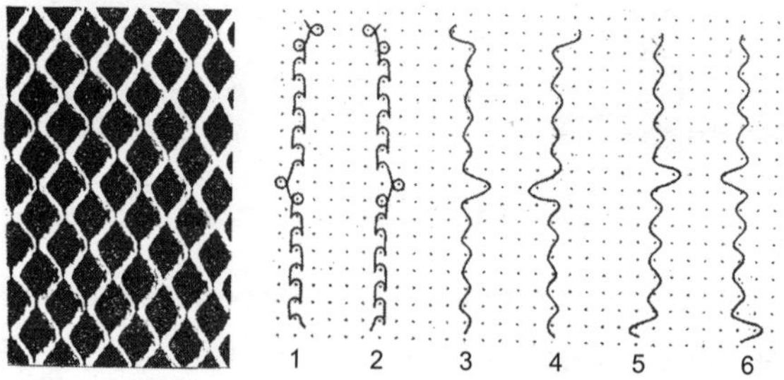

Fig. 21.14

Fishing nets are the main products of this classification. The traditional hand knotted fishing nets are replaced by the specially developed eight guide-bar high speed Raschel machines. To improve the strength of the mesh bar chains and the connection, inlay yarns are made to follow the pillars of the loops and to take part in the binding of each two mesh bars (Fig. 21.14). The inlaid yarns lie almost straight in the construction.

Raschel machines with six to eight guide bars are used for the production of fish nets. Production of fine net structures using fine yarns, heavy duty nets, nets having special selvedges on both sides and extremely wide net fabrics are the possibilities with the high speed Raschel machines.

Sand fly (Fig. 21.15) is a more popular diamond shaped net structure with staggered rows of opening occurring at every other course. Two part-set threaded guide bars with Atlas stitch lapping movements move in opposite direction for every two courses to produce the diamond shape openings. The chain notations are front bar: 1-0/1-2/2-3/2-1 and back bar: 2-3/2-1/1-0/1-2

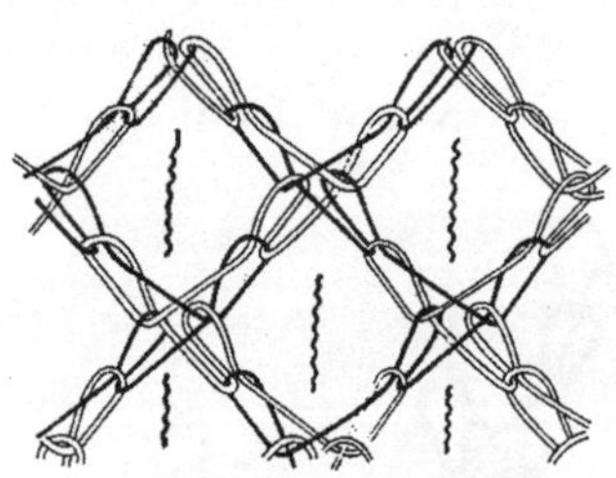

Fig. 21.15

21.5.2 Net Curtains, Drapes and Wall Hangings

Another interesting area of the Raschel machine products is the net curtains. These fabrics are produced on standard Raschel machines. Different techniques such as inlay, weft insertion, fall plate and tulle are used for the production of net curtains with attractive patterns and mesh openings. Similar to the net fabrics mentioned earlier, the production of curtain fabrics also adapt the principle of interconnection of pillar stitches by the above said techniques.

Net curtain structure by inlay technique is very similar in construction to that of marquisette nets as discussed in section 21.5.1.1. One set of yarn knits the ground structure of pillars, while the other sets connect and distort these wales into different mesh openings. The net design shown in Fig. 21.16 is constructed with four guide bars, in which the front guide bar produce pillars while the rest of the three guide bar make the inlay bridging as shown in the lapping diagram.

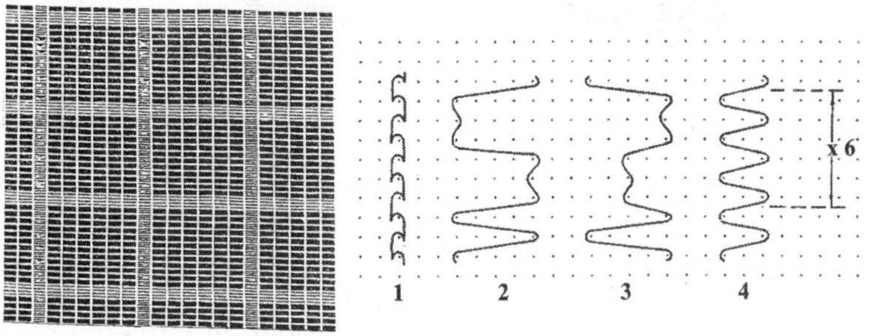

Fig. 21.16

Weft insertion technique is used to produce novel and fancy curtains fabrics. The machine is equipped with a special mechanism to insert straight weft yarns to connect the pillars so that dimensionally stable net fabrics are obtained. Both mechanical and electronically controlled weft insertion magazines are available. The yarns are laid in the fabric horizontally from selvedge to selvedge and perform as weft yarns of the woven fabric. The weft yarns are drawn from the creel cones, stretched across the machine, placed parallel on the hooks of the moving chains and are presented to the knitting zone behind the needles backs. After being placed into the fabric, the weft yarns are trimmed at the selvedges. A net-curtain fabric knitted with weft insertion technique is shown in Fig.21.17 in which the pillars trap the weft yarns which make the fabric stable.

Fig. 21.17

Machines equipped with fall-plate mechanism produce a wide variety of net-curtains designed with superimposed patterning effects on the technical back of the fabric. The technique is more popular with multibar machines rather than the standard Raschel machines. The net-curtains fabric produced with fall plate laps on a marquisette net ground is shown in Fig. 21.18. The structure is produced with four guide bars. The fully threaded back bar and third bar work in conjunction by making inlay and pillar stitches, thus produce the square structure. The two part-set front guide bars which make traverse laps in opposite direction are operated by the fall plate to produce the superimposed diamond pattern.

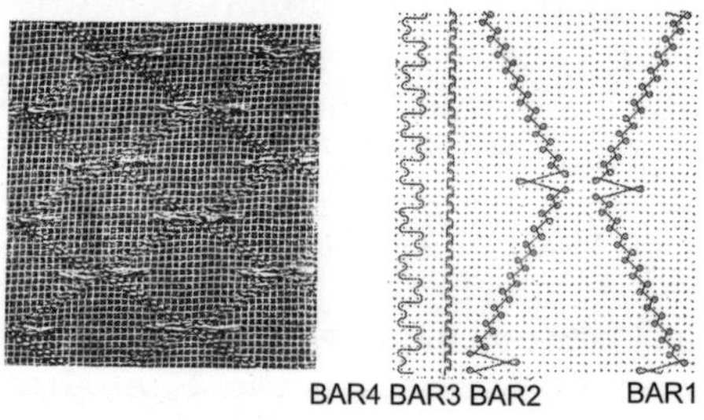

Fig. 21.18

Tulle nets (Fig. 21.19) are used as ground structure in the production of curtain fabrics. For tulle construction, only two fully threaded guide bars are used, the front bar knits the pillars while the back bar introduces in lay threads. The front guide bar chains for three courses and then shogs one needle space to chain on the adjacent needle. The loop which is formed prior to the cross-over is pulled side ways by the neighbouring underlap for the hexagonal shape of the net opening. The pillar and inlay stitches formed by the respective guide bars are continuously associated with each other and virtually follow the same paths. The great advantage of this machine type is that any pattern can be produced on this ground. The ground structure shows a perfect uniformity especially around the motifs.

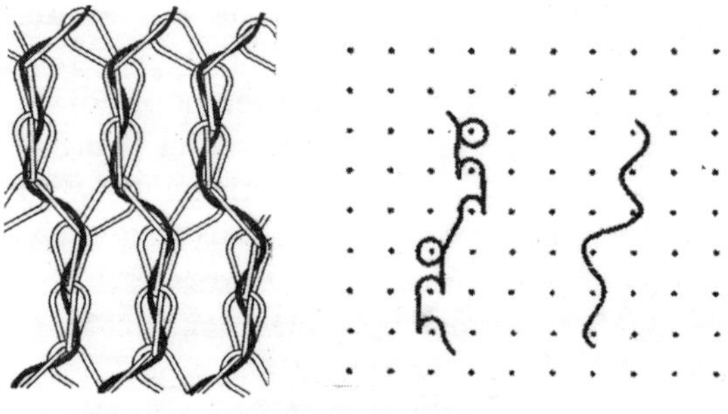

Fig. 21.19

21.5.3 Outer Wear and Upholstery Fabrics

Outer wear and upholstery fabrics can be produced on standard Raschel machines by incorporating weft inserted yarns. The machine equipped with weft magazine unit produce fabrics according to the design effect and repeat size. The ground construction is knitted with two or three guide bars (Fig 21.20) and the design effect is made by weft inserted yarns. By suitably feeding the weft inserted yarn of different nature, such as synthetic/woolen, filament/boucle, course/fine and coloured yarns, fabrics can be produced based on the end uses and application.

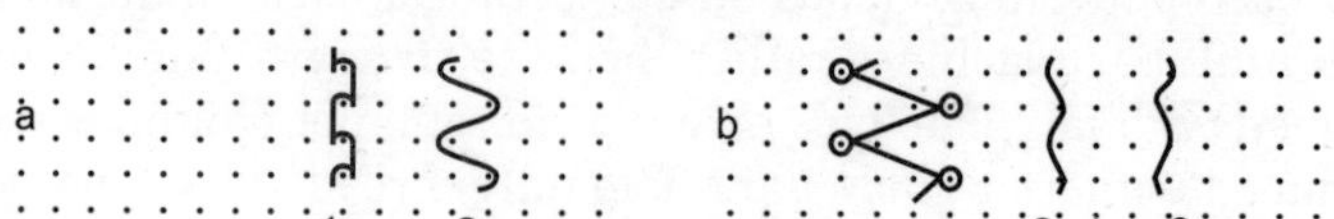

Fig. 21.20

21.5.4 Elasticated Fabrics

Elastomeric yarns laid into the knitted structure produce elasticated fabrics on standard Raschel machines. Both plain and net structures are produced with the elastromeric yarn threaded in the back guide bar power-net (Fig 21.21) is such a construction in which the lapping movement of the four guide bars produce a structure similar to that of a fish net. The front two guide bars knit with continuous filament yarn to produce the basic net structure and the rear two guide bars inlay the natural rubber or elastomeric yarns to give the elastic properties to the fabric in warp direction. All the guide bars are threaded in a 1 in 1 out arrangement.

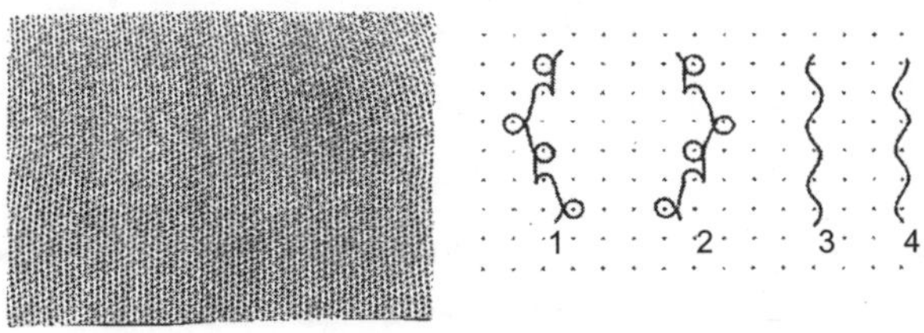

Fig 21.21

21.5.5 Dress Laces

As mentioned earlier, the standard Raschel machines are more popular in the manufacture of variety of laces. Laces produced by hand made and bobbin lace machines are only in very small quantities due to their low productivity and more uneconomical. Similar to the net curtains, the production of lace involves the construction of ground structure and pattern designing over it. However the net curtains machines use 2 to 6 guide bars where as the lace machines employ 78 guide bars. The advent of multi bar machines, and developments in synthetic yarn types and fineness a variety of ground construction lead to enormous scope in the production of interesting lace designs. The modern multi bar Raschel machines use compound needles with a gauge of 24NPI. In a 32 guide bar machine, for example, the fully threaded front two guide bars produce the ground and the remaining light weighted patterning guide bars which are nested in the back displacement lines produce the lace designs. In lace machines, the patterning yarns are traditionally fed off small pattern beams which are placed on machine frame. Since the guide bars of the machine make only the pillar and inlay stitches, the patterning mechanism is usually simpler than that used on curtain machines.

One group of the ground structure is the Tulle net (Fig. 21.18) in which the fully threaded front guide bar chains on adjacent needles for every three courses and the fully threaded back guide bar make inlay connection so that the hexagonal shape of net openings are obtained. In the other group of ground structure, the fully threaded front guide bar constantly chains and the back guide bars make inlay connection (Fig. 21.22). By using very fine yarn, attractive transparent effect can be achieved.

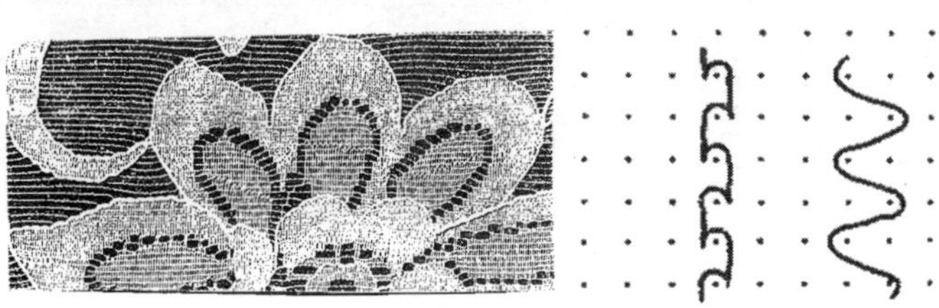

Fig. 21.22

Dress lace products are of two types. In the first type, full width lace fabrics are produced to the width of the machine and are later cut and sewn by the making up industry. The second type are the narrow width banded laces which are used to decorate products made of other fabrics. Lace edgings (*i.e.,* laces sewn to the edges of the other fabrics), lace inserts (*i.e.,* laces sewn between two fabrics), galloons (*i.e.,* laces which are scalloped or curved on both sides) and cut-outs (*i.e.,* motif is cut out of the lace fabric) are the classification of the banded laces. The band laces are produced side by side to the machine width as continuous fabrics. The finishing operations are carried out in the fabric form after which the bands are separated. The separation is carried out by the following methods :

(*i*) by dissolving the thread that connects the adjacent bands in a suitable solvent.

(*ii*) By using very fine yarn to connect the bands during knitting and ripping apart them after the finishing operation.

(*iii*) By using draw threads which inlays alternately into the selvedges of the two adjacent bands and separating the bands by pulling the draw threads.

(*iv*) Galloons and other bands with scallopped edges are separated by cutting.

Plate No. 21.4 Shows the Multibar Jacquard Raschel machine for the production of curtains, elastic laces and rigid laces.

Plate No. 21.4 *Multi Bar Jacquard machines for Curtains and Laces.*

22

Yarn Preparation, Yarn Feed and Fabric Take-up

22.1 YARN PREPARATION FOR WARP KNITTING

Preparation of warp beams is the main task of yarn preparation in warp knitting. The warp beams used for warp knitting are in cases different from those of weaving. Use of synthetic yarns and absence of sizing due to moderate tensions applied on knitting yarns make the yarn preparation simple. The cotton and other combed yarns should preferably be combed types with moisture content not exceeding 10%. They should not be over waxed and the twist should be similar to and not higher than that of thrown yarns used in warp knitting. Steaming would stabilize the twist. During warping, the yarn should be oiled and water-clear odourless oils with a non-ionic emulsifier are the best. The successful use of staple yarns in warp knitting mainly depends on knot free or minimum frequency of knots. Warpbeam quality is very important as differences in yarn thickness, tension, twist and other factor can result in faulty fabrics. Warpbeam can be prepared independently by the knitters themselves or the prepared warpbeams can be obtained from the yarn manufacturer/suppliers. Two methods of warping can be used to prepare warpbeams for warp kintting machines, viz., indirect and direct warping.

22.1.1 Indirect or Mill Warping

This method is similar to that of section warping of weaving. Here the yarn packages are wound onto an intermediate cylinder (*i.e.,* mill) and then back wound onto the warpbeam. A parallel group of yarn ends from the creel is wound on the mill to the required number of revolutions. When the required length is wound, the yarns are cut, the mill is moved sideways, the yarns are re-attached to the mill and a new section is wound adjacent to the first. After the required sections are wound on to the mill (*i.e.,* required number of ends across the whole width of the knitting machine), the whole sheet of yarn is back wound in one operation onto a warpbeam. Though very small creel is sufficient to produce the warp, it is a slow process require greater number of sections on the mill.

22.1.2 Direct Warping

Here the yarn ends are wrapped in one operation from the yarn packages onto the warpbeam. Unlike indirect warping, with moderate creel size and to simplify warphandling, section beams are warped. Each section has its own flanges and a number of section beams are placed on a common shaft to make up the required number of ends for the warp. It is important that all warping sections placed on a common shaft must be of exactly the same length, warped with the same number of revolution, have the same oute- circumference and the individual ends of yarns must be under the same tension. Section beam widths of 65 inches and flange diameter upto 40 to 50 inches are available. Warping assembly of the direct warping system is shown in Fig. 22.1.

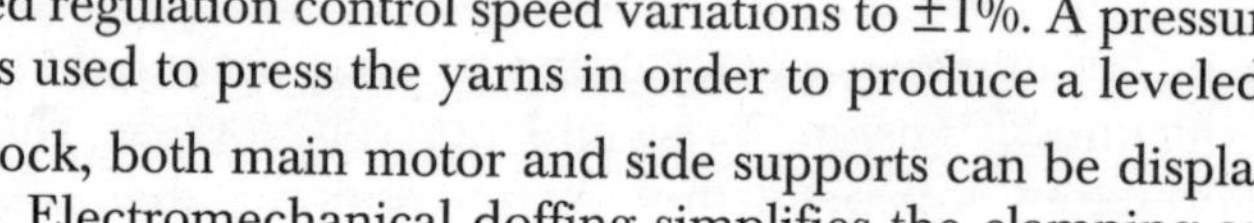

Fig. 22.1

The head stock (1) turns the beam to wound the warp onto it. Higher take-up speeds upto 1200 m/min with auto speed regulation control speed variations to ±1%. A pressure roller, fits precisely between the flanges is used to press the yarns in order to produce a leveled warp surface.

Within the head stock, both main motor and side supports can be displaced to accommodate different beam width. Electromechanical doffing simplifies the clamping and doffing operation.

A front reed (2) spaces the yarns to fit the width of the beam. The reed is slightly traversed from side to side to prevent yarns from forming ridges and to fill the space between the flanges evenly.

Oiling device (3) with emulsion level control, apply lubricants to the ends.

A yarn store (4) allows the beam to be turned backwards and the broken ends can be easily repaired. Electronically controlled braking system ensures short braking distances, not exceeding 6 m even at maximum speed.

Tension rollers (5) iron-out the tension differences in the warpsheets. Yarn tension can be leveled by passing the yarn sheet over tension compensating rollers. A DC motor drives the tension roller in synchronization with the warping machine and according to the yarn tension which is constantly measured by a built-in-gauge.

A back reed (6) gather the yarns coming off the creel.

A creel (7) accommodating the yarn packages. It is a rigid frame with pegs onto which the yarn packages are mounted. Different types of creels viz. magazine creel, carriage creel and rotary creel are used in modern warping which depend mainly on yarn type, size of yarn package and the size of warp section. The magazine creel has two pegs in which reserve cones can be accommodated. With extremely large packages, the spaces between the pegs must be much greater and so the magazine creel becomes too large and the thread lines become too long. In carriage type creels, the creel is loaded off the machine and is then positioned in place when the other carriage, with the previous yarn packages, is removed. This type of creel is especially suitable for yarn producers as the same carriage can be used in the spinning department. In rotary creels, one side of creel is loaded during warping operations so that when a change is required, the creel sections are turned through 180° and the new yarn packages can be tied to the previous yarns ends. For loading on these larger creels, motor driven platforms that travel between the creel sections or outside of them can be used. To supply the yarn ends to the warping machine at a required even tension, the creel is equipped with yarn tensioners, which are either the simple cymbal and post arrangements or the sophisticated tension controllers with hydraulic tension peak damping devices.

22.1.3 Warping for Pattern Beams

The above discussed warping techniques are used to prepare the section beams for the ground bars of the warp knitting machines. Warping of pattern yarns on pattern beams for multiguide bar machines is a different operation which requires different equipment. As only a small number of pattern ends are used, they are wound over a shaft as small individual beams (resembling yarn cheeses) across the full width of the shaft. Without flanges, the pattern ends are wrapped over the shaft in a cross winding movement, much like yarn cheeses. These cheese beams do not exceed 30 cm. of diameter. Warping equipment for this type of beam preparation is shown in Fig. 22.2. It consists of a relatively small creel, a traversing reed to cross wind the yarns and friction rollers to rotate the shaft at a constant winding speed, irrespective of the diameter of the warp. The yarn tension is controlled by simply passing it around four tubes, which are adjustable in their position. The warping speed is adjustable between 50 and 200 m/min.

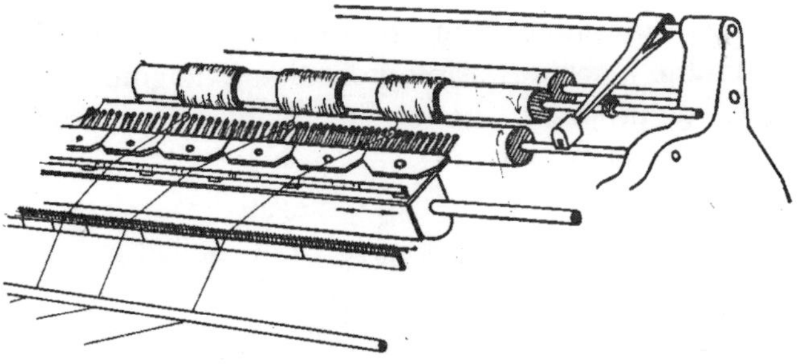

Fig. 22.2

22.2 LET-OFF AND TAKE-UP IN WARP KNITTING

In warp knitting, tension control during knitting and accurate yarn feeding are at most important for effective and faultless knitting. The take-up rollers pull the fabric from the knitting zone and the tension built in this is taken to the back yarn ends. Depends upon the level of tension created and the amount of yarn fed, the take-up system will be driven faster or slower.

During knitting, as unequal quantities of yarns are consumed on the guide bar traverse differently for different courses, four stages of tension variation takes place. At the start of a course, the tension increases as the guide swing through to the hook side of the needles and make the overlap. The tension then reduces as the guides return to the front of the needles and as the needle rises to allow the newly formed laps to slide down the needle hooks and on to the stems. As the needle bar descends to allow the needles for knock over, the tension on the warp thread again increases, followed by a reduction in tension as the needle bar rises back to the starting position. These variations in tension, if not properly controlled, would give rise to imperfect fabric, as the warp threads would not necessarily follow the movements of the guides. Therefore the rates of feeding by the let-off mechanisms are to be controlled and adjusted accurately and also it is necessary to provide a form of take up motion which will recover the warp as the tension reduces. The spring loaded tension rail performs this function and ensures perfect control over the warp threads at all times.

22.2.1 Warp Let-off Mechanisms

The warp let-off motion must work in conjunction with a suitable tensioning device which preserves the tension of the warp threads during the knitting cycle. This usually takes the form of a spring loaded bar which controls the irregularities in tension caused by the movements of the knitting elements. The initial rise in warp tension simply depresses the tension rail until the action of the springs has been counter balanced. Any further increase in warp tension causes the warpbeam to be pulled round to supply a certain amount of warp. The amount of warp supplied depends upon either the amount of weight applied to the friction band around the beam or the delivery of a predetermined yarn requirement *i.e.,* run-in. The former one is self compensating and release the warp on demand, which is a negative let-off type. The later one is positive let-off type, in which the warp-beams are positively driven and their surface speed is monitored to cater a uniform rate of let-off at different reducing beam diameters. In negative let-off mechanisms, a restricting force is applied to the warp beam, its free turning is limited and hence the necessary yarn tension is created. The positive let-off mechanisms are suitable for high speed machines, in which uniform loop structure is important. A constant yarn feed together with constant yarn tension is achieved by positively driving the warp beam instead applying the brake. With the machines using more than one guide bar, the yarn consumption by each guide bar is (*i.e.,* run-in) important and the relative amount of yarn fed from each warp i.e., run-in-ratio is also equally important.

22.2.1.1 Negative Let-off Mechanism: Two systems of negative yarn let-off are available. In the first type, viz., the cord let-off, a constant restricting force is applied to the warp beam and its free rotation is limited. The second one is an intermittent type brake let-off in which the frictional restraint on the beam is removed at times when the warp is required to move forwards, allowing the beam to turn freely.

22.2.1.1.1 Cord Let-off Mechanism (Fig. 22.3): The cord let-off mechanism is used on multibar Raschel and Tricot machines. The mechanism is used to control the light weight pattern beams which supply the partly threaded guide bars. They are negatively turned as a constant frictional force is applied over the beam. The beams at one end has a three spoked star, on which small weights are placed and positioned in order to ensure balanced rotation, so that the beam will not turn without being pulled by the yarns. At the other end, a cord brake, tensioned by braking weights provides a braking force by the friction between the pulley and the cord. It is necessary to reduce the braking weights from time to time as the beam gets smaller in diameter.

Though the mechanism is quite simple and inexpensive, it is sensitive to climate variations and depends solely to the worker to maintain the yarn tension by changing the braking weights.

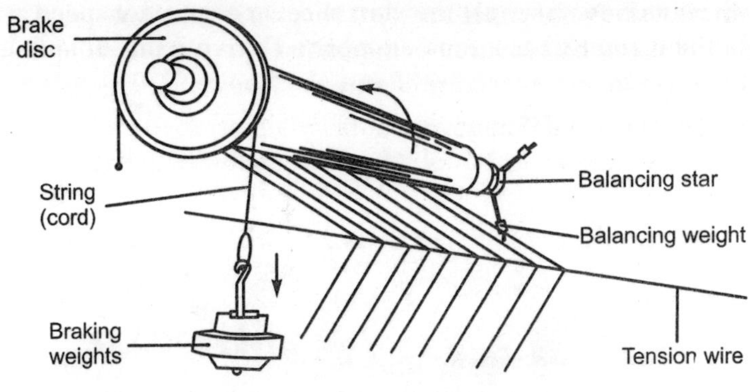

Fig. 22.3

22.2.1.1.2 Brake Let-off Mechanism (Fig. 22.4): This is an intermittent negative brake type let-off mechanism which is employed on slow speed machines. The mechanism, when set correctly, releases the necessary amount of yarn during each knitting cycle. The mechanism consists of a V-belt that held firmly by springs (4) in the groove of a V-belt pulley (3). The warp threads pass over a tension rail, which is mounted on a shaft by the flat leaf springs (L). When the yarns are knitted into the fabric, the tension increases and the tension rail turns slightly. A cam (2), mounted on the end of the tension rail shaft, lifts the push rod and releases the V-belt from the groove in the V-belt pulley (3). The pulley can now rotate freely and the required amount of yarn can be released freely by rotating the warp beam via. a connecting chain. The yarn which is now under a lower tension, allows the tension rail shaft to rotate backwards, the cam (2) releases the pushrod, the V-belt is again set firmly into the groove of the pulley and the warp beam is stopped. The spring (1) which opposes the turning movement of the tension rail shaft, is used to adjust the tension at which the yarn release the belt. When the guide bars swing for their overlap position, the yarn tension reaches its peak and the break releases its hold to operate the system.

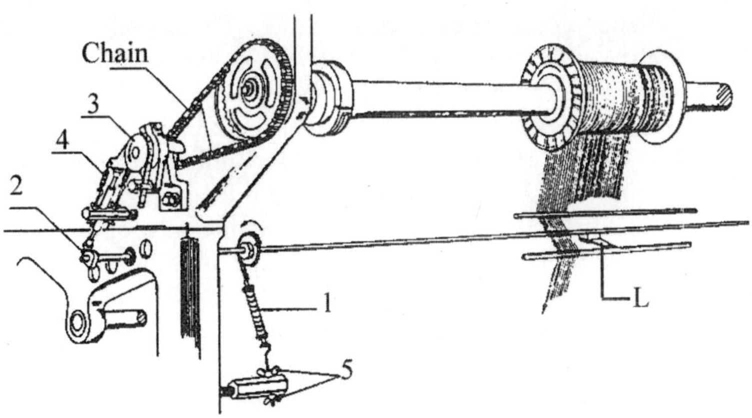

Fig. 22.4

22.2.1.2 Positive Let-off Mechanism: The positive let-off mechanism is suitable for high speed machines where a constant yarn speed together with constant yarn tension is important. As the warp beam is positively driven, it supply the yarn sheet at a constant speed to the knitting zone. Each warp beam of the machine has a separate positive drive arrangement. The positive let-off mechanism are of two types viz., mechanical and electronic.

22.2.1.2.1 Mechanical Positive Let-off Mechanism

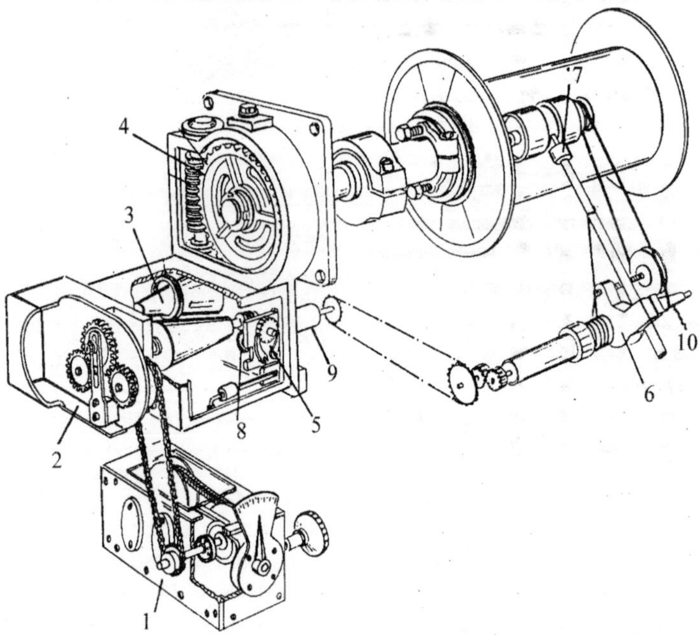

Fig. 22.5

This automatic let-off mechanism is most commonly used in high speed warp knitting industry (Fig. 22.5). The control gear box (1) is driven from the main shaft of the machine, from where the drive is transmitted to change gears (2) and cone box (3). The preselected yarn speed, which depends on the fabric construction, fabric quality, yarn count etc. is adjusted by the control gear box and the change gears are necessary to determine the run-in-ratio of each driven beam. In the cone box, the bottom cone drives the top cone by a steel ring that moves laterally. The top cone drives the warp beam via. a worm and worm assembly (4). Rubber -covered measuring rollers of the measuring device (6) are constantly in contact with the beam surface and are driven by its rotation. The measured speed is transmitted into the regulating box (5) in which it is constantly compared with the desired preset speed. When the linear speed of the warp decreases due to the decreasing circumference of the beam, the alteration is immediately detected by the measuring rollers. The regulating box moves the steel ring on the cones so that the driving ratio between them changes and the rotating speed of the warp beam increases to equal the desired speed. A greater detail of the regulating device (5) is given with reference to the Fig. 22.6. Within the regularity device, the speed of the lower cone in the let-off box is measured against the speed of the measuring cone. When the two speeds coincide, the correct warp velocity is being achieved.

If, however, there is any discrepancy in the speeds, the regulating device causes the warp velocity to be increased or decreased until the balance is restored.

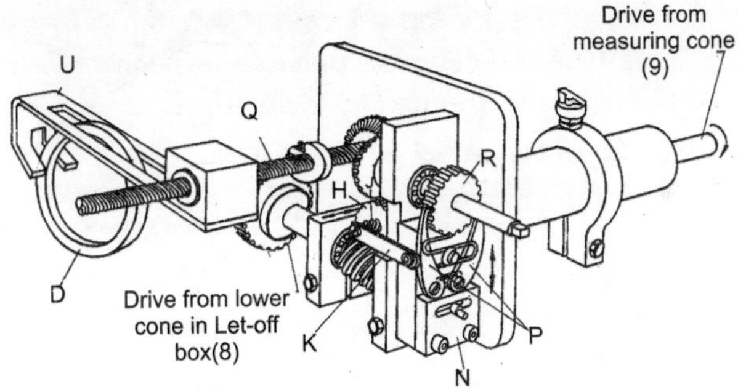

Fig. 22.6

The controlling worm wheel H is driven from the lower cone in the let-off box via the worm drive, and the speed is therefore constantly related to the speed of the machine. The measuring spindle J, which is provided with a stepped-thread screw is fitted into an internal thread in the controlling worm wheel, and is driven indirectly from the warp via, the measuring rollers and measuring cone (10) etc. While the drive to the measuring spindle (J) is exactly the same as the drive to the controlling worm (H), the former remains centrally situated in the latter, but as soon as there is a discrepancy in the speeds, the measuring spindle moves out of its central position and begins the regulating process. Two pawls P are located in bearings on a slide, which is being continuously raised and lowered by an eccentric. A spring between the two pawls pulls them against a regulating pin on the control arm K which is connected to the measuring spindle J. Thus, when the measuring spindle moves out of its central position, it causes the pawls to be moved in the same direction (by means of the control arm and regulating pin), so that one pawl is brought into contact with the ratchat wheel R, and turns it in one direction, at each vertical motion of the slide N. The ratchel wheel turns the threaded spindle of the cone drive and, through the ring holder U, which is mounted on the threaded shaft, causes the steel ring D to be repositioned between the cones in the let-off box. The process continues until the warp velocity has been increased (or decreased) sufficiently to allow the regulating spindle and hence the pawls to return to their central position.

Once the machine has been set to a desired rate of run-in, the operation of the regulating device will normally be repeating at regular intervals, by bringing the same pawl into operation to increase the speed of the warp beam shaft to compensate for the reducing diameter of the warp.

22.2.1.2.2 Electronic Positive Let-off Mechanism : This microprocessor controlled yarn feeding device is increasingly used in more number of modern warp knitting machines. The system consists of a computer, a signal amplifier, main shaft measuring device, DC beam driving motor and beam speed measuring device. The computer is given the information such as empty beam diameter, full beam diameter, number of warping revolutions and the desired run-in-ratio. From the above information, the computer calculates the correct rotating speed of the beam throughout its length. A signal amplifier correctly interprets any slightest difference in the computer signal and also negotiate any slightest reduction in warp diameter. The system offers efficient and controlled yarn feeding over the mechanical regulating device (Fig. 22.7).

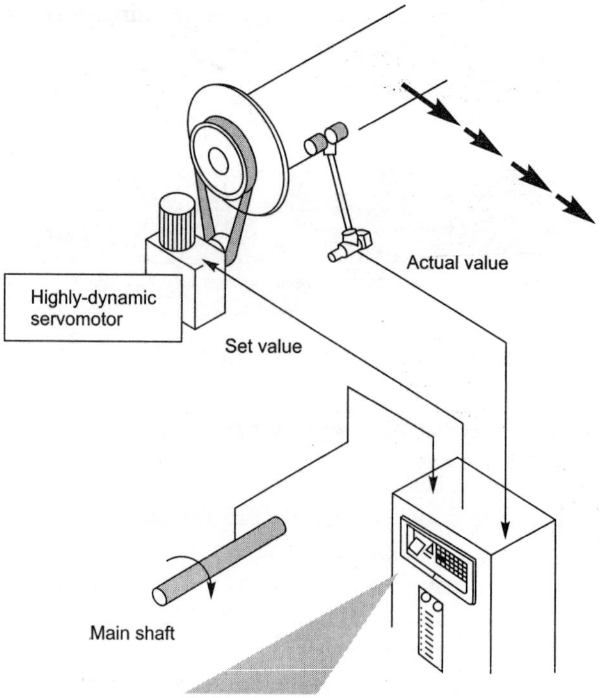

Fig. 22.7

Although the average yarn tension is controlled by the let-off mechanism, the regular relative movements of the different knitting elements cause fluctuations in the yarn tension. These tension changes, which take only fraction of seconds, cannot be compensated by the regulating devices of the let-off mechanism. To overcome this problem, a tension rail is used in which the yarn sheet passes over the smooth surface of a bar and the tension rail which is mounted on flat springs. The low inertia of the tension rail allows it to react rapidly to tension fluctuations.

22.2.2 Fabric Take-up and Batching Mechanisms

Unlike weft knitting, in warp knitting, the fabric take-up determines, to a great extent, the quality of the fabric being produced. The take-up mechanisms used for Tricot and Raschel machines are different due to some basic requirements. In Tricot knitting, as the positive let-off mechanism set the yarn content for each loop, the take-up affects mainly the shape of the loop and yarn tensions. Whereas in Raschel knitting, with negative let-off mechanisms (due to pattern beams), the fabric take-up mechanism determines the fabric quality and appearance. Also as the Tricot take-up mechanism mainly to collect the fabric, it is positioned further away from the knitting zone, while the Raschel take up mechanism is positioned as close as possible to the needles, thus to pull the fabric very tightly downwards.

In Tricot take-up mechanism Fig. 22.8 the drive from the main shaft is transmitted by change gear and worm gear arrangements to the take up rollers. The surface of the take-up rollers is covered with a special material to have high friction coefficient and to eliminate fabric slippage. The take-up rollers speed can be altered by changing the ratio of the change gears. The fabric collection batching roller is frictionally driven.

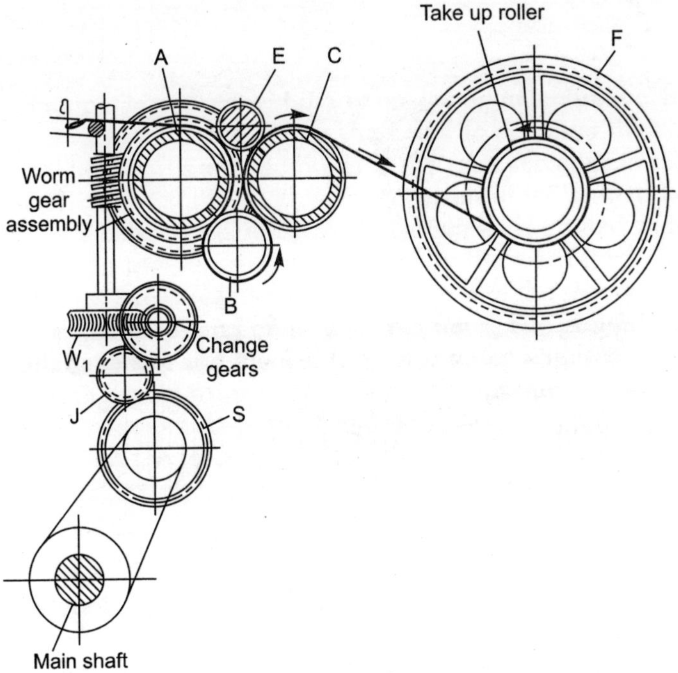

Fig. 22.8

In Raschel take-up mechanism Fig. 22.9 the quality rollers C & D are driven from the main shaft via. change gears. The quality rollers are geared at a 1:1 ratio. The quality roller C is mounted close to the trick plate and the fabric passes under roller C around tension roller F and under quality roller D, before being directed downwards over the guide roller G to be wound on to the batching roller H. The fabric batching roller is frictional driven by roller M and is therefore driven at a constant linear speed sufficient to windup the fabric at a reasonable tension. The quality rollers C and D are covered with high friction material, while the tension roller F, which is loosely mounted between the quality rollers, is usually felt covered. The roller F is pulled into close contact with quality rollers under the tension of the fabric, which increase the area of contact between the fabric and the rollers and also ensure no slippage of the fabric.

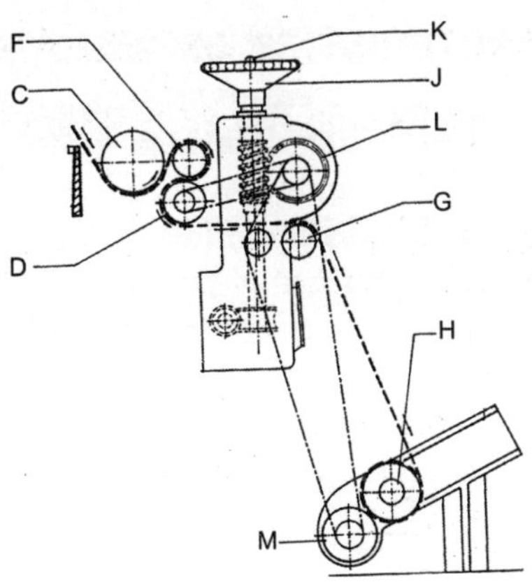

Fig. 22.9

23

Warp Knit Fabric Geometry and Calculations

Warp and weft knitted fabrics, although seems similar when their loop construction is viewed from their face side, are quite different in their mechanical behaviour and properties. In weft knitting, the fabric stability to washing could be obtained only by the finishing treatments to the relaxed dimensions and easy stretching and fabric flexibility are the generally desirable properties. Where as the warp knitted fabrics should not generally be finished to the relaxed dimensions and in this aspect they are deviating from the weft knits. Most of the yarns used in the warp knitting industry are thermo plastic and the fabrics can be stretched considerably even from their relaxed dimensions. Therefore, they are to be heat set to give a generally adequate washing stability. If the fabrics were confined to the relaxed dimensions, the washing stability could be improved, but the mechanical properties of the fabrics *e.g.*, the ease of stretching make it unsuitable for the desired end-use. This deviation is known as a parameter called "Structure ratio" which provides a basis for setting up specifications for the knitter and the finisher. In this chapter, dimensional characteristics, geometry and production calculations of warp knitting are discussed in brief.

23.1 DIMENSIONAL CHARACTERISTICS OF WARP KNIT FABRICS

A two bar warp knitted structure is shown in Fig. 23.1. With reference to the diagram,

The Course spacing, $$C = \frac{1}{\text{courses per cm}} = \frac{1}{c}$$

Wale spacing, $$W = \frac{1}{\text{wales per cm}} = 1/w$$

Where, c = courses/cm, w = wales/cm

And d = yarn diameter in mm

l = stitch length in mm

Like weft knitting, in warp knitting also the stitch length is proportional to the wale and course spacing.

i.e., $$\frac{1}{w} \propto l \text{ or } \frac{1}{w} = \text{constant} \times l$$

$$\frac{1}{c} \propto l \text{ or } \frac{1}{c} = \text{constant} \times l$$

Therefore, $$\frac{1}{w} \times \frac{1}{c} = \text{constant} \times l^2$$

i.e., $$\frac{1}{wc} = \text{constant} \times l^2$$

and $$\frac{1}{wc} l^2 = \text{constant}$$

or, $$wc\, l^2 = \text{constant}$$

Let, $wc\, l^2 = \text{K}s$, where Ks is a constant, then

$$wc = \frac{\text{K}s}{l^2} \qquad \text{... (23.1)}$$

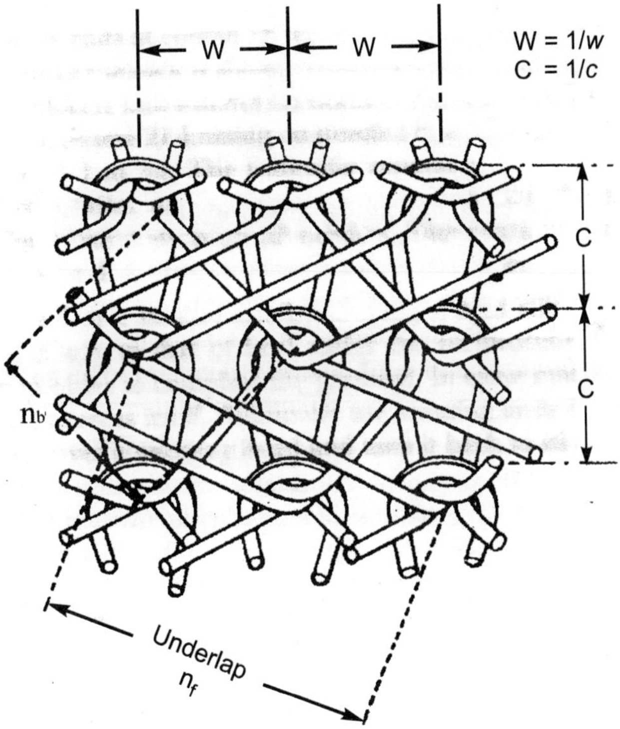

Fig. 23.1

The following are the important dimensional parameters of warp knitting, which influence the quality and yield of the warp knitted fabrics.

23.1.1 Rack

A rack is a set of 480 courses of stitches produced by the guide bar.

23.1.2 Run-in

It is the length of yarn in inches (or centimeters) fed to the needles per rack of 480 courses. The run-in will vary according to the lapping motion of the guide bar which is being supplied. In a two guide bar knitting, a motion of 1-0/4-5 will require a greater run-in than a motion of 1-0/1-2. By feeding different amounts of yarn into knitting zone, the loop size is changed. A longer run-in produces a slacker fabric with big loops while a shorter run-in produces small and tight loops.

23.1.3 Quality

The length of cloth knitted during one rack measured in inches (or centimeter) is called the quality.

23.1.4 Run-in Ratio

It is the ratio of the run-in of different guide bars. When the knitting takes place with more than one guide bar, the relative amount of yarn fed from each warp is important. The run-in ratio is different for different fabric construction and qualities. In a two bar machine, based on the relative lapping movements of the front and back guide bars, the run-in ratio can be determined by assuming 2 units for overlaps and one unit for underlaps and one unit for every additional needle space covered by the guide bar during its underlapping, as shown below for some structures.

Structures	*B. Bar*	*B. Bar units*	*F.Bar*	*F. Bar units*	*Run-in ratio*
(*i*) Locknit	1-0/1-2	6	2-3/1-0	8	3:4
(*ii*) Sharkskin	3-4/1-0	10	1-0/1-2	6	5:3

23.1.5 Stitch Length (l)

The stitch length in warp knitting can be considered as the sum of the loop length and the underlap. The underlap varies with the type of knitted structure and can be expressed in terms of the number of needles traversed by the guidebar, i.e., the shogging movement.

The stitch length can be estimated as $l = \frac{\text{run-in/rack}}{480}$. For two bar fabrics, there are two run-in values and therefore two stitch length values.

23.1.6 Structure Ratio (E)

It is the deviation of fabric dimensions of a given construction to its relaxed states. The value E will be reduced if the fabric is stretched. The structure ratio (E) can be calculated as,

$$E = \text{Stitch Length/Geometric mean of wale and course spacing}$$

The geometric mean (G) of the wale and course spacings is given by,

$$G = \left(\frac{1}{w} \times \frac{1}{c}\right)^{\frac{1}{2}} = \frac{1}{\sqrt{wc}}$$

Structure ratio, $$E = \frac{l}{G} = \frac{l}{\frac{1}{\sqrt{wc}}} = l\sqrt{wc}$$

or, $$E = \frac{l}{\sqrt{wc}}$$

Where, W = wale spacing, C = course spacing, w = wales/cm, c = courses/cm.

23.1.7 Areal Density

The areal density of a single bar fabric is determined as,

Areal density = Stitch density × Stitch length × Yarn linear density

$= S \times l \times T \times 10^{-1}$ g/m^2

Where S = Stitch density/cm^2, *i.e.*, cpcm × wpcm

l = Stitch length in cm

T = Yarn Tex

The areal density of a two bar fabric is given as,

Areal density $= S \times T \times 10^{-1} (l_f + l_b)$ g/m^2

Where l_f and l_b are the stitch length values of front and back guide bars.

23.1.8 Tightness Factor

It is the ratio of the fabric area covered by the yarn to the total fabric area. It refers the looseness or tightness of the fabric. The tightness factor is defined as $K = \frac{\sqrt{tex}}{l}$.

For a two bar, full set fabric, the tightness factor is,

$$K = \sqrt{\text{tex}} \cdot \left(\frac{1}{l_f} + \frac{1}{l_b}\right)$$

Where l_f and l_b are the stitch length values of the front and back guide bars.

23.2 GEOMETRY OF WARP KNITS

The warp knitted structure consists of a set of open or closed loops known as overlaps and connecting section where the yarn has moved backwards and forwards between the knitting needles called as underlaps or cross links. The structure simply consists of single loops and interconnecting cross links. Depending upon the magnitude and direction of the interconnecting cross links, the loops in the structure lean over to various directions.

The geometrical study of knitted fabrics is useful for understanding their structural behaviour. A good understanding of the fabric geometry may be helpful for the knitter for his production plans such as yarn requirements, prediction of production rate and products. Many studies have been carried out by the research workers on warp knitted fabric geometry with respect to loop shape and size, length of cross link, yarn tension, frictional forces, yarn diameter etc.

Grosberg[1] attempted to establish theoretically the run-in of warp knitted structures through the geometry and compared with the actual measured values (Fig. 23.2).

He calculated the run-in through the simplified geometrical model by establishing the loop length, the length of cross links in the loom state fabric and in the relaxed fabric. The following notations are used:

C – distance between courses (course spacing)
W – distance between wales (wale spacing)
b – vertical length of loop
a – a proportionality constant.
x – distance measured horizontally
y – distance measured vertically
n – the number of wale spaces that the underlap i.e., cross-link connects the loops
$l\,(l_f, l_b)$ – loop length (frontbar & back bar)
d – yarn diameter

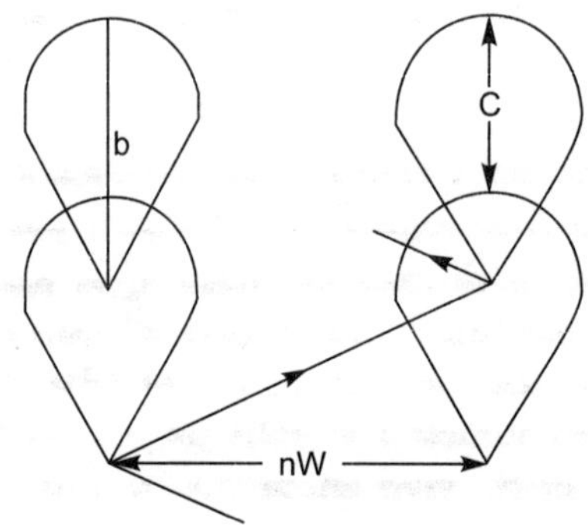

Fig. 23.2

The warp knitted structure is considered into two parts, viz., the loop and the cross link. It is important to consider all the forces acting on both the loop and the underlap.

Though the underlap on the machine is straight , it curves as a part of a circle when the fabric relaxes. The loop however, does not alter noticeably. It is also assumed that the ratio of the loop length to loop height is always constant. Grosberg assumed that the yarn is an elastic unit and that the loop shape is created by bending forces applied to the yarn at the base of the loop. Fig. 23.3 shows the simplified model of forces acting on a newly formed loop. Assuming the take down tension is negligible, the various forces acting on the loop are the forces in the underlap and the reaction produced on the previous loop. These forces are shown as solid lines in Fig. 23.3. The shape produced by such forces will be very similar to the elastica produced by two opposed forces as shown in Fig. 23.4 The bending moment formed by the simple positioning of forces is proportional to the perpendicular distance from the line of action of forces. By reducing through standard mathematical methods, the length of the loop is given as,

Loop Length $= 2.55\ b.$

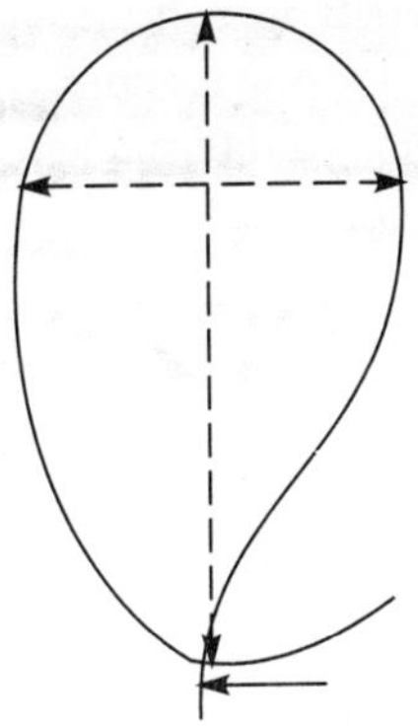

Fig. 23.3

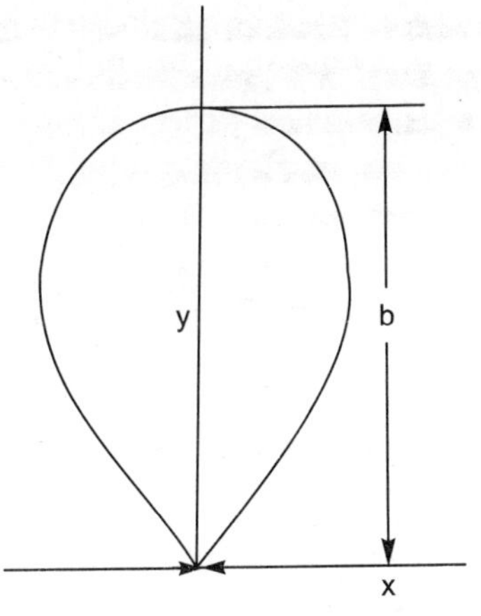

Fig. 23.4

As mentioned earlier, the cross link is straight in the loom state. By neglecting the yarn diameter, the length of the cross-link is estimated as, the length of the cross link $= \sqrt{C^2 + n^2W^2}$.

Therefore, the length of one stitch is given as the total of loop length and underlap length.

i.e., stitch length, $\quad l =$ loop length + length of cross link

$$= 2.55\ b + \sqrt{C^2 + n^2W^2}$$

The force acting along the underlap would correspond to a tension in the underlap and thereby keep the underlap straight. However, on relaxation, the force tend to disappear. Therefore, it is supposed that on relaxation, the underlap has a constant bending moment along its length and, consequently, forms part of a circle. The length of the radius of this circle would depend on the bending moment involved, which would be very difficult to find. The ratio of the length of the arc to the chord connecting the two loops is found for this purpose. This ratio is expressed in terms of the angle subtended at the centre of the circle by the chord. This angle most probably lies between 30° and 60° from the vertical.

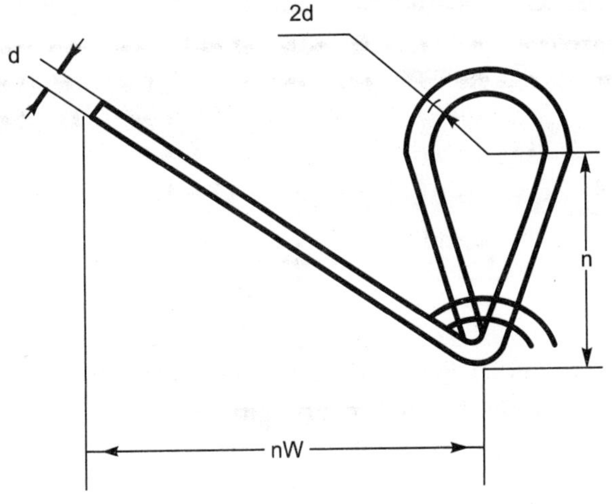

Fig. 23.5

G.L. Allison[2] attempted to relate the stitch length to fabric geometry through a loop configuration of warp knitting. He split the warp knitted cell into four sections. The loop's head as a semi circle of $2d$ radius, two straight lines for the loop arms and a third straight line for the underlap as shown in Fig. 23.5, and the yarn length calculations are,

Loop's head $= \pi \times 2d$

Loop's arms $= 2\sqrt{4d^2 + C^2}$

Underlap $= \sqrt{C^2 + n^2W^2}$

where d is the yarn diameter, C is the course spacing, W is the wale spacing and n is the underlap's needle spacing.

The stitch length was calculated by the sum of the above parts.

i.e., Stitch length, l = loop head + loop arms + underlap

$$= \pi 2d + 2\sqrt{4d^2 + C^2} + \sqrt{C^2 + n^2W^2} + 2d$$

The term $2d$ is a correction for the yarn passing at the loop's base.

23.3 CALCULATIONS IN WARP KNITTING

Warp knitting calculations can be attempted with respect to warping calculations and fabric production calculations. Machine gauge, effective width of the machine, running speed, number of guide bars, number of needle bars, yarn linear density, run-in, run-in ratio and quality are the data required for the production calculations of warp knitting. Where as the effective width of the knitting machine, machine gauge, total number of threads, lengths of warps, dimensions of the sectional beam, yarn linear density and yarn type are the information required for the warping calculations.

23.3.1 Warping Calculations

1. **The number of threads per guide bar can be calculated as,**

   Number of threads = effective width of the machine × machine gauge

2. **Warping time per sectional warp beam:** For a given warping speed, for a known warp length of each sectional beams, the running time per sectional warp beam is calculated as,

   Running time per sectional warp beam = warp length/warping speed × efficiency

3. **The production of a warping machine in Kgs/hr. can be calculated as,**

$$\text{Production (Kgs/hr)} = \frac{\text{warping speed} \times 60 \times \text{yarn fineness} \times \text{number of threads} \times \text{efficiency}}{10000 \times 1000}$$

E.g.1: For a warping speed of 1000 m/min, 0.45 machine efficiency, 50 d tex polyester yarn and 1176 threads, the warping production is,

$$P = \frac{1000 \times 60 \times 50 \times 1176 \times 0.45}{10000 \times 1000}$$

$= 158.76$ Kgs/hr

4. **Warp length determination based on run-in-ratio:** For a warp knit pattern produced with two or more number of guide bars, it is essential to ensure that the separate warps run out at about the same time and no major residues occur. The warp length required for different guide bars can be calculated using the run-in-ratio from the lapping notations.

E.g.2: A warp knit construction is produced with two guide bars, the run-in in terms of meters, of the front and back bar are 1.18 and 2.05. The fabric has 25 CP cm and 2000 meters of fabric needs to be produced.

Number of racks the machine has to knit

$$= \frac{\text{Fabric length} \times \text{CP cm} \times 100}{480}$$

The length of warp required for each guide bar is

$$= \frac{\text{Fabric length} \times \text{CP cm} \times 100 \times \text{run-in}}{480}$$

Warping length for each of the front guide bar section is

$$= \frac{2000 \times 25 \times 100 \times 1.18}{480} = 12291 \text{ m}$$

Warping length for each of the back bar sections is

$$= \frac{2000 \times 25 \times 100 \times 2.05}{480} = 21354 \text{ m}$$

E.g. 3: If a warp knitted structure is produced with three guide bars with a known run-in -ratio, the warp length for individual guide bars are calculated, provided the warp length of one of them is known.

Let $L_1:L_2:L_3 = 38 : 31.75 : 9.25$

and warp length of $L_1 = 40000$ m

warp length of $L_2 = \frac{40000 \times 31.75}{38}$

$= 33421$ m

warp length of $L_3 = \frac{40000 \times 9.25}{38}$

$= 9737$ m

Each of the front, second and third guide bar warp sections should have the warp length 40000 m, 33421 m and 9737 m respectively.

23.3.2 Production Calculations

1. **The warp knitting production in terms of running meters per hour (rmph) and kilograms per hour can be calculated as below:**

$$\text{Production (rmph)} = \frac{\text{Machine Speed} \times 60 \times \text{efficiency}}{\text{Courses per cm.} \times 100}$$

Production (m^2/h) = rmph × finished width

Production (Kgs/h) = m^2/h × weight per unit area

E.g: 4. For a 130″ wide, 2400 rpm, high speed Tricot warp knitting machine of 90% efficiency producing 2.5 m wider fabric, having 30 cpcm and 105 gsm,

$$\text{Production (rmph)} = \frac{2400 \times 60 \times 0.9}{30 \times 100} = 43.2$$

$$\text{Production (m}^2\text{/h)} = 43.2 \times 2.5 = 108$$

$$\text{Production (Kgs/h)} = 108 \times 0.105 = 11.34$$

2. **Having known the run-in (meters/rack) value of the guide bar, the rmph can be calculated as,**

$$\text{Production (rmph)} = \frac{\text{rpm} \times 60 \times \text{run-in}}{480}$$

Total yarn consumption of the guide bar in meters/hour

$$= \frac{\text{rpm} \times 60 \times \text{run-in} \times \text{number of threads}}{480}$$

$$\text{Production (kgs/h)} = \frac{\text{rpm} \times 60 \times \text{run-in} \times \text{number of threads} \times d \text{ tex}}{480 \times 10000 \times 1000}$$

E.g. 5. A fully threaded two guide bar fabric is produced on a 28 gauge, 172″ wide, 1800 rpm speed machine. The run-in of 40 *d* tex yarn at front and back guide bars are 2.15 m/rack and 1.05 m/rack respectively.

$$\text{Production (Kgs/hr)} = \frac{1800 \times (2.15 + 1.05) \times 60 \times 28 \times 172 \times 40}{480 \times 10000 \times 1000}$$

$$= 13.87 \text{ Kgs/hr.}$$

23.4 ANALYSIS OF WARP KNITTED FABRICS

The analysis of warp knitted fabrics enables to understand the knitted structure, guide bar lapping, design principles, yarn materials and the type of machines used. Equipments such as piece glass, scissors, rulers, working needles, and tweezers are the basic needs for the analysis. Use of microscope (7 to 30 fold magnification power) and yarn and fabric balance for analyzing the yarn material are preferable. The procedure discussed here may provide brief guidelines of the analysis of the warp knitted fabrics. However it takes time and practice to acquire the skill in the analysis.

To start with, it is better to learn the warp knitted fabric analysis with the so-called training fabrics which are threaded with single-colored threads in every guide bar. With such fabrics, the lapping of the guide bars can be better recognized. As mentioned previously (Chapter 21), the warp knitted fabrics are characterized by their needle loops on the face side and under laps on the backside of the fabrics. In 2 bar fabrics, under normal conditions, the threads of the front guide bar bend to dominate the face as well as the back of the fabric *i.e.*, the underlaps of the front guide bar will appear at the top of the technical back of the fabric. Thus it is possible to identify the yarns of the front and back guide bars.

The warp knitted fabric analysis may consist of the following:

1. Determination of the technical face and technical back of the fabric
2. Determination of the guide bar lappings & drafting the lapping diagrams
3. Drawing of the loop structure
4. Marking the repeat in length
5. Determination of the yarn run-in
6. Numbering the needle gaps
7. Specification of the threading
8. Reading the chain notation
9. Determination of WPI and CPI
10. Determination of yarn run-in ratio
11. Determination of the machine gauge and the possible machine type
12. Examination of the yarn material.

In the following example, the magnified photographic views of a single-face (single need bar) warpknitted fabric are shown for their technical face and technical back, as given in Fig. 23.6.

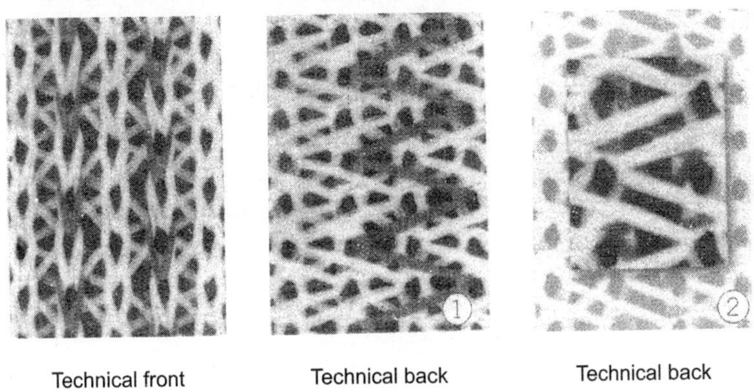

Fig. 23.6.

It is first to determine the technical face and technical back of the fabric. The technical face can be recognized by the needle loops, especially by the appearance of the stitch legs. The technical back can be recognized by the underlaps (also referred as the cross-links, stitch feet and sinker loops). By the magnifying lens, the stitch heads and stitch feets can be identified.

Having recognized the technical face and technical back, now the fabric is observed for its technical back and the construction of the front guide bar can then be determined by means of the cross-links running on top of the surface. The following can be found out in the given example.

- Stitches are formed in every course
- The underlap direction changes in every course
- A wale is situated between the needle stitches.

As a result of the above findings, it can be concluded that a 2 and 1 (2 × 1) stitch was made by the front guide bar.

The next is to find out whether the stitches of front guide bar 2 × 1 construction are open or closed. The thread end of one cross-link which runs into or out of a needle loop is plucked by a hooked needle. By observing the thread movement in the needle loop, one can determine the thread run and thus the type of the stitch (open or closed).

After having established that the front guide bar produce clos type 2 × 1 stitch construction, its lapping diagram can be drawn by showing the needle numbering, repeat and the chain notation. By assuming two units for overlap, one unit for underlap and one unit for every additional needle space covered by the guide bar during its underlapping, the yarn run-in for the front guide bar is determined as 8 units.

The above procedure can be followed to analyze the construction of the back guide bar. From the Fig. 23.6 one can clearly see that the cross-links of the back guide bar runs from one wale to its neighbouring so that it does not jump over any needle loop *i.e.,* no wale is situated between the needle stitches. It also alters direction in every course. By plucking the thread end of one cross-link, which runs into or out of a needle loop, the open or closed type of the stitch can be determined. Thus it is established that the back guide bar carried out a closed type 1 × 1 construction. The direction of lapping movements of the front and back guide bars can be found by observing the cross-links of both the guide bars. Opposite directional movements of the guide bars are known as counter-lapping and same directional movements are known as equal lapping. By observing the cross-links in the Fig. 23.6, it is found that the guide bars are counter-lapped. Also by seeing the arrangement of needle loops on the technical face, one can differentiate the counter and equal lapping. If the stitches are alternately sloped or tilted at various angles at each course, then the fabric is made with equal laps. If the stitches are straight, then it is counter lapped. Now the lapping diagram for the back guide bar can be drawn by showing the needle numbering, repeat and the chain notation. Also, by assuming 2 units for overlap and one unit for underlap, the yarn run-in of the back guide bar is determined as 6 units.

The complete construction is therefore summarized as:

Single-face fabric/Two bar structure.

Front guide bar: Closed 2 × 1 stitch.

Back guide bar closed tricot stitch.

Guide bars counter-lapped.

Run-in ratio is 4: 3

The structure is known as "Locknit".

References

1. Grosberg. P., The Geometry of Warp Knitted Fabrics, J. Text. Inst., V.51, T 39.
2. G.L. Allison, Silk and Rayon Record, March 1958.

24

Speciality Warp Knits

24.1 LAYING-IN

In warp knitting, it is not necessary that all the guide bars need to form the complete loops. Based on certain design and commercial requirements, some guide bars do not knit the yarn into the fabric and just insert horizontally into the fabric. The machine is equipped with a special lapping movement for such purpose. This laying-in principle allows to use yarns which otherwise cannot be knitted. Less fabric weight due to no loop formation, saving of special and costly patterning yarn length, overcoming the limitations of using coarser and thicker yarns are some attractive advantages of laying-in. In the following illustrative diagram (Fig. 24.1), the fully threaded front guide bar produce the ground structure and the partly or fully threaded back guide bar with a different coloured yarn produce the laying-in. The in-laid underlaps are made at the back of the needle and does not enter the hook of the needle.

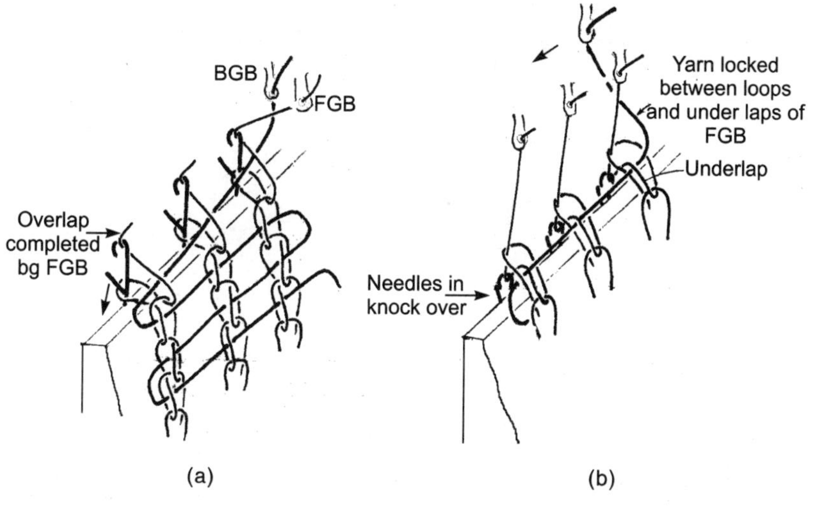

Fig. 24.1

Fig. 24.1a: Front guide bar completes overlap and its yarns are in the needle hooks. Back guide bar swings out at the back and its yarn did not enter into the needle hook.

Fig. 24.1b: Needles are in their knock over and loops are formed with FGB yarns. BGB yarn is laid-in and locked between the loops and underlaps of the FGB.

Fig. 24.1c: Needles move up for their next knitting cycle.

Underlap of the front guide bar is completed.

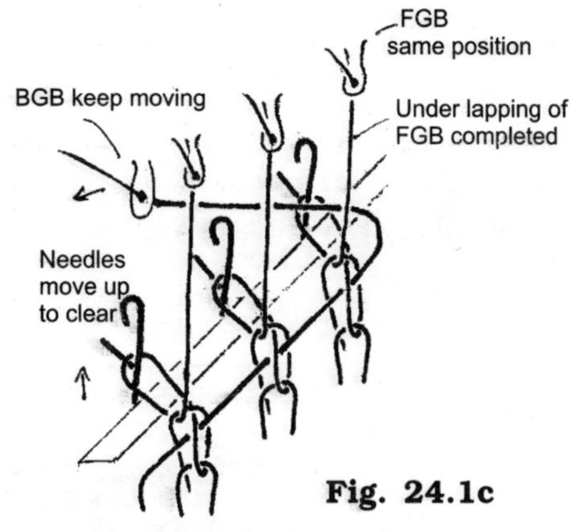

Fig. 24.1c

The inlay yarn runs between the needle's back and underlaps of FGB.

24.2 FALL-PLATE

Similar to the guide bars, the fall-plate is a bar having a thin metal plate extended to the full width at the machine. It is placed between the guide bars and is attached to the guide bar brackets. In addition to the same swinging movements of the guide bars, it also makes vertical upward and downward movements. The mechanism is exclusive to the latch needle Raschel machines. The fall-plate is lowered once for every courses knitted. With this technique, yarn can be incorporated into the fabric, without actually being knitted-in. A simple illustration of the fall-plate principle is shown in Fig. 24.2.

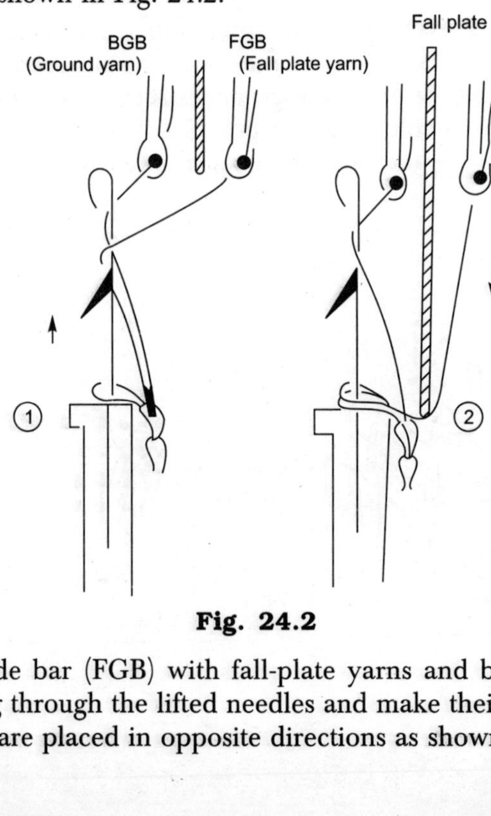

Fig. 24.2

- Both the front guide bar (FGB) with fall-plate yarns and back guide bar (BGB) with ground yarns swing through the lifted needles and make their overlaps. The overlaps of the FGB and BGB are placed in opposite directions as shown in Fig. 24.2a.

- The fall-plate is lowered. As the lap of both the guide bars are in opposite direction, only the fall-plate yarns are depressed and placed over the old loops of the fabric, which has been formed during the previous knitting cycle as shown in Fig. 24.2b.
- As the needles descend, both the old lap and the fall-plate new loop are knocked-over together. The fall-plate loop is positioned on top of the old loop and is appeared on the technical face of the fabric.
- As the needles lift after knocking-over, the fall-plate is also lifted and remains until for the next knitting cycle.

Fancy and heavy yarns can be used as fall-plate yarns. Open or closed lap fall-plate loops produce attractive relief designs.

24.3 CO-WE-KNIT (Fig. 24.3)

The fabrics of this technique, as the name implies have the combined advantages of both the woven and knitted, with respect to their appearance and characteristics. This combined weaving and knitting was developed by Karl Mayer of West Germany in 1960. The laying-in principle is used in which weft yarn is laid in front and behind the warp threads to form a typical woven structure. Non-apparels such as upholstery, furnishing and industrial fabrics are the major applications.

The structure consists of knitted pillars which are held together by laying-in threads under the fall-plate principle. The vertically running warp threads are the blind stitches which are woven in and out of the laid-in threads. The fabric is manufactured on a Raschel machine with four guide bars and a fall-plate displaced between the guide bars two and three. The front guide bar makes pillar stitches vertically on the same needles. The second guide bar makes weft laying-in whereas the third and forth guide bars control the two sets of warps. The motion of the needle bar is modified in such a manner that the needle remains stationary while the guide bars complete their cycle of operations. The warp threads supplied from the third and fourth guide bars give density and stability to the fabric. The laying-in threads of the second guide bar appear as weft and when combined with the pillar stitches of the front guide bar, gives a stable fabric. Coarser and fancy yarns can be conveniently used for design effects.

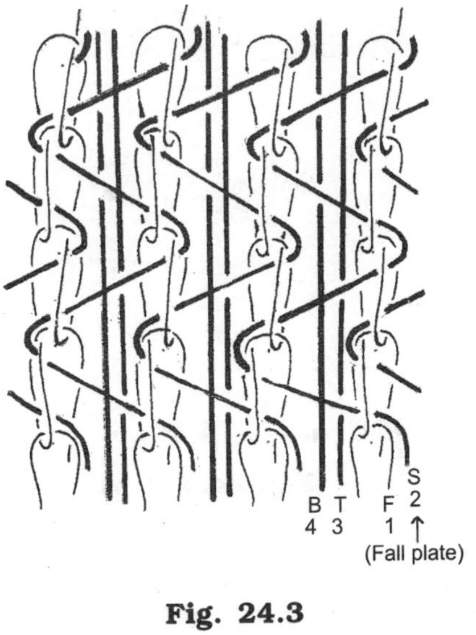

Fig. 24.3

24.4 CUT-PRESSER

Some spring bearded needle Tricot machines, in addition to their regular presser for the opening and closing of the needles, are provided with a separate presser, known as cut-presser. The regular presser processes all the needles collectively to allow each needle to form a new loop.

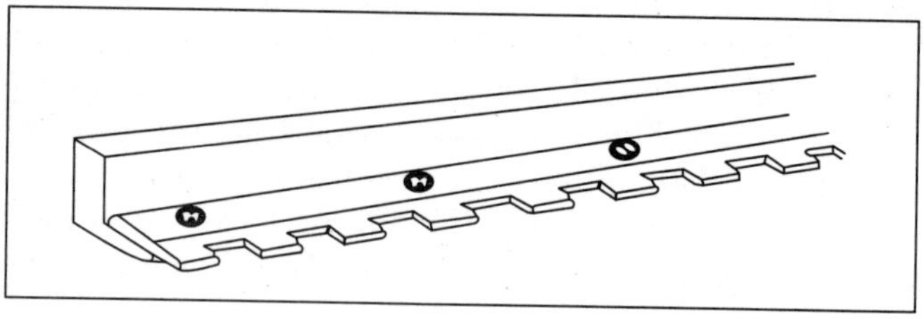

Fig. 24.4

Whereas the cut-presser acts on selected needles at certain intervals according to the designing requirement. The pressing edge of the cut-presser (Fig. 24.4) can be cut in different arrangements in regular or irregular manner (4 × 4, 4 × 6 etc.) to create sequences of pressed and non pressed needles. They have definite relationships with the order of threading of the guide bar.

In a machine working with two or three guide bars, the partly threaded guide bar may work in conjunction with the cut-presser; the fully threaded guide bars make the ground of the fabric by supplying yarns to all the needles and being operated by the regular presser. As they work in conjunction with the partly threaded guide bar, the cut-presser undergo lateral movements also in addition to their forward and backward movements. Like the guide bars, the cut-presser also receives its lateral motion from a chain of links of varying heights. This chain is mounted on the same drum as the guide bar pattern chain.

The pattern principle is such that by feeding different needles with yarns, while others maintain their previous loops, and by pressing groups of needles, while others remain open, a pattern is formed. As the cut-presser make lateral traverse, in order to connect the sections of wales together, the corresponding guide make atlas traverse lapping movements (Fig. 24.5). The fabrics made with the use of cut-presser fall in two main groups such as shell stitch fabrics and spot, or knop effect fabrics. Blouses, dresswear and shawls are some important applications.

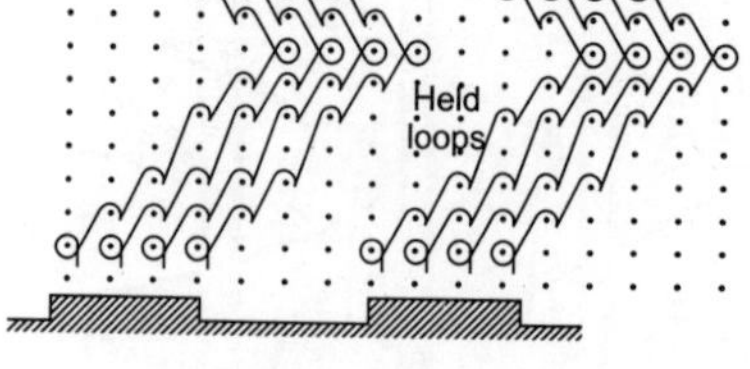

Fig. 24.5

24.5 WEFT INSERTION

The weft insertion technique, in a way is similar to the laying-in technique, in which weft yarns are introduced to the entire width of the fabric. The yarns do not enter the hooks of the needles and do not form loops. These yarns are laid horizontally from selvedge to selvedge, perform as weft yarns of woven fabric, so that the widthwise properties of the knitted fabric are similar to those of woven fabric. As a basic principle, across the completed knitted width, the weft yarn is placed and trapped between the backs of the needles by the warp threads as the guides pass through the needles to form overlap and return to the front of the machine (Fig. 24.6). That is, the weft yarn passes over the loops and under the underlaps. The ground construction may be of locknit or open work net effect and the weft yarn is inserted with the help of weft bits at the rate of one weft per knitting cycle from a weft insertion magazine.

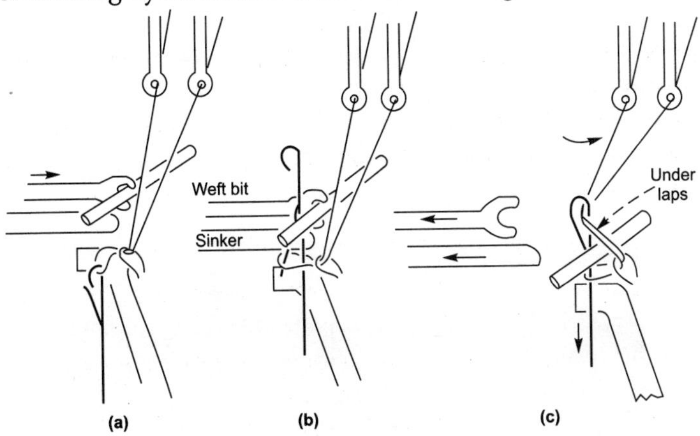

Fig. 24. 6

In magazine weft insertion, eighteen or twenty four threads are drawn-off by the weft laying device from a creel and are laid into stationary gripper rails arranged on both sides. The grippers on one side hold the weft yarn end firmly, until the weft insertion device has laid the yarn lengths into the gripper bar on the other side (Fig. 24.7 *a* & *b*).

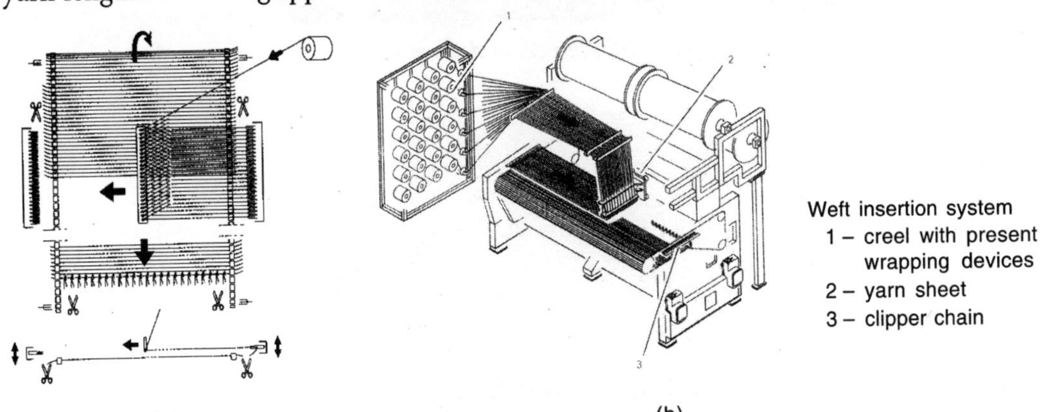

Fig. 24.7

The two gripper bars then lower the yarn sheet into the opened clips, the clips close, and finally the grippers release the threads. The weft laying-in device then returns to the other side, from where the process is repeated. The clips transport the prepared weft threads continuously to the knitting elements. The clips release the full width weft at the moment when the weft bits advance over the lowered needle heads to insert the weft. After being placed into the fabric, the weft yarn must be separated and trimmed at the selvedges by two pairs of scissors placed one on each side of the machine. Both the Tricot and Raschel machines are constructed with weft insertion technique to produce high quality fabrics.

24.6 RASCHEL JACQUARD KNITTING

The Jacquard principle in Raschel Knitting provides effective control over individual guides and hence individual warp threads. The machine operates with two types of guide bars viz. the ground guide bars and Jacquard guide bars. The ground guide bars produce a ground net structure of open pillar stitches using relatively finer yarns and the Jacquard guide bars with relatively thicker yarns, cover the openings of the ground to produce a motif. The ground yarns are fed from the sectional beams and the Jacquard yarns are fed from the cones arranged in a creel. There is normally only one, or occasionally two Jacquard guide bars and the rest are the conventionally controlled bars. Raschal jacquard machine and its knitting elements are shown in Plates 24.1 and 24.2 respectively.

Plate No. 24.1 *Jacquardtronic Raschel warp knitting machine*

Plate No. 24.2 *Knitting elements of Raschel Jacquard.*

24.6.1 Knitting Elements

The knitting zone of the machine consists of latch needles (1), trick plate (2), ground guide bars (3), jacquard guide bar (4), dropper pins (5), an optional fall plate (6) and sinker bar (7) as shown in Fig. 24.8. Specially designed long and flexible guides are fixed to the jacquard guide bars.

These guides are staggered and can be deflected one needle space either to the right or to the left, so that each one, when deflected, is placed behind or in front of the adjacent one and thus in alignment with each other. The gauge of the jacquard guides is twice finer to that of the needles so that there are two guides between adjacent needles arranged in two staggered rows. Yarns of different type or count can be threaded through them.

The jacquard guides are deflected and displaced by means of dropper pins of dropper pin bar. These dropper pins are spring loaded and are either kept lowered in their displacement position (*i.e.,* at their bottom position) or raised out of action by means of the jacquard apparatus placed above the machine. The number of dropper pins is equal to the number of needles in the machine. The different deflections of the individual guides of the jacquard guide bar are important for the production of jacquard designs. As shown in Fig. 24.9, long, flexible guides are shaped to form two displacement lines and dropper pins are placed between each two guides. For the deflection to take place, the selected dropper pin is lifted up; the dropper pin bar is shogged side ways one needle space to the left. This makes the dropper pin which is in lowered place to deflect and displace the respective guide one needle space to the left.

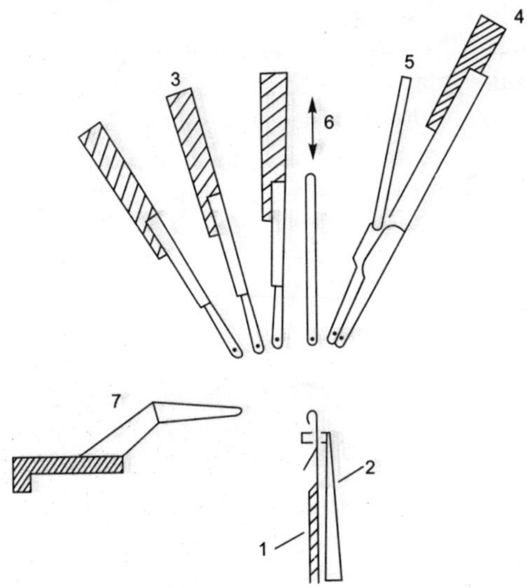

Fig. 24.8

The deflected guide is now placed in front of the adjacent one so that both are in one alignment without touching each other. They swing together into the same needle space. However, the jacquard bars do not swing, otherwise the harnness strings could become entangled. The swinging movement is always produced by the needle bar and the needles. The lateral shogging movement of the jacquard bar affects the harness length only slightly and presents no problem.

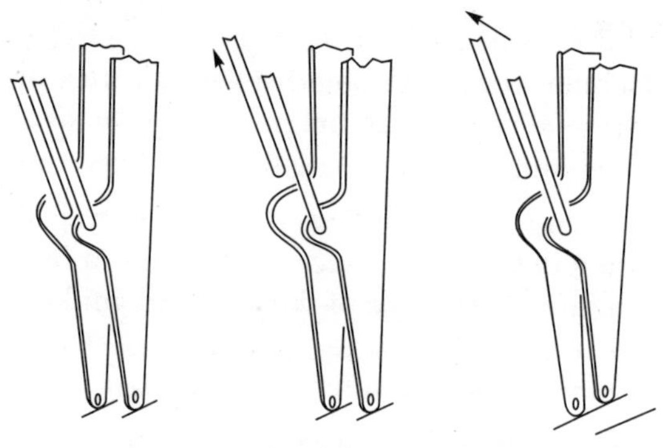

Fig. 24.9

Instead of shogging the dropper pin bar, the deflection of the jacquard guides could be possible by the shogging movement of the jacquard bar also. The deflection could also be obtained by giving relative shogging movements to both the dropper pin bar and jacquard guide bar in the same direction.

As mentioned earlier, the jacquard bars, by their basic inlay movements over two needles cover the openings of pillar ground structure formed by the ground bar. By three different inlay lapping movements of the jacquard guides, three different pattern effects are produced on the ground structure as given below in Fig. 24.10.

(*a*) The jacquard guides follow the pillar stitches of the ground structure so that the ground net is free of overlaps and the fabric is fully opened.

(*b*) The jacquard guides make inlays between two adjacent needles so that the net opening is covered by two underlaps and the fabric is semi-opened.

(*c*) The jacquard guides make inlays between three adjacent needles, i.e., over two needle spaces so that the net opening is covered by four underlaps and this appears fully covered at this area.

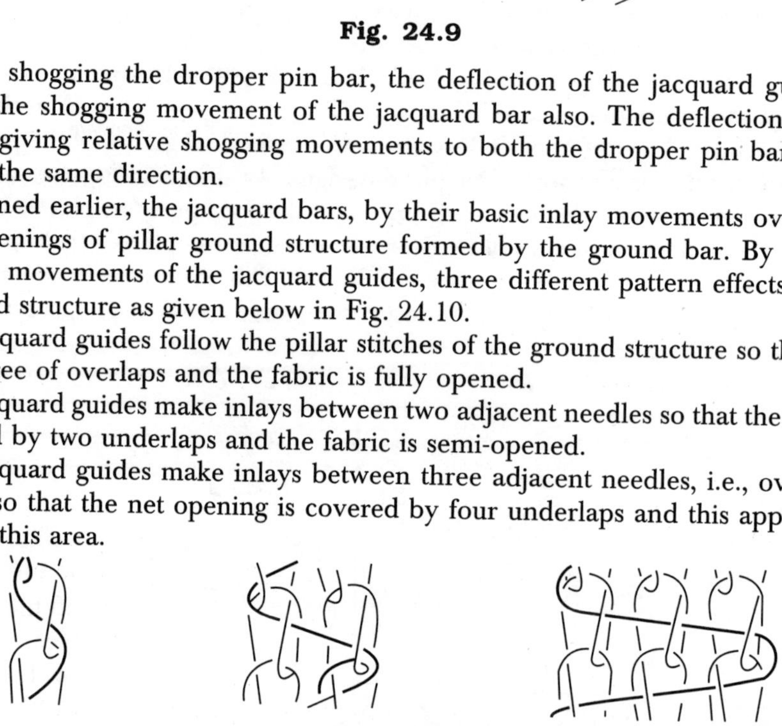

Fig. 24.10

The three different inlay-lapping techniques mentioned above are used for jacquard patterning effects and are known as "three-needle technique". For patterning representation on point papers, three different colours are used to represent uncovered full net openings, semi-solid openings and fully covered solid areas.

24.6.2 Jacquard Mechanism

Figure 24.11 shows the simplified working principle of the Raschel Jacquard. The jacquard frame, mounted overhead, consists of a perforated jacquard card which is placed on a perforated plate and is shifted to a new selection at every knitting course by means of a transport wheel. Selection pins are resting over the card and according to the pattern, some of the selected pins fall into the holes punched into the cards. As the selection pins are connected to the horizontal selection rods they also thus lowered. The pressing frame moves forward and pushes the selected horizontal rods to the right and the non-selected rods escape. Lifting hooks, which are connected to the dropper pins by means of cords are controlled by the edge blades of a cam operated lifting frame which is mounted at the top.

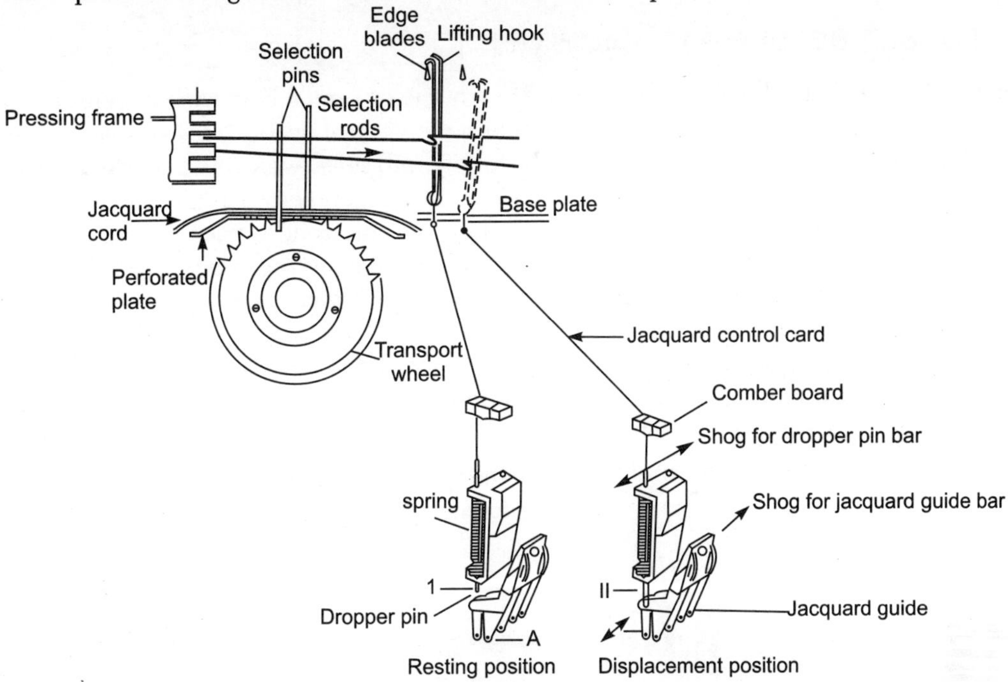

Fig. 24.11

The selected horizontal rods when pressed by the pressing frame, displace their respective hooks to the right so that these hooks evade the edge blades and are not pulled upwards. All the hooks which are not pulled, remain down and rest on the base plate, and in turn not pull their respective dropper pins. Whereas the lifted hooks, pull the cords and lift the dropper pins placed in the dropper pin bar out of the jacquard guides. The dropper pins, which are not pulled, remain between the jacquard guides due to the springs. In general, a hole in the jacquard card causes the selection pin to fall, displaces the lifting hook out of the edge blade, not pulls the cord upwards and then keeps the dropper pin in between the jacquard guides and deflects them as the dropper pin bar shogs. The unselected selection pin causes the edge blades to lift the hook so that the dropper pin is lifted out of the jacquard guides and hence no deflection to the guides. The dropper pin bar and jacquard guide bar are shogged by their respective pattern wheels.

One of the greatest advantages of jacquard technology is the production of attractive large motifs for decorations. Although curtains, elastic and rigid laces, florals, table cloths, underwear fabrics are the various warp knit jacquard products, curtains are the biggest single products.

The new generation, electronically controlled Raschel Jacquard machines are developed, which result in extremely large pattern variety in the production of curtains and laces. The dropper pins are activated for their lifted and lowered positions by means of electromagnets. Each electromagnet operates one harness cord and one dropper pin so that the number of magnets is the same as the number of dropper pins, the number of guides in the jacquard bar and the number of needles in the needle bar. A novel method of guide deflection is offered by the Piezo-electric jacquard system. 16 or 32 piezo-elements are cast together and attached to the jacquard guide base. When an electric current is applied, the piezo crystals contact or expand and thus bend the jacquard guides out of their normal position and displace them side ways.

24.7 DOUBLE NEEDLE BAR MACHINES

Tricot and Raschel warp knitting machines with two sets of needle bars were developed for the production of double-face fabrics. In such machines, the vertical needle beds are placed back-to-back so that the technical face of the fabric is shown on both sides, while the underlaps are locked within the structure. The fabric is drawn downwards in the gap between them. Guide bars pass between the needles in both the beds as they swing from the front to back of the machine and vice-versa. Compared to single bed knitting, additional swinging movements to the guide bars *i.e.,* the third swinging motions are necessary after each underlap to complete the knitting operation, so that the other needle bar can rise to commence its knitting cycle. Having loops on both sides, the fabrics of two needle bar machines are thicker than those of the single needle bar machines. Outerwear, gloves, mesh, crepe, pile and thermal fabrics are some of the products of these machines. Knitting elements of double needle bar machines are shown in Fig. 24.12.

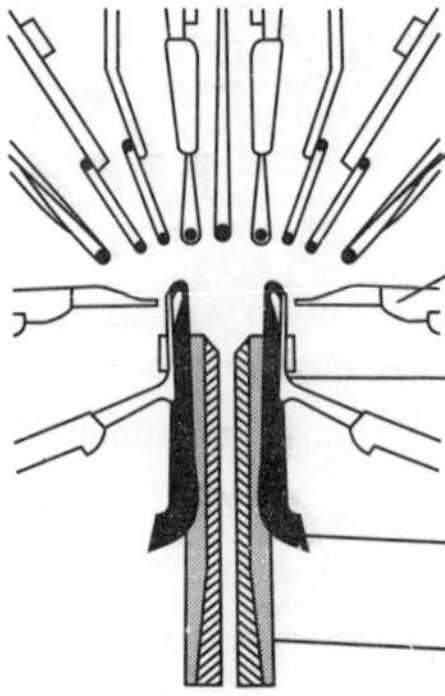

Fig. 24.12

24.7.1 Simplex Machines

Simplex machines are bearded needle warp knitting machines equipped with two sets of needles. The two needle bars are mounted back-to-back so that the angle formed between the needles is about 45°. The knitting elements consist of two needle bars, two sinker and presser bars and two guide bars. The knitting elements of each side lie directly behind and in alignment

Plate. 24.3 *Simplex machines*

with the elements of the other side. The needles of both the bars are positioned in the same gating arrangement and the guide bars swing in between the needles of each needle bar and produce one double sided fabric. A double needle bar simplex machine is shown Plate No. 24.3.

Fig. 24.13 shows the simplex knitting cycle

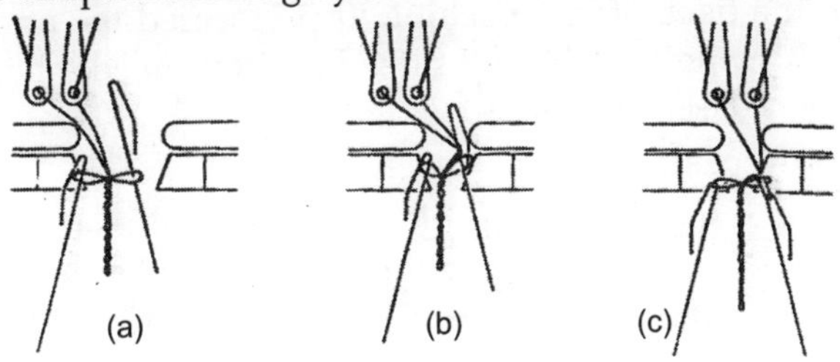

Fig. 24.13

(*a*) The back needle bar completed its knitting action and the front guide bar rises sufficiently so that the old underlaps slide down into the needle stem. The guide bars swing towards the back of the machine.

(*b*) As the guide bars swing towards the back of the machine, they wrap yarns over the front needle beards. The front needle bar is lifted and the overlaps slip into the needle stem. Further, as the front needle bar is lowered, the new overlaps are trapped into the needle hook. The presser closes the beard and the needle descends through the old overlaps.

(*c*) The front bar needles descend further to knockover. The guide bars make their underlap shog behind the front needle bar and then swing towards the front of the machine to allow the back needle bar to rise for second part of the machine sequence.

Identical sequence occurs on the back needle bar to complete the machine cycle.

The traditional cotton glove fabrics are produced from the simplex machines. Today, by using a variety of synthetic fibres, ladies underwear and outerwear with bigger mesh openings are also produced with simplex machines.

24.7.2 Two Needle Bar Raschel Machines

The machine consists two sets of vertically mounted latch needles facing back-to-back. The arrangement of knitting elements and the knitting action of the Raschel is less complex to that of simplex machines. The needles of both the bars are in the same gating and the fabric is drawn downwards in the gap between them. The two needle bar Raschel machines offer much scope for modification in order to knit a wide variety of products ranging from packing sacks to artificial blood vessels. Unlike simplex, more number of guide bars are used in double needle bar Raschel machines with special attachments such as fall-plate, weft inlay, creeping motion etc.

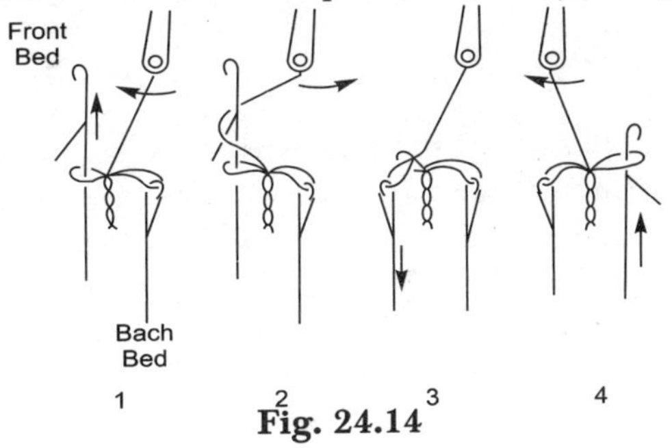

Fig. 24.14

Fig. 24.14 illustrates the knitting sequence for the double needle bar Raschel machines. The diagrams show the loop formation with respect to the front needle bar and for the back needle bar, identical operations are carried out.

(*a*) The guide bar is at the back of the machine. It performed the underlap shogging for the front needle bar and swing to the hook side. The front needle bar ascends for clearing.

(*b*) The guide bar make overlaps to the front bar needles and then swings back. The front needle bar starts to descend.

(*c*) The front needle bar descends further, old loop closes the latch and the new yarns are trapped into the needle's hook. The guide bar swing for the third time towards the front needle bar. Underlap shogging movement for the back needle bar can start.

(*d*) Front needle bar knock over and completes the new loop formation. The back needle bar starts moving up for their clearing position.

As mentioned earlier, there is a good scope for knitting special structures with double needle bar Raschel machines. Different fabric types and effects are produced by the patterning mechanisms:

- By using 2 guide bars, two separate fabrics, one on the front needle bar and the second on the back needle bar can be produced so that each of the guide bars is knitting on one needle bar only (Fig 24.15).
- By using a fully threaded middle guide bar, a sandwich of the two fabrics can be produced, connected by yarns. The fabric is then cut and separated into two plush fabrics by special equipment (Fig. 24.16).

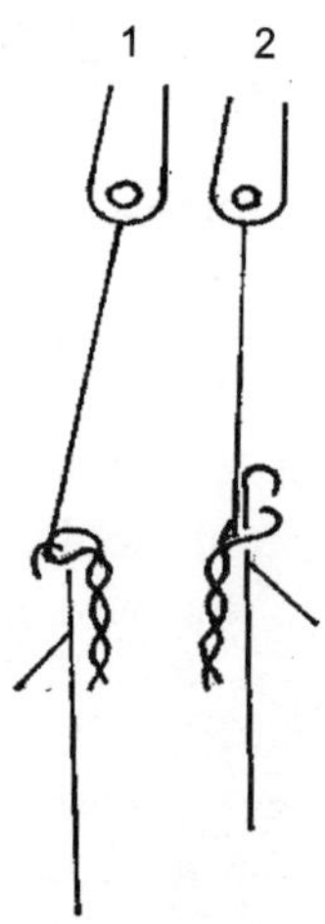

Fig. 24.15

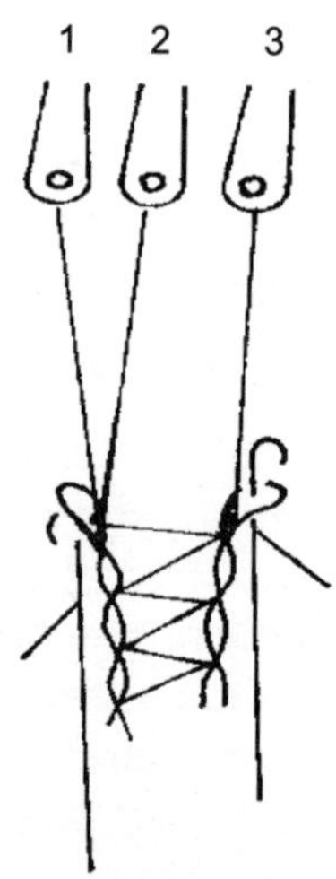

Fig. 24.16

- Unlike the tightly and closely constructed woven sacks, in double needle bar Raschel machines, loosely constructed packing sacks can be produced for coal, vegetable, fruit and even rice packing. The machine operates with as many as nine guide bars of which four are fully threaded and five are light weight partly threaded pattern guide bars. The guide bars 3, 4, 6 and 7 construct two separate fabrics of the sack in such a manner that the guide bar 4 knits pillar and guide bar 3 make inlays with the front needle bar. Similarly guide bar 6 pillars and guide bar 7 inlays with the back guide bar. The guide bar 5 is used to close the bottom of the sack by joining the two fabrics at one side. Guide bars

2 and 8 are threaded with very thick draw threads in their single guide fingers which are used to pull and close the sack opening after filling. Guide bars 1 and 9, which are also threaded with one guide finger, cover the draw thread every few courses.

- Tubular warp knitted fabrics can be produced on a double needle bar Raschel machines using 4, 6, 8 or even more guide bars in such a way that the side connections are identical to the body structure and a seamless tube is formed. Of the four guide bars, one fully threaded guide bar produces the front fabric on the front needle bar, the other fully threaded guide bar produces fabric on the back needle bar and the inner two guide bars produces the connection on each side. Packing tubes, tubular packing nets etc. are the products of this category.
- Pile fabrics are produced on double needle bar Raschel machines with six fully threaded guide bars. Two guide bars form the front fabric, two guide bars form the back fabric and the middle two guide bars threaded with the pile yarns knit on both needle bars to produce floating yarns (Fig. 24.17). After the knitting operation, the structure is separated into two pile fabrics by cutting. The distance between the needle bars can be set apart upto 60 mm to produce different heights of pile. Using different yarn types and different printing techniques, attractive pile effects can be obtained. Upholstery fabrics and automotive textiles are the major end uses of these pile fabrics.

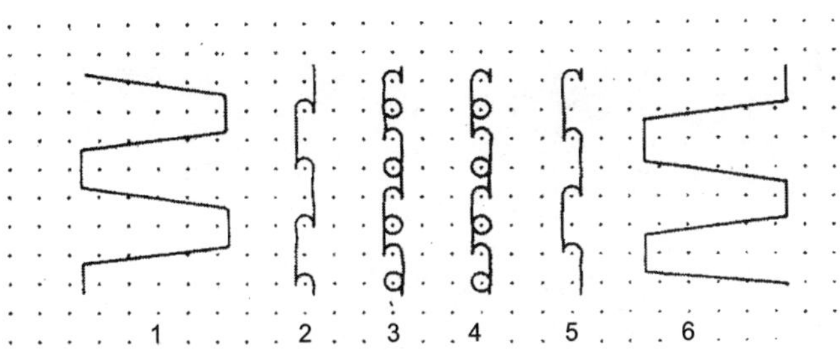

Fig. 24.17

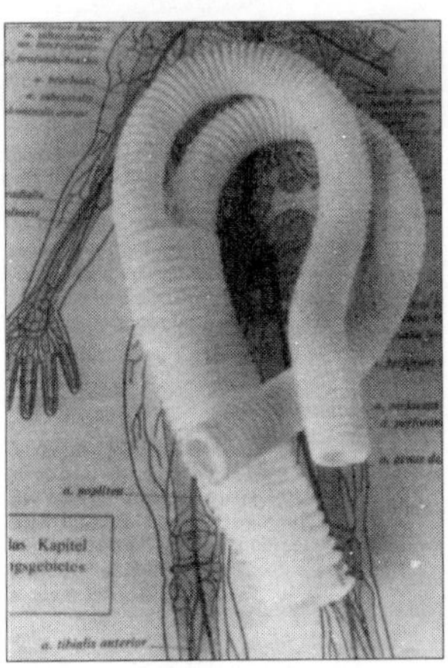

Fig. 24.18

Another pile effect, viz. pointed pile is produced by replacing the front needles by a point or pin-bar around which the pile yarns are overlapped. In order to secure them firmly, the pile yarns may be overlapped in the base fabric of the back needle bar or inlaid on yarns. The point pile lie at right angles to the base fabric and on some machines the loops are cut to produce the floating fibres. The structures are particularly suitable for floor coverings and carpeting.

- Branching tubular fabrics such as artificial blood vessels are the other important products of double needle bar Raschel machines. Use of fully threaded ground guide bars and nesting of pattern guide bars, as a total of 16 are used to produce the different sections of the tube. Two pattern drums, operating alternatively are used to produce different parts of the product. Different sizes of arteries with respect to their tube diameter and length can be produced (Fig 24.18). Another type of branching tubular fabric such as pantyhose can also be produced with the same technique.

A double needle bar Raschel machine is shown in Plate No. 24.4.

Plate No. 24.4 *Double needle bar Raschel machine.*

25

Warp Knitted Technical Textiles

25.1 INTRODUCTION

The production of technical textiles is a rapidly developing trade in the textile industry to which the knitting sector also contribute to a large extent. Warp knitting is by far the most versatile fabric production system in textiles. Warp knitted structures can be produced elastic/stable, with an open or closed structure; they can be produced flat, tubular or three-dimensional. The new concept of fabric engineering was conceived during the evolution of technical textiles as a natural development of the trial and error methods. The commercially available fabrics failed in many cases to satisfy the specifications set by the producers and the designers and textile engineers were called upon to create entirely new structures. The flexibility of warp knitting techniques makes them attractive both to the designer and the manufacturer of technical textiles. There is increased applicability of the warp knitted structure to the technical textiles in recent years. Advantages with respect to the mechanical properties, high productive knitting machines and use of synthetic yarns opened new ventures for warp knitted technical textiles. In this chapter, a review of the applicability of warp knitted structures to various sectors of technical textiles is highlighted.

25.2 APPLICATION SEGMENTS

In recent years, the importance and significance of the technical textiles is enhancing to large extent. Technical textiles are broadly segmented into the following main applications:

GEOTECH : Geo-textile and Civil engineering
HOMETECH : Technical components of furniture, household textiles and floor covering
MEDTECH : Hygiene and medical
MOBILTECH : Automobiles, Shipping, Railways and Aerospace
INDUTECH : Filtration, Conveying, Cleaning and other industrial uses
BUILDTECH : Building and Construction
PROTECH : Personal and Property protection
PACKTECH : Packaging
SPORTTECH : Sport and Leisure
CLOTHTECH : Technical components of footwear and clothing

25.3 WARP KNIT STRUCTURES FOR TECHNICAL TEXTILES

The warp knitted fabrics used as technical textiles can be produced with the following constructions for their increased stability and end use requirements.

(*i*) *Pile/Velour/Fleece*: With single needle bed, 2 bar machines, pile can be produced either by raising the long underlap of the front guide bar, or by producing pile loops by supplying warp at a faster rate to the back guide bar than it is required for the conventional structures so that the surplus warp forces its way through to the surface. The pile fabrics produced on double needle bed Raschel are of two types, viz., cut pile and point pile.

- The cut pile is achieved by knitting a separate base fabric on each needle bed, but joining the two together by the lapping movement of the pile which is later slit to produce the two cut pile fabrics.
- Point or looped pile is produced by replacing the front bar needles by a point bar or pin bar around which the pile yarns are overlapped. The loops lie at right angles to the base fabric or on some the pile loops may be cut by rotating cutting blades.

(*ii*) *Conventional Structures–Directionally Structured Fabrics (D.S.F)* (Fig. 25.1): By using standard stitches and their combinations, lengthwise, widthwise and both way stability of the structures can be obtained.

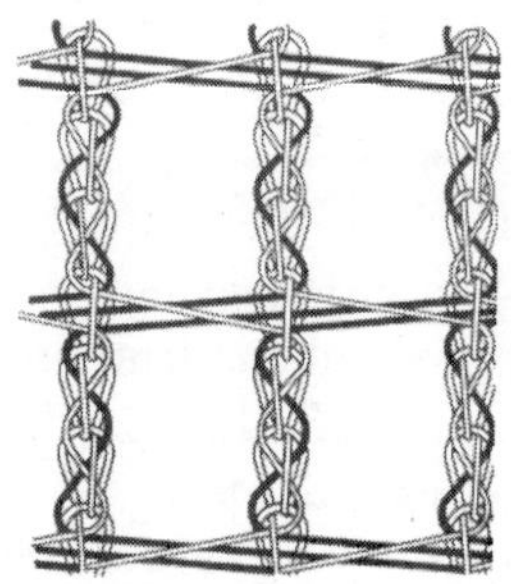

Fig. 25.1

(*iii*) *Three-Dimensional Structures* (Fig. 25.2) : Conventional knitted structures can be stitched together in layers to form a three-dimensional structure. These are also sandwiched spacer textiles, which are produced with two needle bar machines. By using 4 guide bars lapping over the 2 needle bars, a sandwiched structure is produced. The space between the needle bars can be kept upto 60 mm.

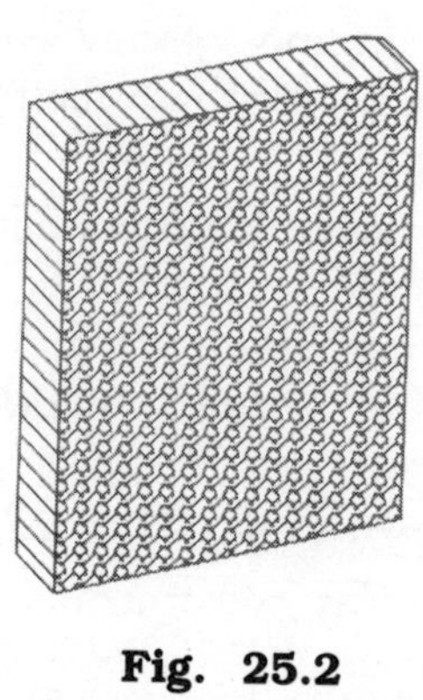

Fig. 25.2

(*iv*) *Directionally Oriented Structures D.O.S* (Fig. 25.3) : D.O.S is unique fabrics in which straight and parallel set of yarns are inlayed into the structures at almost any desirable angle such as weft/warp monoaxial direction, monaxial, perpendicular biaxial and multiaxial. The principles of laying in and weft insertion are conveniently adopted for the above technique. Compared to the mono and biaxial, the multiaxial structures withstand forces in all directions. In warp knitted multiaxial structures, four sets of yarn are inserted in different direction. Apart from warp and weft, two diagonals are placed at the required angles. Also, such structures withstand the shearing forces in all direction and the high tensile strength of the fibre is utilised.

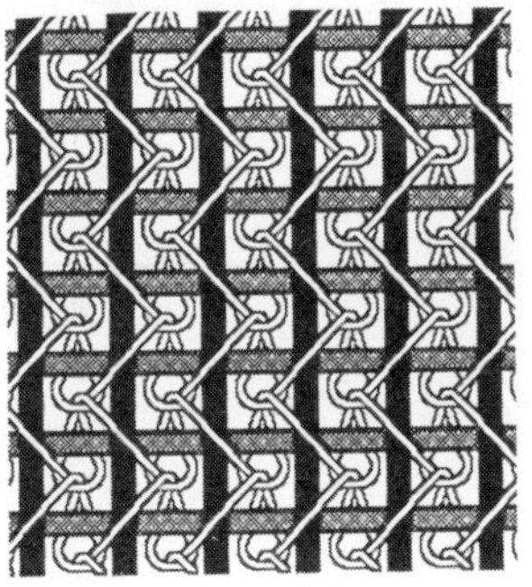

Fig 25.3

(*v*) *Composite Structures* (Fig. 25.4) : Composite structures are fabrics, which are composed of two or more dissimilar components to create a single structure that performs much well. Reinforced non-woven fleece is a popular composite structure. The fleece can be placed either at the back of the fabric or in between warp and weft yarn.

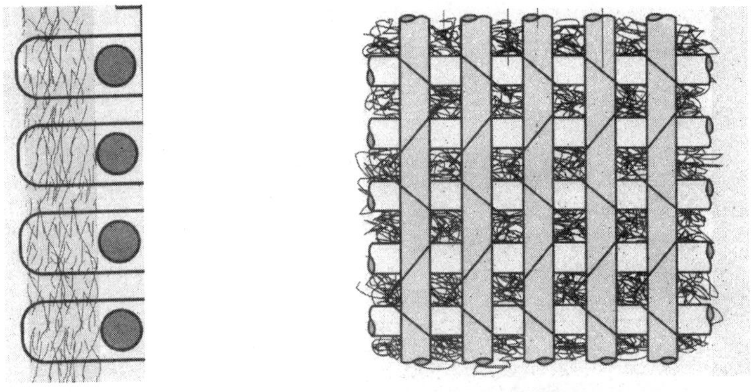

Web on the reverse side of composite fabric

Biaxial reinforced non-woven composite

Fig. 25.4

25.4 GEO-TEXTILES

Quite large quantities of warp knitted fabrics are used for geo-textiles application such as building foundation, drainage layers, railway construction, marine wave defence, landscape protection etc. With respect to the properties such as tensile properties, tear strength, hydraulic properties, warp knitted geo-textiles are advantageous when compared to others. Fabric/soil shear and Fabric/soil friction characteristics of warp knitted geotextiles ensure better performance of them.

(*i*) *Vertical Walls*: Warp knitted D.O.S geo-textiles can be utilized as reinforcing elements to strengthen soil and thus prevent the collapse of steep embankment faces and vertical retaining walls. The uncrimped yarn arrangement in the fabric ensures that the stress generated in the soil will be transmitted directly to the high tenacity polymer without any straightening of the textile structure resulting in minimized movement of the geostructure. Fig. 25.5 depicts the fabric construction and structure for vertical walls. Three guide bar RS3MSU Raschel machine with magazine weft insertion is used for the production for such structure.

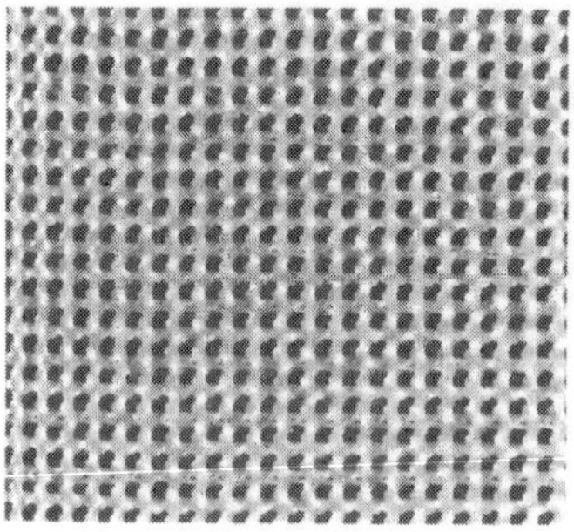
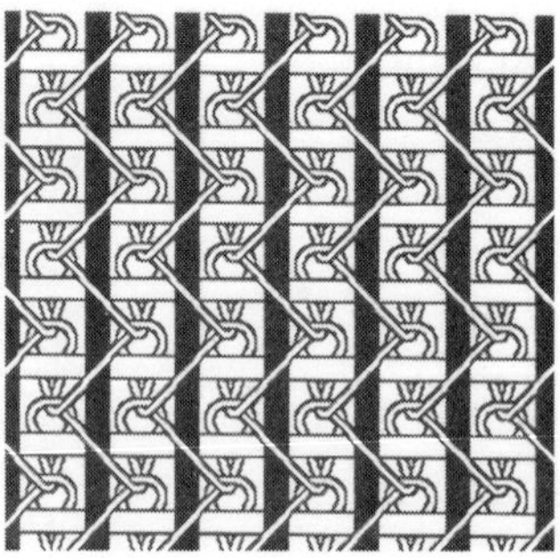

Fig. 25.5

(*ii*) *Drainage and Filtration*: In such applications, the geo-textile does not filter the soil, but supports the exposed soil surface preventing it from break down. A simple combination of pillar stitch and inserted weft produces fine net. The absolute uniformity of the opening enables the civil engineer to select the correct fabric for the soil particle distribution. Fig. 25.6 depicts the fabric construction and structure for drainage and filtration. Three guide bar RS3MSU Raschel machine with magazine weft insertion is used for such construction.

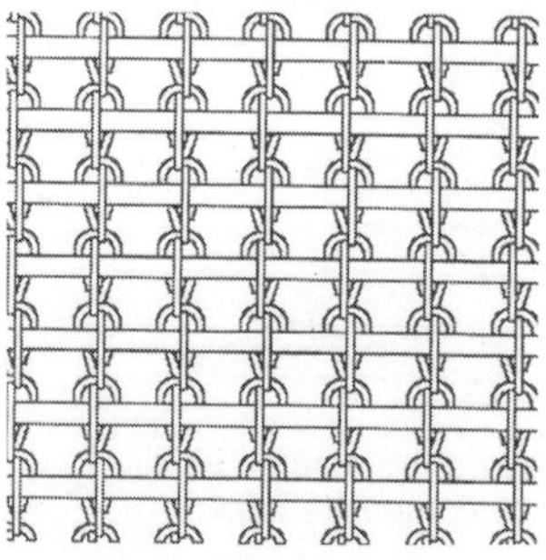

Fig. 25.6

(*iii*) *Fin Drains*: It is a composite polymer drain in the form of a vertical sheet. It is permeable to water and is designed to conduct water along its length. Three-dimensional cuspated honeycomb textiles as the permeable core in the drain can be used. Fig. 25.7 depicts the fabric construction and structure for fin drains. For such construction, the basic Tricot machine is used.

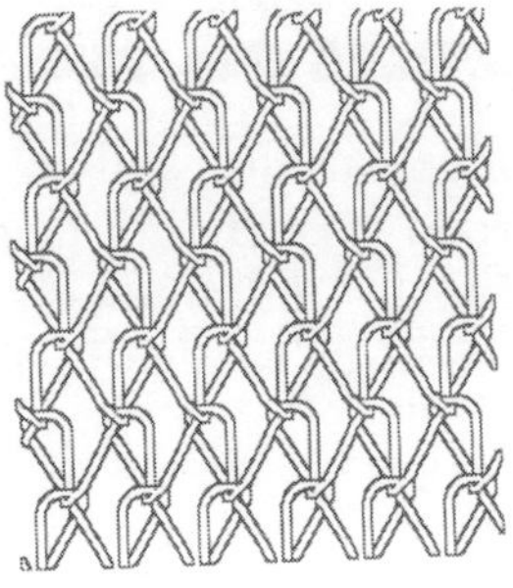

Fig. 25.7

(*iv*) *River and Marine Erosion Protection*: Geo-textiles can be used beneath the stone or concrete facing of river banks and canals, to prevent erosion where water currents and waves would wash out the fine soil particles. The use of geo-textiles can by means of heavier grade materials can be extended into erosion control structure in tidal estuaries and even on coastal zones facing direct attack by large waves. The composite D.O.S structure is used for such applications. Fig. 25.8 depicts the fabric construction and structure for river and marine erosion protection. Three guide bar RS3 MSU-V high performance Raschel machines for re-inforced nonwoven fabrics are used for such construction.

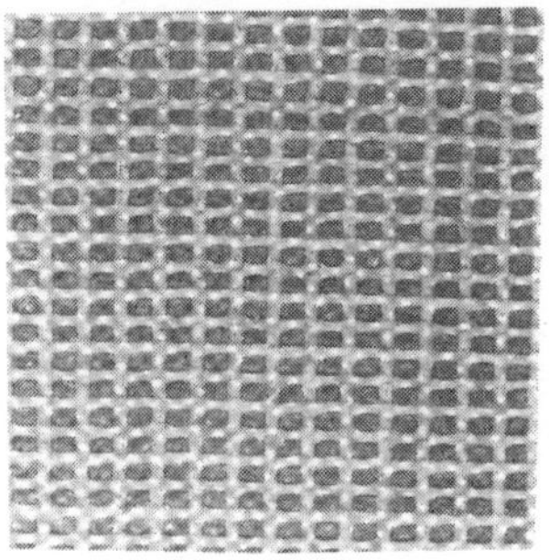

Fig. 25.8

(*v*) *Embankment Support*: With embankment construction, local, underlying cohesive soils may be protected from failure by geo-textile reinforcement. A very special fabric is used in embankment reinforcement construction. Fig. 25.9 depicts the fabric construction and structure for Embankment supports. Raschel RS3MSU-V machine is preferred.

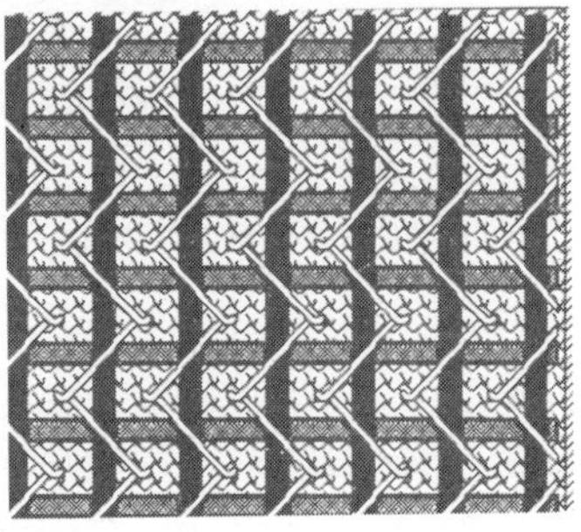

Fig. 25.9

25.5 HOME TEXTILES (Fig. 25.10)

This mainly includes curtains and table cloths, furnishing fabrics, decorative fabrics, and lightweight net curtain. RJWB 4/2F Jacquard Raschel machine, with its piezo jacquard system offers the user virtually unlimited design possibilities, without any restriction on repeat. "Cliptronic" is a two piece jacquard bar fabric, together with individual yarn selection offer a high degree of versatility, productivity and technical perfection.

Fig. 25.10

25.6 MEDICAL TEXTILES (Fig. 25.11)

Mattress covers, bandages, carrier materials for sensors, wheelchair cushions for bedsore prevention, prothesis padding, pads and working clothes for operating room, orthopedics applications, prothesis for arms and legs, branched artificial blood arteries are some of the applications of warp knitting fabrics in medical sector. Fine and course technical spacer fabrics produced with double-needle bar Raschel machines have great application in medical textiles. Multiaxial high tech Raschel knitting machine-MALIMO with several weft insertion systems is used for the production of medical textiles.

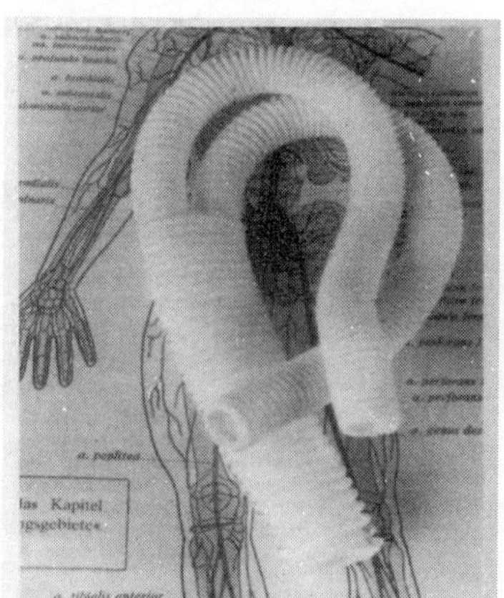

Fig. 25.11

25.7 AUTOMOTIVE TEXTILES

The application areas of warp knitted textiles in the field of automotive industry are seat covering fabrics, decorative cladding, upholstery backing etc. The fabric constructions such as

(*i*) flat with slight relief or raised or surface patterning
(*ii*) fabrics, plain or with surface pattern
(*iii*) spacer fabric with or without decorative side
(*iv*) fleece fabric with light surface pattern effects, and
(*v*) compact/voluminous fleece effect

are used for the above automotive applications. Tricot machines and double plush Raschel machines are used for the above constructions. The decorative cladding includes covering for around the seat base and back, door panels, support pillars for the door and windows and parcel shelves. The seat covering and upholstery sectors were previously coated with foam fabric, which was covered on the reverse side with a smooth base fabric. The demand for single material sourcing replace the foam fabrics with warp knit constructions.

25.8 INDUSTRIAL TEXTILES (Fig. 25.12)

Warp knitted structures are applied for reinforcing substrates for grinding wheels and belts, Conveyor belts, reinforcing substrates for synthetic materials, preforms used in the construction of vehicles, airplanes, ships and tanks. In case of machine construction, fast moving parts are made from warp knitted fabrics. In power engineering sector, they are used for rotor blades for wind power plants, wind channel blowers. For ship construction, sailer yachts, fast patrol boats, passenger ships, boats for detecting mines, submarine boats are some of the applications. In aero spatial industry, they are highly used for the manufacture of antennas, paraboloidal-type reflector, pressure vessels, and satellite structures.

Fig. 25.12

25.9 BUILD TEXTILES

This mainly comprises of fabrics for building industry, plaster base, waterproof sheetings for roofs, light weight supporting frameworks, air-inflated tents, stadium roofing, and flat-roof coverings and face insulation. For these applications, warp knitted structures are successfully applicable.

25.10 PROTECTIVE TEXTILES

This includes flame and weather protection, protection against the sun, camouflage nettings, insect-repellant mats and sun-protective blinds. Tricot machine and Raschel machines can be used for the manufacture of such structures, which are fabrics and nets with varying mesh openings and constructions.

25.11 PACKING TEXTILES (Fig. 25.13)

Horizontal sacks as continuous sack runs can be used for various packing purposes used in fully-automatic packing lines. Specially designed sack producing Raschel machines are used for this purpose. High-performance Tricot machine is used to produce stable fabric for high quality suitcases and bags. This textile is extremely dimensionally stable, abrasion resistant, water-repellant, offering highest stability to sharp pointed and pushing through objects.

Fig. 25.13

25.12 SPORT TEXTILES

MALIMO Multiaxial is a high tech knitting machine with several weft insertion systems produce multiaxial fabric layers for various applications such as tennis racquets, skis, snowboards, surfboards, sport boats etc. Spacer fabrics are widely used in sportswear such as surf jackets, submerses for diving suits, functional jackets, and protection areas in sportswear, breathable shoe fabrics. Spacer fabric with two dense faces or one face with semi-open structure is used for sport shoes. It is also used as inner shoe soles or inlay shoes for sports and casual shoes.

25.13 CLOTHING TEXTILES

Very interesting application in this field is the lace manufacture. The new generation of jacquard lace machines has even surpassed the pattern possibilities and has set new standards in terms of the technology and design varieties. The Jacquard Raschel machine type RJWBS 4/2F offers the end user numerous application possibilities in lace edging and all over lace with relief-like patterning in combination with various ground structures such as powernet, tulle etc.

Flat Bad Knitting

26

Flat Bed Knitting

26.1 INTRODUCTION

As mentioned earlier, flat knitting is an important division of weft knitting industry. The use of flat bed machines and flat knitted fabrics, were much limited in India few decades before, and also they are hand operated. However, they are indispensable as for the present knitting industry is concerned. The first flat bed machine was demonstrated in 1862 and patented in 1865 by Rev. Isaac Wixon Lamb, an American. Now a days high productive power driven machines are available.

In this chapter, a brief description of some important aspects of the flat knitting machines such as knitting elements, knitting operation and fabrics produced are discussed briefly.

The flat knitting machines are advantageous due to their less capital cost, economical running of small lots, superior fabric quality and more versatility. The facilities for loop transfer and racking are easier in flat knitting machines. These machines can be complimentary to the standard, intermediate and full jacquard machines, as collars, sleeve cuffs and body rim trims for fully fashioned sweaters can be conveniently produced on these machines. While the fronts and backs and sleeves can be produced on regular machines.

It is interesting to mention that the fabrics which can be produced on circular knitting machines can be easily produced on V-bed flat knitting machines also, but some of the fabrics which can be produced on V-bed machines are difficult and sometimes impossible to get produced on circular weft knitting machines.

26.2 MACHINE TYPES AND CLASSIFICATION

The flat bed machines are divided into three groups as given below:

(*i*) V-bed machines.
(*ii*) Horizontal opposed flat bed machines.
(*iii*) Straight-Bar multisectional full fashioning machines.

Latch needles are used with the first two categories where as spring bearded needles are used with the third group.

The hand operated machine width range from about 80-120 cm and the power driven automatic machines from about 66-200 cm or more.

The following chart gives further information on the classification of flat knitting machines.

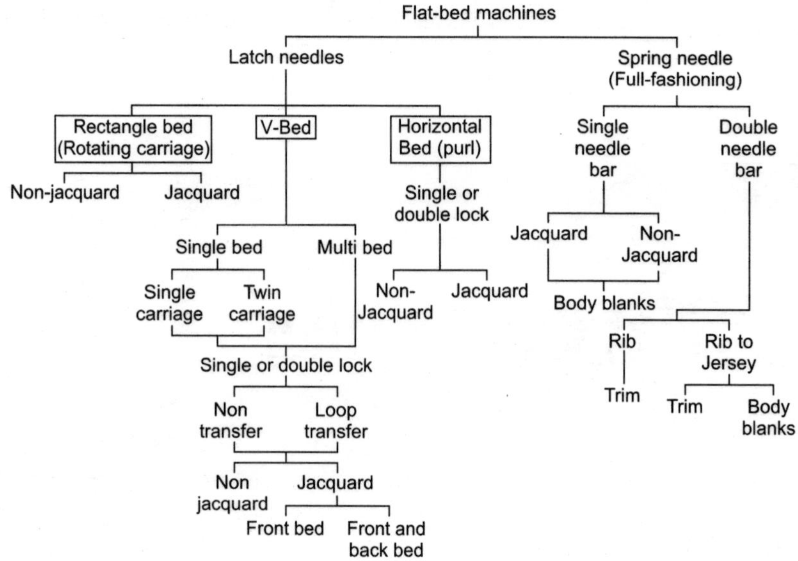

26.3 KNITTING ELEMENTS

26.3.1 Needle

Majority of the flat knitting machines use the latch type of needles, which is punched from a flat strip of extremely high quality steel. The latch needles used in flat knitting machines have an additional attachment, known as transfer spring, which enables the needles to transfer its loop to another needle and thus facilitate elaborate patterning (Fig. 26.1).

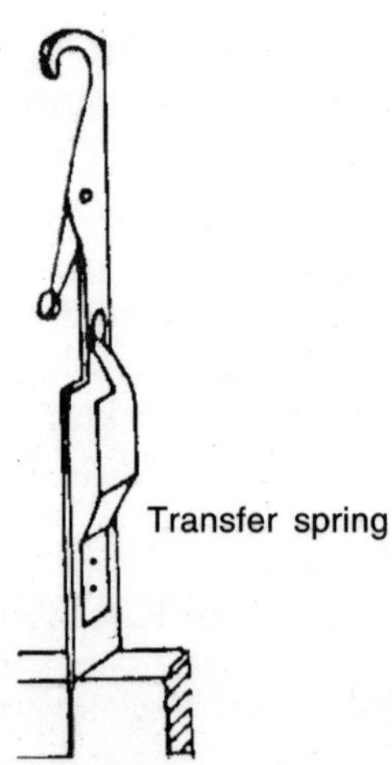

Fig. 26.1

26.3.2 Needle Beds

The V–bed machines have two rib gated, diagonally approaching needle beds, set at between 90 and 104 degree to each other and at 45° angle to the horizontal plane, giving an inverted V-shape appearance (Fig 26.2).

The needles are placed in parallel tricks formed in a flat metal plate. This plate is called the needle bed.

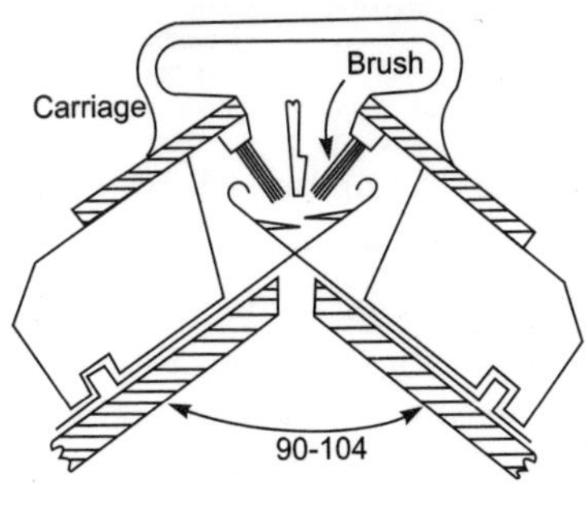

Fig. 26.2

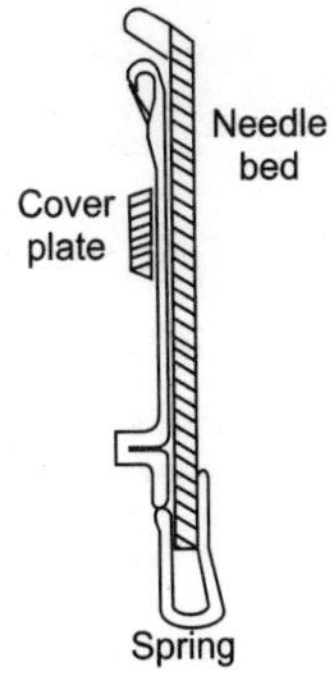

Fig. 26.3

The needle beds are attached to a stationery stand or table. The needle hooks face out from the needle bed and the needles can move freely up and down in their tricks. A plate called a needle cover bar prevents the needles being tilted by the fabric take down tension. The blade separating two adjacent needles is called the trick wall. At the end of the trick-wall a tooth called a land or "knock over bit" is soldered onto the needle bed. The whole set of bits form the knockover comb. During knitting, the edges of the knock over bits restrain the sinker loops as they pass between needles and thus assist in the knock over of the old loops and in the formation of the new loops. Flat machines do not normally employ holding down sinkers as the take down and the loops on the needles in the other bed help to hold the old loops down on the needle stems (Fig. 26.3).

26.3.3 Cover Plate

A thin metal blade located in a slot across the top of the needle bed tricks is called the cover plate. This prevents the stems of the needles from pivoting upwards out of the tricks as a result of the fabric take down tension drawing the needle hook downwards, whilst allowing the needles to slide freely in their tricks. The plate can be withdrawn sideways out of the needle bed to allow needles to be removed.

26.3.4 Latch Brushes

Latch brushes are attached to the cam-plates of both needle beds to ensure that the needle latches are fully opened. The supports of the brushes are adjustable to ensure precise setting of the bristles relative to the needles.

26.3.5 Carriage Guide Rails

The cam carriage either slides or runs on ball bearings or wheels along the guide rails. It is propelled either by hand or from a motor driven continuous roller chain. The carriage reciprocates over the needle beds, carrying cams and yarn, producing one course of knitting on each side way movement from right to left to right.

26.3.6 Cams

To operate the needles up and down from the stationery needle bed and to knit a course of loops, a set of cams viz., raising cam (clearing cam), guard cam and stitch cam are necessary (Fig. 26.4).

A triangular shaped raising cam (1) push the needle butts upwards when it is moved from left to right (Fig. 26.4*a*).

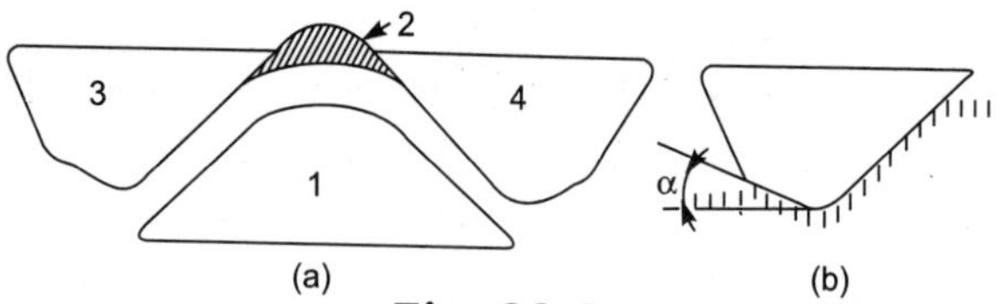

Fig. 26.4

When the needles are lifted up, the latches are opened up by the brushes and the clearing of the loops takes place.

As the retractable stitch cam (i.e., knock over cam) comes in contact with the butts, the needles slide downwards. The needle hooks draw the new loops through the old ones, while the latch close the hook. Actually the cams are not triangular exactly, their upper part are rounded to make smooth running of the needles. The level of stitch cam is adjusted by translating the entire cam up or down a slot.

The guard cam (2) is set between the two knock over cams (3 and 4) thus limits the butt displacements to the positive action of both cams.

Immediately after knock-over, the loop tension will cause the needles to be raised to different heights when they are guided at their lowest position by the knock over cams. In order to achieve the maximum uniformity of the loops, the needles are controlled at this point by rounding the lower point of the cams and by damping the needle upthrow with a gentle slope, at an angle "α" to the horizontal (Fig. 26.4*b*).

With V-bed machines, symmetrical camming arrangement in each needle bed enables a similar action to be achieved in both direction of the carriage transverse. Only one cam of each type carries out its function during a traverse and the other acts as a guard cam, forming cam track for the butts (Fig. 26.5).

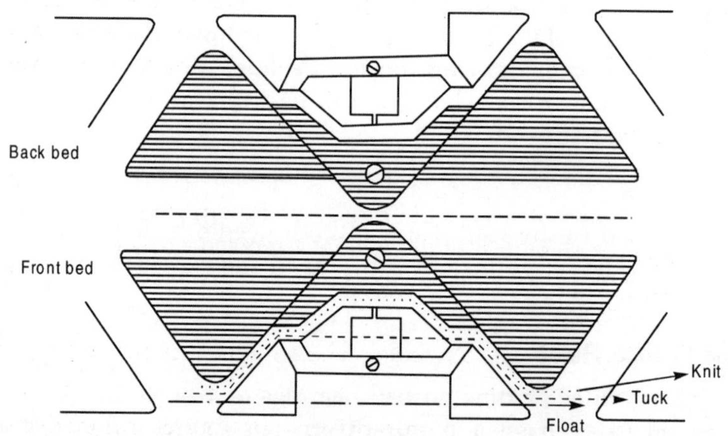

Fig. 26.5

26.3.7 Security Spring

A spring is placed at the lower end of the need,le bed and its purpose is to maintain the needle butt at a fixed height. They are held in position by the covering bar. When the spring is pushed into the high position, it moves the needle butt into the level of the cam track and at this active position the needle knits. Conversely, when the spring is in the low position, the needle butt is brought below the level of the base of raising cam. At this inactive position the needles cannot knit.

26.3.8 Fabric Take-down

Take-down tension is applied to the fabric, either by suspending weights or take down rollers. The take down action is designed to exercise a tension on the loops held on the needles to ensure correct knock over. Otherwise excessively loose loops may not close the latches, and may not be completely knocked over.

26.4 YARN PATH IN FLAT KNITTING (Fig. 26.6)

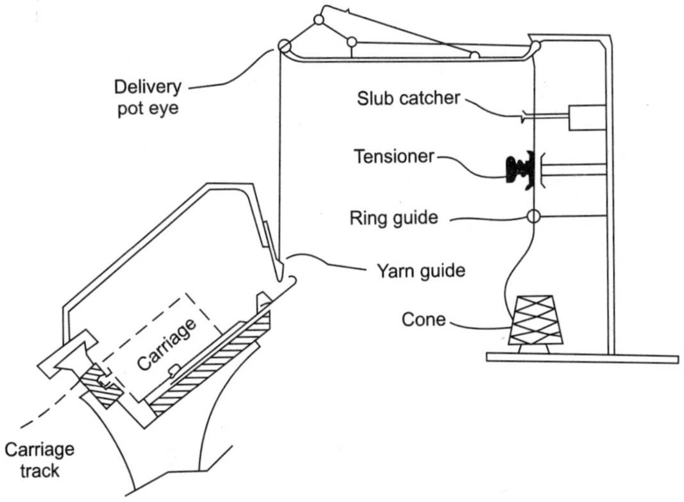

Fig. 26.6

The flat machine consists of a bobbin support stand at the back of the machine. The yarn from the supply cone passes through a ring guide, tensioner, knot/slub catcher, spring arm tensioner, delivery pot eye and finally to the yarn guide at the end of the yarn carrier.

26.5 KNITTING CYCLE (Fig 26.7)

The needles are actuated by the cams, which are mounted together with yarn feed on a carriage. The carriage reciprocates over the needle beds, carrying cams and yarns, producing one course of knitting on each side way movement from right to left and vice-versa. The cam actuates the needles by coming in contact with the butts. The needles are placed in slots called 'tricks' with their butts outwards.

Rest position: In this position, the needles take straight line motion and they are in level with the edge of the knock-over bits.

Clearing position: The needle slide upwards in the slots along the edge of the rising cam. When they get to the top, the latches are opened up by the brushes and the needles are above and clear of the loops.

Feeding position: As the needles come in contact with retractable cam, the new yarn is fed to the needle.

Knock-over position: The needle hook draw the loops through the old ones while the latch closing up the hook, allows the old loop to slip over the needle end, and the downward pull of the weight placed on the fabric end draws it down.

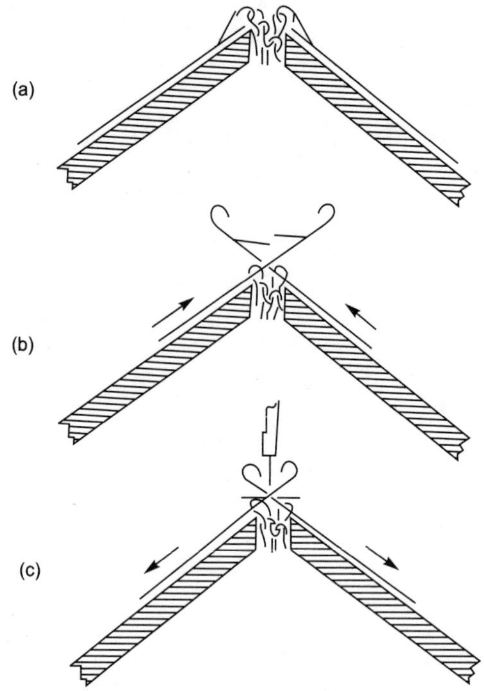

Fig. 26.7

26.6 FLAT KNIT FABRICS

26.6.1 Single Jersey Fabrics

The V-bed is the flat bed counter part machine of the circular cylinder and dial machines. It can be used for the manufacture of both jersey and rib. Only one needle bed is needed for the single jersey structures. For that, the back needle bed may not have any needles and a flat piece of fabric or a fabric with a single thickness is obtained. By using both the beds, tubular fabrics can also be produced (Fig. 26.8). When the carriage travels from left to right, the yarn is knitted on the front bed only. Then, when the carriage return from right to left, only the back bed needles knit the yarn. The cycle is repeated. The same yarn is thus knitted on each bed, the wales of the selvedges AD and BC will be joined.

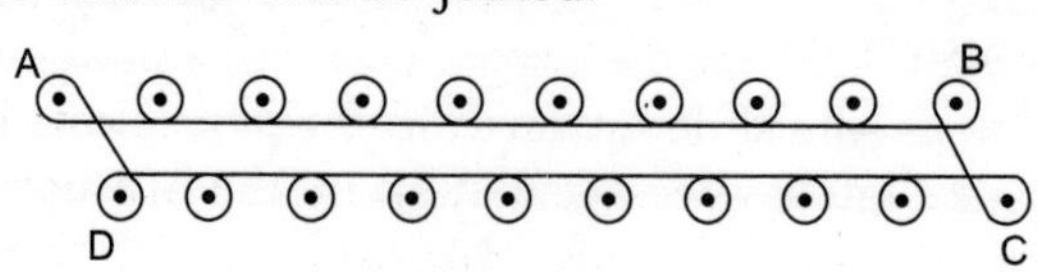

Fig. 26.8

Tubular fabrics can be made on V–bed knitting machine by using needles in both the needle beds. The cams would be then set in such a manner that, when the carriage is moved in one direction, the front cams will operate the front needles and back cams will be put out of action.

And when the carriage is moved in opposite direction, the front cam will be put out of work. By continuing the operation in this manner, a tubular knitted fabric will be produced, closing-up both sides of the fabric.

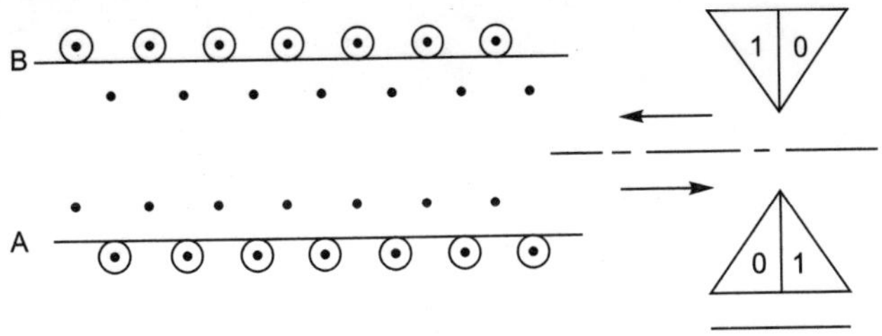

Fig. 26.9

As shown in Fig. 26.9, in A, the carriage traverses from left to right, the raising cam of the front bed is in action (1); the back bed raising cam is out of action (0). The front bed needles form loops while these on the back bed are in active. In B, the carriage moves from right to left. The front bed raising cam is out of action (0) while the back bed cam is in the active position. In this case, the front bed needles are inactive, while the back bed needles knit. In the diagram, 0 represents the cam out of action and 1 represents the cam in action.

Drop stitch patterns are produced when the front bed needles are cleared of their loops by a carriage traverse without a yarn carrier (Fig. 26.10).

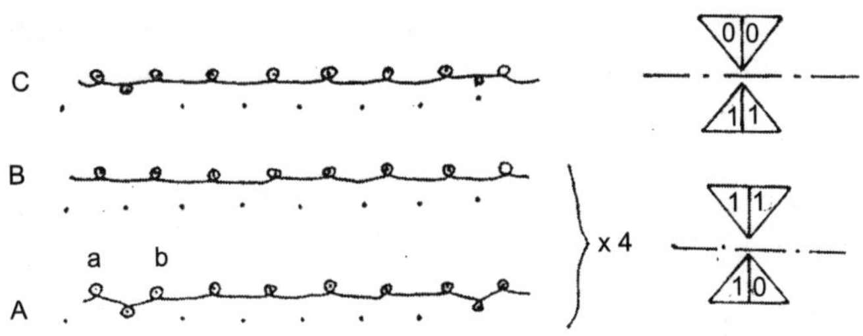

Fig. 26.10

During knock over, the loops of the front bed are cast-off and drop. The yarn that was fed to the dropped loops is redistributed among the adjacent stitches which become enlarged, thus forming an opening between the wales of the stitches formed at "*a*" and "*b*".

26.6.2 Rib Fabrics

Rib fabrics can be produced by operating front and back cams simultaneously for both front and back needles. The cams are exactly opposite to one another so that they push the needles from both the needle beds at the same time and draw them down at the same time. The needles are so set that the needles of the front plate come at a point in the middle of the spaces between the needles of the back plate. By manipulation of yarns and needle movements, a variety of fabrics differing in appearance and feel can be produced.

26.6.2.1 Half Cardigan Rib (Fig. 26.11) : In this rib structure, greater number of tuck stitches are added to make the fabric heavy, wide and soft. This structure comprises one cardigan course alternating with a 1 × 1 rib structure course. For each double traverse of the carriage, two successive stitches are knitted on the front bed, while held and tuck loop are formed at the back bed.

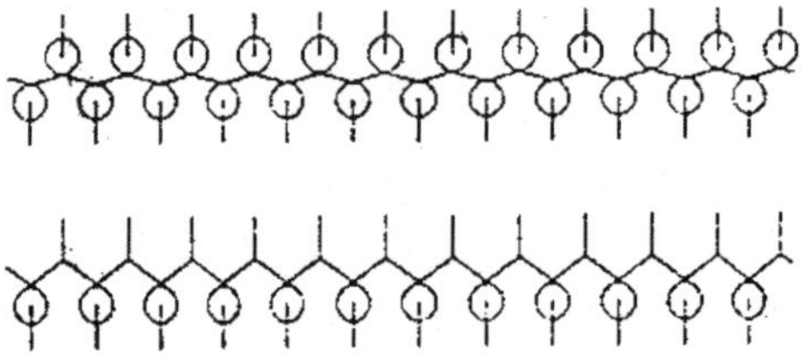

Fig. 26.11

26.6.2.2 Full Cardigan Rib (Fig . 26.12)

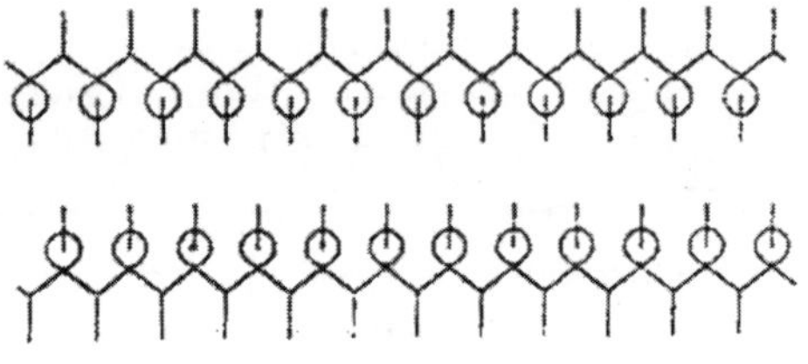

Fig. 26.12

In full cardigan, more tuck stitches are introduced which makes the fabric widen, heavier and less flexible than the half cardigan. The full cardigan is symmetric on both sides.

Fig. 26.12 shows that all the loops on each course will have a tuck loop superimposed upon them, to produce the structure.

26.6.2.3 Racked fabrics (Fig. 26.13) : The racked effects are obtained from the lateral displacement of one needle bed with respect to the other. Such effects cannot be obtained on single jersey fabrics.

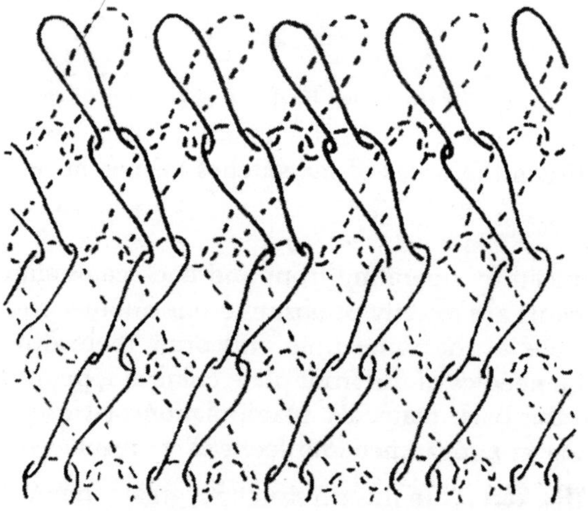

Fig. 26.13

Of the simple asymmetrical structures, full and half cardigans are the most common found structures on which the racked effects are produced. Racked effects are produced by using a tuck rib fabric as a basis and moving one set of needles, one needle distance sideways to the right or left, in relation to the other. The sideways movement must take place when the tuck loops are on the needles.

A stitch becomes inclined because the racking movement of one needle bed will displace the needle relative to another needle, which draws the loop from the same yarn in the same row.

Racking is more effective when tuck loops are used. Hence, it is always made on cardigans. Racking is also more pronounced if the racking is extended to two needle spaces, but this puts strain on the yarn.

26.7 FLAT KNIT DEVELOPMENTS

The development of power driven automatic flat knitting machine is a breakthrough in the flat knitting industry. Although the basic principles remain the same, the automatic power flat knitting can be programmed to knit through a garment length sequence at high speed and offer greater patterning facilities.

26.7.1 Needle Selection

Needles on either beds can be selected individually by both mechanical and electronic controls, for extensive jacquard patterning.

In mechanical selection, there are selectors in the tricks beneath each needle and their tails are supported by a jacquard steel which extends across the full width of the needle bed. The jacquard steels have punched holes and several steels are connected to form an endless chain loop that runs over a prism. As the prism turns, the jacquard steel on its upper surface thrust the protruding tails of the selectors, thus produce a simultaneous selection at every needle trick. The prism can dwell to repeat a selection or rack forward/backward by one or two positions. A punched hole in the jacquard steel allows the selector tail to sink through it into the groove in the prism and thus be unaffected by the thrust of the prism, so that its needle is left at the inactive level. Unpunched portion causes the corresponding selector to be pushed upwards in the trick, aligning its butt with a raising cam in the carriage so that eventually the needle above it will be lifted possibly to knit. If the needle bed is rocked, the jacquard apparatus moves side ways to keep in alignment with it.

In electronic selection, the jacks are no longer selected by punched blades but by an electro-magnetic device operated by a perforated tape or matrix. The selection may be operated on all the needles simultaneously at the end of a traverse or on successive needles during the carriage traverse. These electronic systems facilitate the production of jacquard patterns and permit rapid sampling of new designs.

26.7.2 Press Foot

Press foot of rectangular cross section is securely fixed to the carriage and moves across the machine with it. The presser foot precedes the rising needles. The wire of the presser foot slides underneath the intersection of the needles of the opposing needle beds and presses against the stitches laced between the two needles. The main objective of the presser foot is that the take down roller tensioning is no longer necessary to control the loops.

26.7.3 Seamless Garment Knitting

The seamless garment knitting techniques on V-Bed machines offers variety of advantages in knitting process by eliminating the cutting and sewing operations. Saving in cost and time, and higher productivity provide benefit to the manufacturers as well as end users. Seamless garments on a flat bed machine can be created in three separate tubular forms, by knitting one wider tube for the body part and two narrower tubes for the sleeve parts. On the flat machine, the three tubes are knitted on a pair of front and back needle beds. The flat machine knits and transfers loops between the front needle bed and back needle bed with different yarn carriers, one for body tube and two for sleeve tubes. The three tube knitting continues to the under arm point. At the under-arm point, the two carriers knitting the sleeve parts are taken out of the knitting zone. The remaining carrier, that knitted the main body part knits together the three tubes into one tube. The tubes are joined at the under arm, shoulder and neck points. In this manner, the seamless garment knitting is accomplished. The number of garment tubes knitted on the machine depends on the desired type of the knitted product, i.e., the sweater may require three tubes whereas the seat cover requires one. This three dimensional knitting technique has been extended to areas other than apparels, such as upholstery, automotives, industries and medicals. Even in apparels, products such as gloves, hats, socks, trousers and skirts can be produced by this technique. Three dimensional seamless seat upholstery enhances appearance and enables the users to feel comfort. Automotive seat covers made by seamless knitting process eliminate the cut and sewing operations, bring down warranty costs and adds quality. Tubular knitted structures such as bandages, orthopaedic supports and medical compression stocking are developed by their technique. Incorporation of high performance fibres and additional sensors or electronics could provide further opportunities for seamless knitted products in health care applications.

Panel production is still the most common form of flat knitting. Whether a panel or a garment is made, it is possible to knit into the fabric packets, closed pouches, collars, button hole panels, facings, overlapping collars, bows and most any garment component imaginable.

Socks Knitting

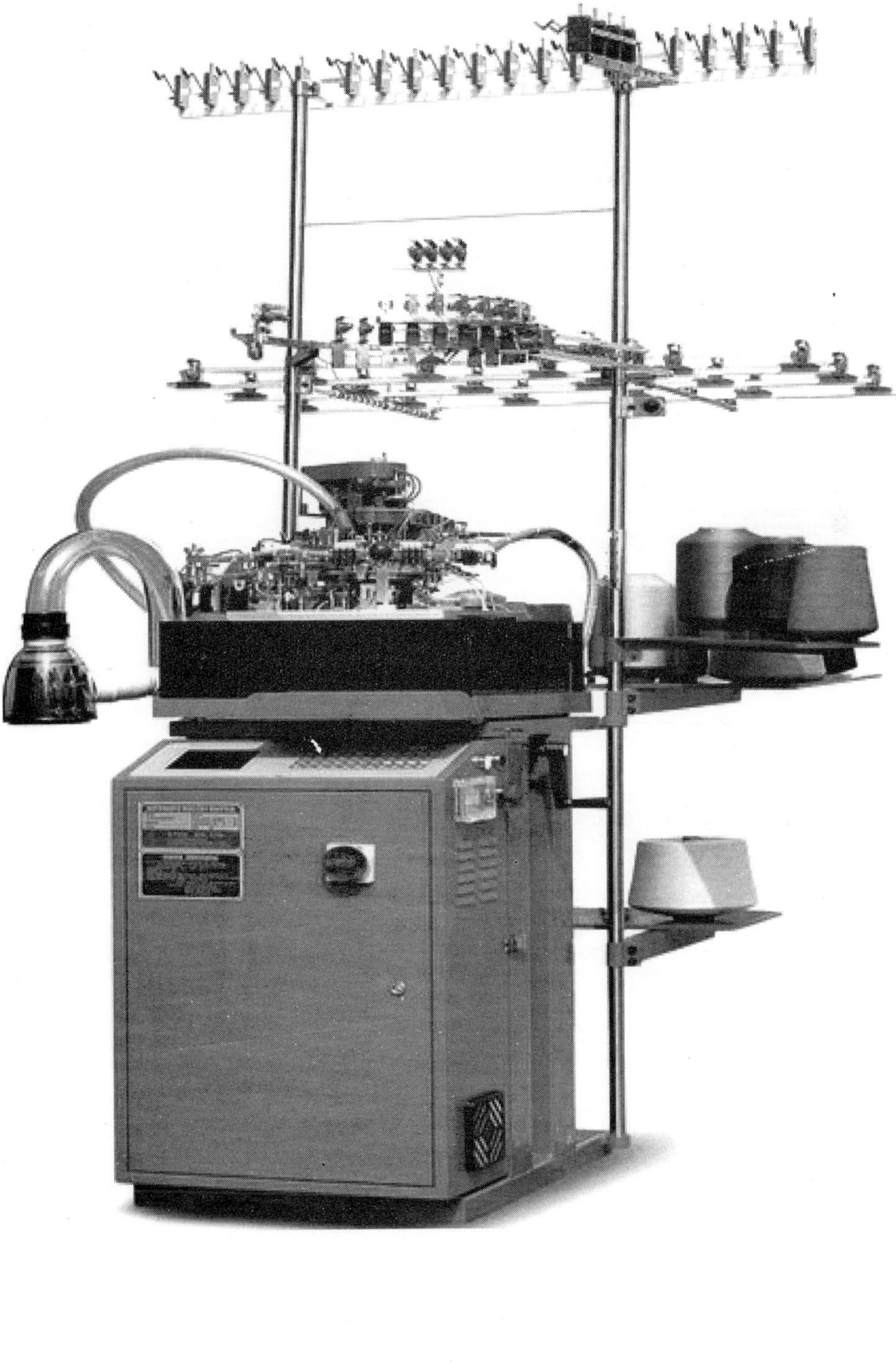

27

Hosiery Socks Knitting

27.1 INTRODUCTION

The term hosiery is derived from the word hose and has always referred to a series of articles which are used for the complete covering of the legs such as stockings, socks, tights, panty hose etc. Production of child's hand knitted foot covering was the origin of this seamless circular knitting which was estimated to be the fifth century A.D. Lee hand operated stocking frame established the principle of bearded needle knitting. The invention of the latch needle and its introduction to circular knitting enabled the development of small diameter hand operated sock machine with single rotating cam system which can be oscillated for heel and toe pouch knitting. The industrial revolution led to the development of power operated circular sock-machines both of single and double cylinder types. Later, each part was automated in turn, and the control of the number of courses from a chain which was racked at each course was taken place. Studs on the chain enabled the operator to stop the machine in order to make necessary mechanical changes manually at the different parts of the socks. Around 1900 B.C. every change of motion were almost automated.

The double cylinder machines use the double ended latch needles which are operated on both the top and bottom cylinders to change the character of the stitch. The earlier double cylinder machines work under rotary cam and stationary cylinder principle, and significant improvements happened only after the development of the revolving cylinder machines in 1912. The present day double cylinder machine is capable of producing a wide range of fancy effects. The objective of this chapter is to provide a brief idea on single cylinder socks knitting machines with respect to their shaping principle, drive mechanism, timing and control and fabric construction.

27.2 SOCKS GEOMETRY

The sock is engineered for a specific purpose of producing a specific product. The overall geometry of the socks is shown in Fig. 27.1 with its different portions. The first portion of the sock is the starting welt which is the top portion. It is a roll welt produced on a 1 × 1 rib basis and consists of a start course made in 1 × 1 rib followed by several courses knitted on the plain needles only, while the rib needles float.

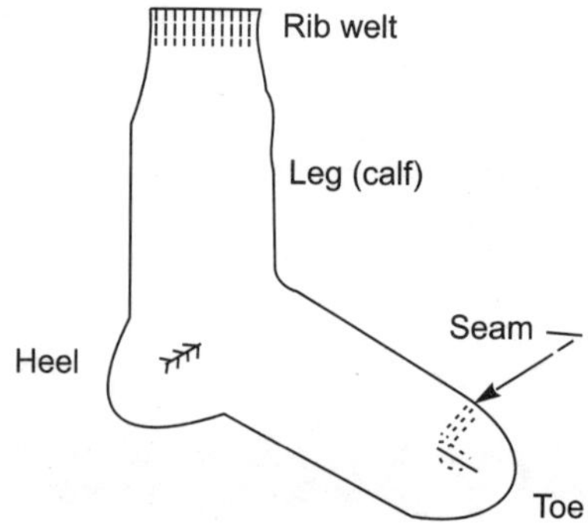

Fig. 27.1

Following the welt is the calf which usually has some spandex laid-in for increased elasticity. The next portion is the body that runs from the calf to the ankle. This is followed by the heel. The heel area of the sock usually has a thicker yarn or additional ends having good abrasion resistance to withstand the constant rubbing as the shoe is worn, put on or taken off. Following the heel is the foot portion. After knitting the foot, the toe portion is knitted with toe yarns. The sock is then completed with a short straight seam produced coursewise across the knitting and then pressed.

27.3 SOCKS SHAPING

The heel is the primary area of the sock and the sock turns right angle at the heel. The sock must undergo a shaping process at the heel and to perform this shaping operation, the knitting machine works in a reciprocating mode. It means that the machine oscillates back and forth past selective needles. Miss stitches are not formed as the yarn never passes the needles which are never raised. The oscillatory or reciprocating knitting is complex and is carried out at slow speed. During the first part of the heel, the fabric is narrowed by one needle at each course and the width of knitting is gradually reduced until only one third of the pouch needles are left in action. At this stage, the widening operation is introduced so that the second half of the heel is completed. And then the circular knitting recommences.

In the shaping of the heel, the machine knits all the needles of courses one through eight as shown in Fig. 27.2. The heel is made by wales 1 through 14 (rest 14 are on the other side of the sock). From course 9 onwards, the machine starts the reciprocating mode and narrows each course by one wale *i.e.,* at course 9 the machine knit stitches on needles one through 13, at course 10 knitting on needles one through 12, at course 11-knitting on needles one through 11 and so on.

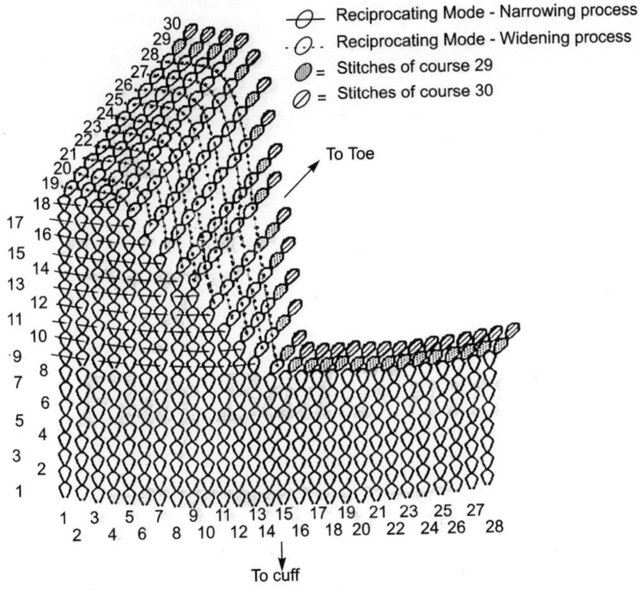

Fig. 27.2

The narrowing process ends at course 18 where the knit stitches are formed at needles one through 4. Only one side of the sock is shown in the Figure. At course 19, the widening process begins and the machine forms knit stitches on needles one through 5, at course 20, knitting on needles one through 6, at course 21-knitting on needles one through 7, at course 22, knitting on needles one through 8 and so on. The widening process ends at course 28 where the knit stitches are formed at needles one through 14. At course 29, the reciprocating mode is terminated and tubular knitting starts on all needles. This continues until the toe is reached. The machine is knitting the foot portion of the sock at this time.

The toe area is shaped to a smaller size by knitting a tight stitch. The toe shaping is also carried out in a similar way to that of heel under the reciprocating mode. After the toe is completed, it is sewn in order to complete the garment. In some machines, the toe closing can be done in the knitting process itself. Automatic toe seaming units can automatically turn the hose inside out, seam it over a seaming head and turn it back to its correct side.

27.4 YARN PASSAGE (Fig. 27.3)

There are two general supply of yarns to the knitting machine, viz., the feeding of main yarns and elastomeric yarns. The main yarn supply is checked as close as possible to the supply point so that in the event of a yarn failure, as much time as possible is allowed for stopping the machine. After passing through a cymbal tensioner, tension hook stop motion and guides the yarn reaches the latch needles. On yarn breakage, the fall of the hook lever will complete an electric circuit which will energise a magnetic coil to operate a switch to stop the machine drive. Similar kinds of stop motions are provided with elastomeric yarns also. Needle detector placed close to the needle bed guards against broken needle hooks and other needle damage.

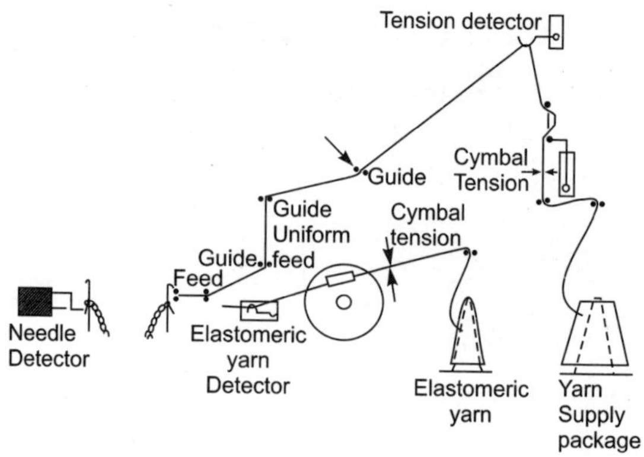

Fig. 27.3

27.5 DRIVING MECHANISM

The drive to the sock knitting machines are of two type, *i.e.,* the cylinder is driven with simple circular motion and complex oscillatory motion. Accordingly the machine is driven faster and slower respectively. During the formation of welt, body and foot, the knitting speed is maximum with simple circular motion and no mechanical change is taking place. During heel and tow knitting, cylinder is given slow speed oscillatory motion, approximately 380° in each direction.

The drive from the main shaft to the cylinder is through bevel gears as show in Fig. 27.4. The driving shaft A is fixed to the left bevel gear and is driven by means of a key and slot in the clutch C. When the clutch is moved to the right, it has circular motion from the rotating pulley through the driving shaft and when moved to the left it has oscillatory motion from the gear B. Oscillatory knitting is mostly performed at slow speed through gear B driven by a quadrant Q which in turn is driven by pitman L and big gear D. The big gear D is driven by E and the clutch C disengages gear E and engages the recess in gear B. The clutch received its horizontal lateral movement from the timing and control mechanism. The oscillatory knitting taken place at approximately two-thirds of the speed of circular knitting.

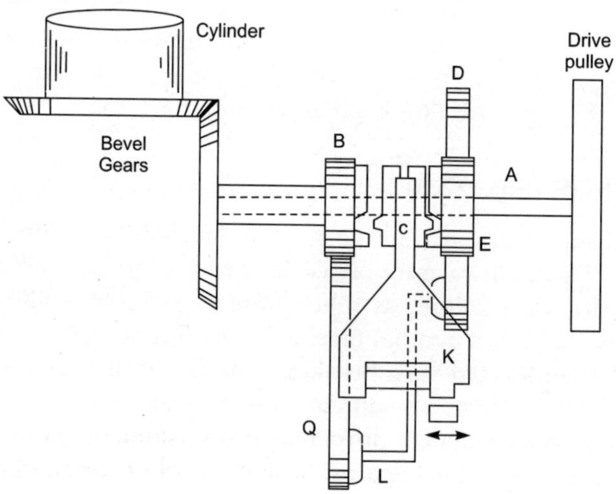

Fig. 27.4

27.6 TIMING AND CONTROL MECHANISM (Fig. 27.5)

The timing and control of the hosiery machine resolves itself into the counting of courses and this normally involves the use of an endless chain made up of detachable square links to enable the size of the product to be changed quickly and conveniently.

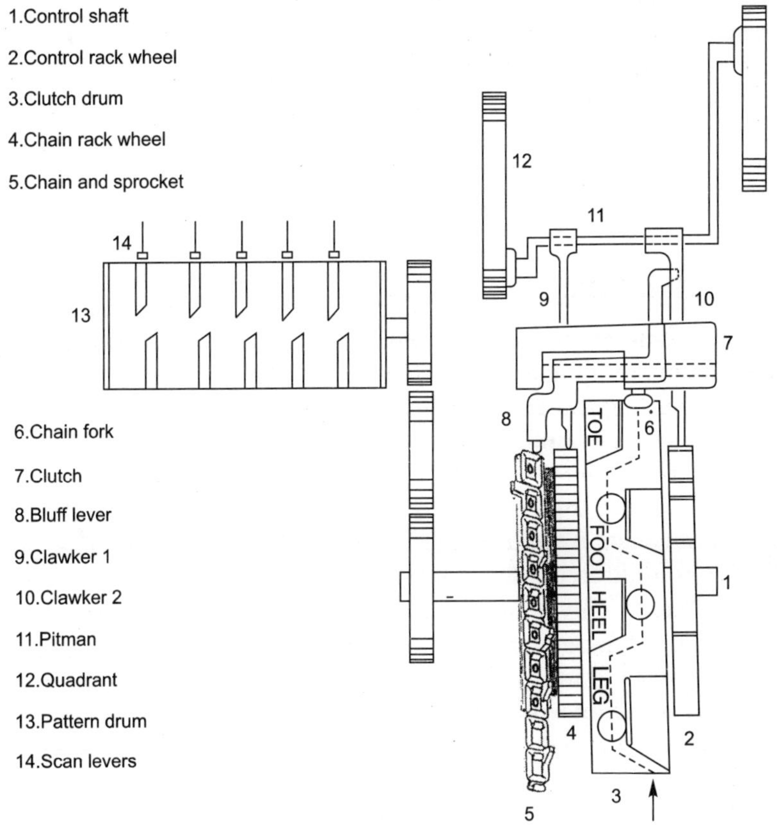

Fig. 27.5

The number of courses produced per chain link can vary according to the type of the machine and the number of racks on the chain rack wheel required to move the chain one link. The single cylinder chain is mostly racked once for every four machine revolutions and since there are three racks per chain link, this gives twelve machine revolutions for every chain link. Double cylinder machines normally have two racks per chain link and each rack represents one machine revolution, therefore one chain link will represent two cylinder revolutions. The chain consists of some links having lugs or studs, which are the movement links and some links are plain, as shown in Fig. 27.5.

The plain links are used purely for providing time between major changes whereas the movement links with projection or studs initiate mechanical changes, usually by causing the control shaft to be racked. The movement links may be of three heights, viz., low links, medium height links and high links to present different time contacts of the clawker-2 with the control rack wheel.

On the control shaft, the following are mounted.

(*i*) **Chain Sprocket Wheel and the Chain Rack Wheel:** The endless chain loop is mounted over the sprocket wheel which is compounded with the chin rack wheel. The chain rack wheel is racked by means of a clawker 1, which is continuously reciprocated forward and backward by the quadrant.

(*ii*) **Clutch Drum:** The clutch drum controls directly the clutch to cause the cylinder to rotate or reciprocate. The shift of the clutch is effected by a clutch fork actuated from a series of cams located on the clutch drum. The cams are shaped and spaced in such a manner that by riding the clutch fork over the clutch drum, mechanical changes for the production of different sections of the sock such as welt, leg, heel, foot and toe are taking place. The circles indicate the path of the clutch fork in one complete revolution or one complete hose.

(*iii*) **Control Rack Wheel:** The control rack wheel is irregularly notched and is the means whereby the control shaft is advanced. The control rack wheel is rotated by a clawker-2, which is continuously reciprocated forward and backward by the quadrant. The clawker-2 can be held away from engagement with the control rack wheel teeth by a bluff lever which is positioned underneath it. Movement links with a stud which projects upwards from one limb cause the bluff lever to be pivoted and lowering the clawker-2 on to the control rack wheel. Movement low links introduce the clawker-2 for the shortest time, for example for introducing the heel pickers; medium height links are for yarn and speed changes; and high links introduce the clawker-2 earlier and keep it in action for the longest rack for complicated mechanical changes such as heel.

(*iv*) **Pattern Cam Drum:** This large diameter cam drum is driven from the control shaft. One or more such drums may be employed on the machine. The length of the drum is divided into a number of racks with long narrow cams screwed into position. The cams have a tapered leading and trailing edge to transmit smooth movements through their corresponding levers or rods which scans over them. The cams function for speed changes, knitting cam changes, pickers, the verge, take down splicing and pattern drum racking.

27.7 COLLECTIVE WORKING MECHANISM

The quadrant pitman L on each revolution reciprocates the clawker 1 forward to rack the chain rack wheel which in turn moves the chain over the sprocket wheel. Major mechanical changes take place whenever the bluff lever is pivoted by the movement links of the chain. The other end of the bluff lever is lowered, thus introduce the clawker 2 to rack the control rack wheel as it reciprocates forward by the quadrant pitman L. The control rack wheel turns the clutch drum and shift the clutch by the clutch fork to the left or right according to the required mechanical changes such as welt, leg, heel, foot and toe. Thus the cylinder can now be driven for rotary or reciprocatory mode. Respective yarn feeding can be made by the pattern cam drum through the scanning levers which are in contact with the cam tracks of the drum. During rotary knitting mode, the cylinder is driven faster and all the machine needles undergo the looping operation. Two stitch cams "*a*" and "*b*" govern the needle movements as shown in Fig. 27.6.

During oscillatory knitting mode, approximately half of the total needles are raised into a high inactive cam-track by the introduction of a switch cam as shown in Fig. 27.6. The switch cam operates only on the long knitting butts allocated specially to them so that they retain their loops from the last course of rotary knitting. During narrowing, the leading needle in each direction of oscillation is lifted up to join the other needles in the inactive track by the action of one of the two side pickers as shown in the Fig. 27.6. These side pickers are alternately in action according to the direction of oscillation, and they operate throughout the oscillatory motion. During widening, a down picker is introduced which lowers two needles at a time, thus cancelling the effect of the up picker and putting an extra needle into operation.

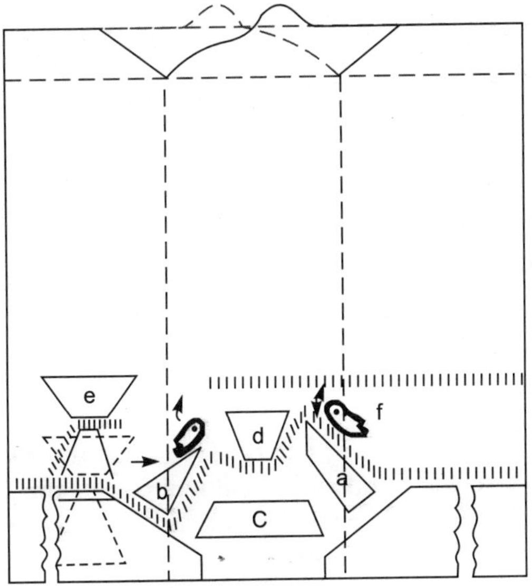

L – Line of Sinker
a, b – Stitch cam
d – Control cam
e – Switch cam
f – Up picker

Fig. 27.6

Each of the side pickers has an L-shaped recess and they are positioned facing upwards at the approach into the cam system so that in either direction of oscillation, the knitting butt of the leading heel needle is caught by the recess. Fig. 27.7 shows the engagement of the up pickers to that of needle butts.

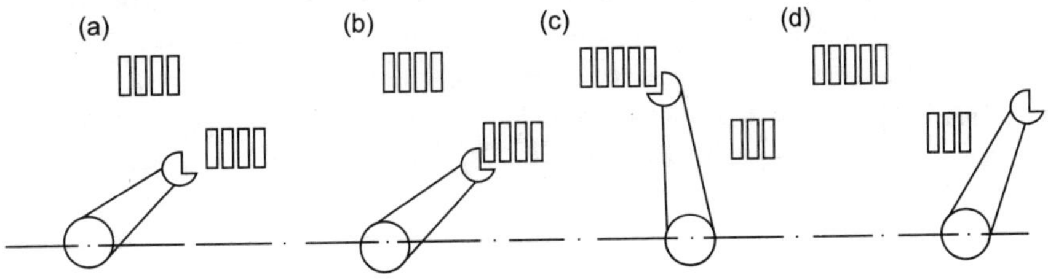

Fig. 27.7

The continued movement of the cylinder causes the picker to be moved backwards and as its movement is restricted, it pivots upwards in its holder to place the butt into the high inactive track. The spring attached to the picker then pulls it down again. The down picker, during the widening process brought into action and moves down two butts at a time from the inactive track. It has a recess on each side of its undersurface so that two butts can be accommodated in each direction of oscillation.

The action of the up pickers is automatic and continues as long as the machine reciprocates. A U-shaped shield, which receives its lifting and lowering from the pattern drum, introduces the down pickers during the widening process.

Half the needle knit in the heel section with narrowing occurring at each side until only one-third of the needles is left in action. As each needle is lifted out of action, the yarn is automatically wrapped over it in the form of a tuck stitch which makes the heel joint stronger. Widening then takes place until all the heel section needles are brought back into operation when circular knitting recommences. A toe pouch is knitted in a similar manner in which the seam will be on top of the toe.

27.8 NEEDLE CAM, SINKER CAM AND STITCH ADJUSTMENT

Two stitch cams govern the needle movements as shown in Fig. 27.6. These cams, in addition to depressing the needle at the stitching points, function also as clearing cams when the needle butts ride over them.

The sinker cam consists three sections ***a***, ***b*** and ***c*** to control the movements of the sinkers as shown in Fig. 27.8. The back or left hand sinker cam 'a' moves the sinkers for rotary knitting. The front or right hand sinker cam '*b*' moves the sinkers when the cylinder is reversed for heel and toe. The centre sinker cam '*c*' withdraws the sinkers on both the forward and backward motions of the cylinder. The relation of the needle points (*d*) and sinker points (*e*) is illustrated by circles as shown in the Fig. 27.8. Though in most of the circumference, the sinker points follow a circle just inside the needle circle, at the stitching point they are withdrawn and follow an arc just outside the needle circle. Relative movements or timing of the sinker and needle are determined by the sinker adjusting screws '*f*' and '*g* ', which are projecting from the sinker caps.

The sinker bed is fixed to the head of the needle cylinder so that any raising or lowering of the cylinder will affect the loop length, as the entire stitch cam assembly is raised or lowered. Scanning levers which ride over the cam tracks of the pattern drum raise or lower the cylinder through adjustable set screws. Raising the cylinder results in shorter distance below the sinker and produce short stitches whereas lowering the cylinder will carry the needles further downward and produce a longer stitch. Separate tracks on the pattern drum are responsible for the adjustment of loop length for calf, heel, toe, ankle and foot etc. For the knitting of calf, graduated stiffening is operated from a rotary eccentric cam which is racked independently of the control shaft. It allows the cylinder to be gradually lifted so that the loops gradually become smaller and the leg tube is narrowed.

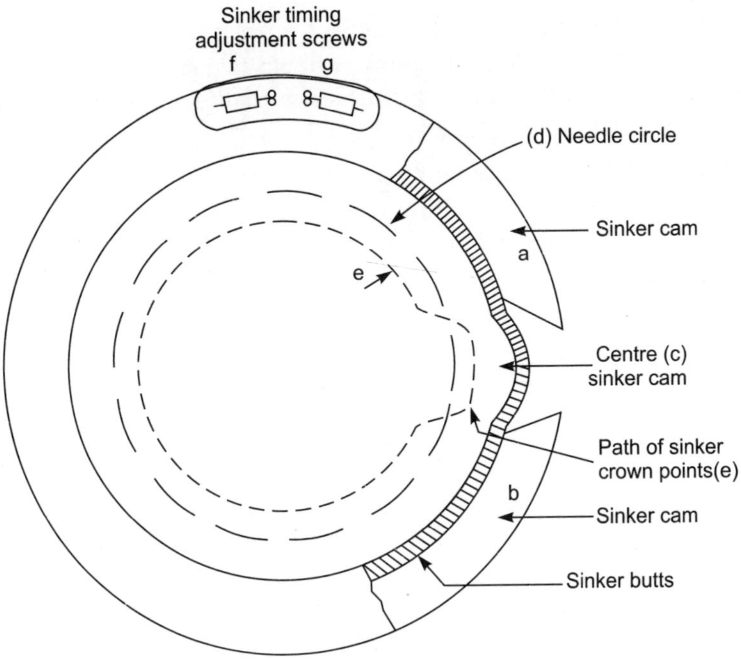

Fig. 27.8

27.9 FABRIC CONSTRUCTIONS

Apart from the appearance, the hosiery fabrics should be functional and should have some special properties. In this regard, some widely used hosiery fabric constructions are discussed below:

(*i*) **Run-resist:** This structure is capable of giving the ideal fabric stretch and flexibility combined with a unique attractive appearance. The popular run-resist structure is 1 × 1 cross tuck as shown in Fig. 27.9. The structure is a four course repeat.

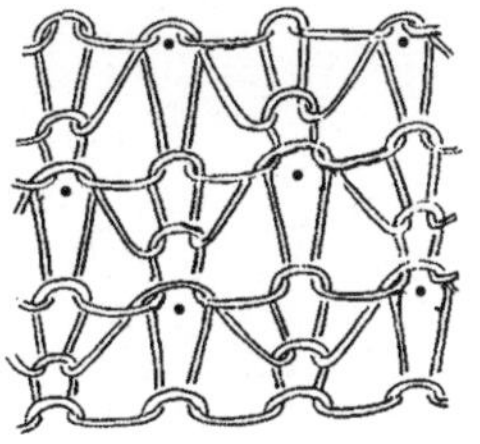

Fig. 27.9

(*ii*) **Texture effects:** Influence of yarn texture is apparent in the appearance of fabric production. The effect is produced by the use of high bulk textured yarns in the same fabric as flat nylon. If the textured yarn either floats or tucks, the flat nylon loop becomes elongated to give a lace or eyelet effect as shown in Fig. 27.10.

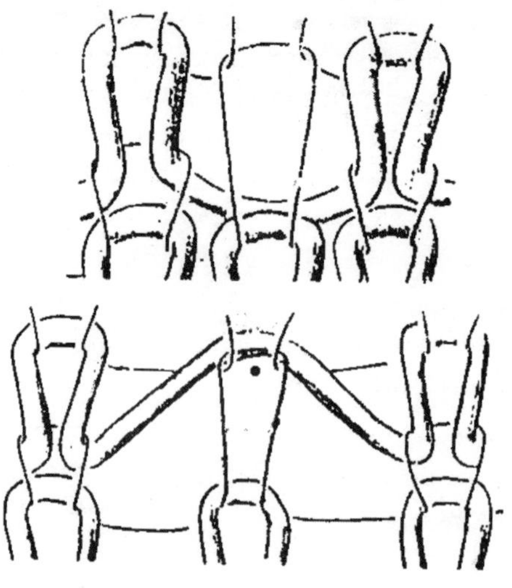

Fig. 27.10

(*iii*) **Plating:** By feeding two yarns together with different characteristics, appearance and functional characteristics can be improved. For example, nylon can be plated over cotton to give the comfort of cotton with the wear characteristics of nylon.

(*iv*) **Terry:** This is a type of plating called as plush plating. Two yarns are used in each loop and these are fed so that they are directed to the needles at different angles. Elements are used to control the two yarns so that one forms a regular loop and the other forms a greater length of stitch. The longer loops project over the ground platform of the regular loops.

(*v*) **Tuck:** By drawing three loops into the needle hooks before clearing (*i.e.*, 3 course tuck), then all of the surrounding loops change their configuration which leads to the production of a lace-knop effect. The adjacent wales are moved away and the centre wale is shortened. This type of construction is widely used in hosiery production.

27.10 DEVELOPMENTS

Developments in socks knitting were first aimed at higher speed of production which in terms of increasing the number of feeders and/or using faster speeds. Increasing application of Lycra and other elastane yarns for improved fit and comfort is considered as the other important development. The increased use of elastane yarns has also given rise to the wider use of special elastic yarn feeding devices.

Single cylinder machines equipped with a needle carrying dial or double cylinder machines having superimposed top and bottom cylinders are the two basic methods of socks knitting. A six feed single cylinder machine will produce a sock in little more than half a minute. Though the single cylinder machines are more simple and flexible, both the single and double cylinder machines received their popularity through their electronic needle-by-needle selection techniques. Versatility and patterning capability have been the most significant feature of the latest generation of sock machines in both categories. Diameters from 3¼ to 5" and gauges ranging from 48 to 240 NPI allow the production of all genres of socks for infants, school children, sports and leisure, men's classic dress socks and heavy weight work and aerobics socks. The terry sports sock either in plain terry or with motif patterning upto four colours has been the prominent product of single cylinder machines. Single cylinder machines are now able to knit jacquard designs in upto seven colours or more. Plating, stripes and embroidery type designs can also be incorporated. Single cylinder machines, also explore the more unusual aspects of sock design. Character application and 3-D effects have led to a riot of colour and humour in infant's and children's socks.

The double cylinder machine continues to find its major use in men's classic dress socks with attractive jacquard designs. Electronic needle selection and quality style change has almost totally superceded mechanical methods.

Introduction of automatic toe closing on the knitting machine simplify the whole production cycle, so that the products can go directly to boarding and packing. This saves the investment in toe closing machinery, saves labour, power and overhead.

REFERENCES

1. Ajgaonkar D.B. Knitting Technology.
2. George A. Tay. Fundamentals of Weftknitted Fabrics.
3. Gottlieb N. Production and Properties of Warpknitted Fabrics, Textile Progress.
4. Harry Wignall Hosiery Technology.
5. Iyer/Mammel/Schach. Circular Knitting.
6. Kudriavin L. Laboratory Practices in Knitting Technology.
7. Maille I.T.F. (i) Flat Knitting Machines.
   (ii) Circular Knitting Machines.
8. Paling D.F. Warp Knitting Technology.
9. Raz. S. (i) Flat Knitting, The New Generation.
   (ii) Warp Knitting Production.
10. Reichman C. Knitted Fabric Technology.
11. Resifeld A. Warp Knit Engineering.
12. Smirfitt J.A. (i) Introduction to Weft Knitting.
    (ii) Production and Properties of Weftknitted Fabrics, Textile Progress.
13. Spencer D.J. Knitting Technology.